Signals and Communication Technology

More information about this series at http://www.springer.com/series/4748

Pedro Amado Assunção · Atanas Gotchev
Editors

3D Visual Content Creation, Coding and Delivery

Springer

Editors
Pedro Amado Assunção
Instituto de Telecomunicações
and Politécnico de Leiria
Leiria
Portugal

Atanas Gotchev
Department of Signal Processing
Tampere University of Technology
Tampere
Finland

ISSN 1860-4862 ISSN 1860-4870 (electronic)
Signals and Communication Technology
ISBN 978-3-030-08552-0 ISBN 978-3-319-77842-6 (eBook)
https://doi.org/10.1007/978-3-319-77842-6

Printed on acid-free paper

This Springer imprint is published by the registered company Springer Nature Switzerland AG
The registered company address is: Gewerbestrasse 11, 6330 Cham, Switzerland

Contents

Chapter 1
Introduction

Pedro Amado Assunção and Atanas Gotchev

Three-dimensional (3D) audiovisual content is nowadays the driving force of many multimedia applications and services as well as development of different support technologies. The recent evolution of 3D media technologies has been quite diverse and progressing in different directions, not only enhancing existing technology but also developing and pushing forward new and richer content-driven applications. The main goals of using 3D multimedia have been maintained over the years as the ability to provide users with perceptual elements (mostly audiovisual) capable of providing an immersion feeling of being part of the scene, interacting and perceiving the 3D nature of the real physical environments conveyed by 3D content. More recently, the search for better technology, more pleasant user experiences and growing consumer markets have been driving a lot of research projects and new results with high potential impact in future evolution of 3D multimedia services and applications.

While immersive multimedia systems have been attracting increasing attention from researchers, industry and consumer market, many technological challenges remain associated with the huge amount of data that has to be dealt with at all stages of delivery systems. Evolution in this field has been essentially accomplished through expansion of audiovisual acquisition and rendering from single to many (virtually infinite) spatial locations, which requires audiovisual scene representations through acoustic wave fields and light fields, rather than single audio and video capturing the scene from a single spatial location. In this context, the ultimate goal of 3D multimedia technologies is to bring higher realism in the visual scenes being communicated and to provide the user with more creative tools for interacting

P. A. Assunção (✉)
Instituto de Telecomunicações and Politécnico de Leiria, Leiria, Portugal
e-mail: amado@co.it.pt

A. Gotchev
Department of Signal Processing, Tampere University of Technology, Tampere, Finland
e-mail: atanas.gotchev@tut.fi

P. A. Assunção and A. Gotchev (eds.), *3D Visual Content Creation, Coding and Delivery*, Signals and Communication Technology,
https://doi.org/10.1007/978-3-319-77842-6_1

with the visual content. Correspondingly, the interest in 3D technologies has been ever strong caused by their potential to enrich the human perception and support the development of novel applications and services in areas such as entertainment, 3DTV, games, medical and scientific visualization. Subsequently, the advances in 3D multimedia technologies open new market opportunities and enhance the user experience. Many research projects have aimed at developing future 3D technologies and reaching the next frontiers. As exciting as the novel results could be, they are always only the current state of the art and a starting point to go beyond.

This book presents recent developments in the field of 3D visual communications, departing from current technologies and analyzing their evolution to reveal the constraints that still limit the ultimate 3D user experience.

Multi-view video and light field representations are characteristic of the current trends in 3D visual technologies, and thus, they are addressed first in order to establish the fundamentals for representing the latest developments in efficient coding and delivery methods and tools. The aim to capture high-quality 3D visual content faces the need to process high amount of captured data. This, subsequently, calls for new efficient representations and effective coding tools. In this context, the book describes advances in multi-view video coding, including both standard-compliant techniques and non-standard ones, dense multi-view and depth-based coding.

Light field imaging is an emerging topic, currently gaining importance in 3D multimedia capture, coding and display. Therefore, the book also presents advanced compression methods aimed specifically at light field compression for storage and transmission, including simulation results and performance evaluation. The impact of network errors and data loss in multi-view video, depth and light field coded streams is further addressed for different packet loss conditions. Advanced error-concealment methods capable of efficiently reconstructing lost data in different types coded streams are presented, including evaluation of their performance in terms of objective quality of the visual information delivered to users.

The book also covers transmission systems, including network technologies and hybrid transport networks, used to support 3D multimedia services and applications. Additionally, recent research results, focusing different networking aspects of 3D delivery systems, are highlighted. For research and engineering, several simulation and emulation tools including testbeds are presented for test and/or performance evaluation, system design and benchmarking. These are particularly useful in research studies or development of innovative solutions for problems affecting the 3D multimedia performance of integrated delivery systems and communications infrastructures.

3D is about immersive and interactive experience; thus, primary factors for its high-quality delivery are the psychological factors in the context of multimedia consumption, the computational models of 3D perception and related quality metrics. The book presents several quality evaluation methods and related metrics for 3D video delivery systems, including monitoring and matching the quality of service (QoS) and quality of experience (QoE). The use of standard methodologies in relation with various quality assessment objectives is discussed. Comprehensive analysis of human factors and their relationship with specific 3D visual technologies, which influence the overall user experience, are further presented.

Another essential element in research and development projects involving the field of 3D video delivery systems is common datasets, publicly available, to allow comparison of results and validation of research advances obtained in different labs worldwide. Following the importance of datasets in this field, the book presents several publicly available datasets, which are relevant for active researchers and engineers dealing with acquisition, processing and coding of 3D visual data, as well as delivery through networks with different types of constraints (e.g. errors, losses, delays, etc.).

Overall, this book includes contributions from many researchers of European universities, companies and research centres, which collaborated together to scientific advances in the field of 3D multimedia delivery systems, within the scope of the European framework for Cooperation in Science and Technology, COST Action IC1105, *3D Content Creation, Coding and Transmission over Future Media Networks* (3D-ConTourNet).

Chapter 2
Emerging Imaging Technologies: Trends and Challenges

Marek Domański, Tomasz Grajek, Caroline Conti, Carl James Debono, Sérgio M. M. de Faria, Peter Kovacs, Luís F. R. Lucas, Paulo Nunes, Cristian Perra, Nuno M. M. Rodrigues, Mårten Sjöström, Luís Ducla Soares and Olgierd Stankiewicz

Abstract This chapter addresses image and video technologies related to 3D immersive multimedia delivery systems with special emphasis on the most promising digital formats. Besides recent research results and technical challenges associated with multiview image and image, video and lightfield acquisition and processing, the chapter also presents relevant results from international standardization activities in the scope of ISO, IEC, and ITU. Standard solutions to encode multiview image and video content and ongoing research are addressed, along with novel solutions to enable further developments in the emerging technologies dealing with capture and coding for lightfield content and free viewpoint television.

M. Domański (✉) · T. Grajek · O. Stankiewicz
Chair of Multimedia Telecommunications and Microelectronics, Poznań University of Technology, Poznań, Poland
e-mail: marek.domanski@put.poznan.pl

T. Grajek
e-mail: tomasz.grajek@put.poznan.pl

O. Stankiewicz
e-mail: olgierd.stankiewicz@put.poznan.pl

C. Conti · P. Nunes · L. D. Soares
Instituto de Telecomunicações and Instituto Universitário de Lisboa (ISCTE-IUL), Lisbon, Portugal
e-mail: caroline.conti@lx.it.pt

P. Nunes
e-mail: paulo.nunes@lx.it.pt

L. D. Soares
e-mail: lds@lx.it.pt

C. J. Debono
Department of Communications and Computer Engineering, University of Malta, Msida, Malta
e-mail: c.debono@ieee.org

S. M. M. de Faria · L. F. R. Lucas · N. M. M. Rodrigues
Instituto de Telecomunicações and Politécnico de Leiria, Leiria, Portugal
e-mail: sergio.faria@co.it.pt

P. A. Assunção and A. Gotchev (eds.), *3D Visual Content Creation, Coding and Delivery*, Signals and Communication Technology,
https://doi.org/10.1007/978-3-319-77842-6_2

2.1 Introduction

Recently[1], both among the research community and in industry, great attention is paid to *immersive multimedia*. The word *immersive* comes from Latin verb *immergere*, which means to dip, or to plunge into something. In the case of digital media, this term is used to describe the technical systems that are able to absorb viewers totally into an audiovisual scene [1–3]. Although *immersive multimedia* may be related to both natural and computer-generated content, in this book, we are going to focus mainly on the natural visual content that originates from multiple synchronized video cameras, and that possibly is augmented by data from supplementary sensors, like depth cameras.

For an immersive system, it is important to reconstruct a portion of an *acoustic wave field* [4] and a *lightfield* [5]. In a classic audiovisual system, audio and video are acquired using a single microphone and a single video camera. This is equivalent to the acquisition of a single spatial sample from an acoustic wave field and a lightfield, respectively. Therefore, the immersive media acquisition means acquisition of many spatial samples from these fields that would allow reconstruction of substantial portions of these fields. Unfortunately, such media acquisition results in huge amount of data that must be processed, compressed, transmitted, and rendered.

Although both video and audio are substantial for the impression of immersiveness, the scope of this book limited to the visual content. Nevertheless, it is worth to mention that significant progress is already made in the immersive and spatial audio technology. The faster development of this audio technology is related to lower bitrates and smaller data volumes for audio than for video. Moreover, the human auditory system is also less demanding than the human visual system. There already exist several spatial audio technologies like *multichannel audio* (starting from the classic 5.1 and going up to the forthcoming 22.2 system), *spatial acoustic*

[1]Written in 2017.

L. F. R. Lucas
e-mail: luis.lucas@ipleiria.pt

N. M. M. Rodrigues
e-mail: nuno.rodrigues@co.it.pt

P. Kovacs
Holografika, Budapest, Hungary
e-mail: p.kovacs@holografika.com

C. Perra
Department of Electrical and Electronic Engineering, University of Cagliari, Cagliari, Italy
e-mail: cperra@ieee.org

M. Sjöström
Department of Information Systems and Technology, Mid Sweden University, Sundsvall, Sweden

objects, and *higher order ambisonics* [6] that are able to produce strong impressions of immersiveness. First, the presentation technology seems to be more advanced for spatial audio than for video. The respective systems comprise the systems with high numbers of loudspeakers but also to the binaural rendering for headphone playback using *binaural room impulse responses (BRIRs)* and *head-related impulse responses (HRIRs)* that is a valid way of representing and conveying an immersive spatial audio scene to a listener [7].

During the last decade, the respective spatial audio representation and compression technologies have been developed and standardized in MPEG-D: MPEG Surround [8], SAOC [9], and MPEG-H Part 3—3D Audio [10] international standards. The spatial audio compression technology is based on coding one or more stereophonic audio signals and additional spatial parameters. In that way, this spatial audio compression technology is transparent for the general stereophonic audio compression. Currently, the state-of-the-art audio compression technology is *Unified Speech and Audio Coding (USAC)* standardized as MPEG-D Part 3—USAC [11].

For the *immersive video*, the development is more difficult, nevertheless, the research on immersive visual media is booming recently. *Immersive video* [2] may be related to both natural and computer-generated content. Here, we are going to discuss mostly the natural content that originates from video cameras and possibly is augmented with data from supplementary sensors, like depth cameras. Such content is sometimes described as *high-realistic* or *ultra-realistic*. The immersive multimedia systems usually include communication between remote sites. Therefore, such systems are also referred as *tele-immersive*, i.e., they serve for *highly realistic sensations communication* (e.g., [12]).

The abovementioned immersive natural content usually is preprocessed by computers before being presented to viewers. A good example of such *interactive* content is spatial video that allows a viewer to virtually walk through a tropical rainforest reach of hidden swamps, poisonous plants, and dangerous animals. During the virtual walk, a virtual explorer is very safe and may enjoy the beauty of nature being relaxed, and without fear. The virtual walker may choose arbitrarily a virtual trajectory of a walk, may choose a current direction of view, may stop and look around, watch animals and plants, etc.

The respective visual content is acquired with the use of many synchronized cameras. Then, sophisticated computer processing of video is needed in order to produce the entire representation of the visual scene. Presentation of such content usually must be preceded by rendering that results in the production of video that corresponds to a particular location and view direction currently chosen by a virtual rainforest explorer. Therefore, the presentation of such rendered video may also be classified as presentation of *virtual reality* although all the content represents real-world objects in their real locations and motions (see, e.g., [13]).

Similar effects may be obtained for computer-generated contents, both standalone or mixed with natural content. In the latter case, we speak about *augmented reality* that is related to "a computer-generated overlay of content on the real world, but that content is not anchored to or part of it" [13]. Another variant is *mixed reality* that is "an overlay of synthetic content on the real world that is anchored to and interacts with the real world contents." "The key characteristic of mixed reality is that the synthetic content and the real-world content are able to react to each other in real time" [13].

Considering the immersive video, we have to refer to 360° video that is currently under extensive technological development. The 360° video allows at least to watch the video in all directions around a certain virtual position of a viewer. More advanced versions of 360° video allow a viewer also to watch video in any direction up and down from its virtual location, as well as to change the virtual location. In popular understanding, the 360° video is even treated as a synonym to the immersive video, e.g., see Wikipedia [14].

The preliminary classification of immersive video [3] was recently discussed by MPEG (Moving Picture Experts Group, i.e., formally ISO/IEC JTC1 SC29 WG11[2]) [15–17]. By drawing conclusions from this discussion some main categories of content may be defined:

1. monoscopic 360° video, where usually video from many cameras is stitched to a panorama,
2. stereoscopic and binocular 360° video that allows a viewer to watch in an arbitrary position with various levels of spatial sensations,
3. 6° of freedom 360° video that provides a viewer the ability to change freely his/her location.

For Class 2, the first generation of 3D video, i.e., the stereoscopic video is the very popular and the simplest case. The last wave of enthusiasm for 3D video was encountered around year 2010 but the lack of user-friendly stereoscopic displays has reduced the interests recently. In this book, we rather consider the next-generation 3D content that allows a viewer to perceive spatial parallax possibly without special glasses that are necessary for traditional stereoscopic displays, like shutter glasses, polarization glasses, or color-filter glasses. Such glass-free systems are still challenging even for a fixed view, nothing to say about 360° video.

The Class 3 is related to virtual navigation that is a functionality of future interactive video services where a user is able to navigate freely around a scene. The systems that provide such functionality are often called free viewpoint television (FTV) [18–23]. The prospective FTV will be an interactive Internet-based system that may output virtual monoscopic video, virtual stereoscopic video or even multiview video, e.g., for watching a virtual view on an autostereoscopic display.

In 360° video, virtual navigation and other types of advanced visual content, the virtual views are synthesized or rendered using a scene representation, or a scene model. The following scene representation types are mostly considered in the references: object-based [24, 25], ray-space [19, 26], point-based [27], and multiview plus depth (MVD) [28]. As the first three types of models are related to quite complex calculations, currently, the MVD representation is used most often and

[2]See Sect. 2.3.

will be extensively considered further in this book. Nevertheless, it is worth to mention that modeling of 3D scenes using point clouds is considered as an competitive and interesting approach, even related to recent standardization projects [16].

The multiview plus depth video format is also vital for the display technology. Although the display technology is also not mature enough for wide adoption of 3D video and for the immersive video and images, the situation seems to be diversified for various display application areas. In particular, the glassless autostereoscopic displays and projection systems are being improved step by step, thus increasing the comfort and quality of spatial (3D) video presentations. Such signage systems may use even 200 views, i.e., they display simultaneously 200 views in order to produce realistic impression of depth [29–31].

2.2 Multiview Video Plus Depth

The complete and general description of a visual scene may be provided using a plenoptic function (POF) [32]. The plenoptic function is usually defined as a function of seven variables, i.e., $POF(x, y, z, \phi, \varphi, t, \lambda)$, where x, y, z represent the coordinates of a point in 3D space, ϕ and φ define the direction of a light ray, t denotes time, and λ denotes the wavelength in light ray. The value of the plenoptic function expresses the "amount of light" (e.g., luminance) of a given wavelength λ, registered at a time instant t at a point (x, y, z), and in the direction defined by the angles ϕ and φ. In order to describe a scene entirely, the plenoptic function should be measured at all points (x, y, z) in some 3D space relevant to the scene, for all wavelengths λ from the visible light interval, and in all directions defined by the angles ϕ and φ possibly from the interval $(-\pi, \pi)$. Obviously, such full description is neither possible nor necessary. Instead, in multimedia technology, we use various simplified representations of 3D scenes already mentioned in Sect. 2.1. Among those types of representation, the multiview plus depth (MVD) representation is the most popular in practical approaches to natural 3D video. More views with the corresponding depth maps we have, more exact is the approximation of the lightfield.

The high number of video views of multiview video results in a huge amount of data that needs to be transmitted over bandwidth-limited channels. This fact motivates the research on compression systems that should be able to drastically reduce the storage and the bandwidth requirements for 3D video data. Practical systems register, process, and transmit only a subset of the required views together with the geometric information of the scene, represented by depth maps. The missing views can then be generated at the receiver side through view synthesis algorithms, based on the transmitted view and depth data. For this purpose, depth maps provide the information related to the distance of each pixel in the video view with relation to the view camera position. Such representation for 3D video, using a small number of video views combined with the geometric information of the scene, is the called multiview video plus depth (MVD) [28, 33] as already mentioned. Figure 2.1 illustrates an MVD system, which uses view synthesis at the receiver side.

An example of a depth map and the corresponding view is depicted in Fig. 2.2.

Depth estimation is still a challenging task. In general, there exist two approaches:

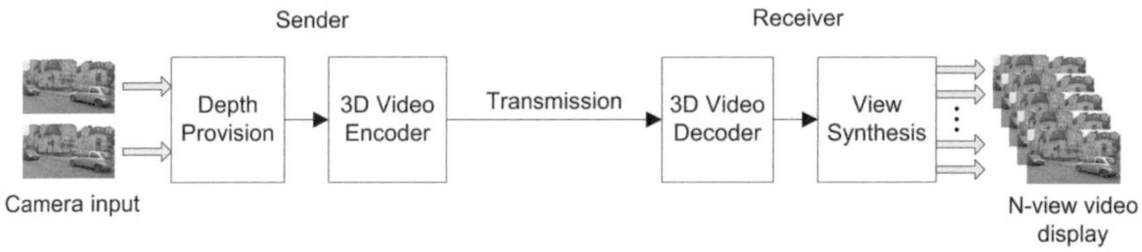

Fig. 2.1 MVD system based on view and depth data with view synthesis at the decoder side [28]

Fig. 2.2 A view and the corresponding depth map from the test multiview sequence *Poznan_Street* [34]

- application of special depth sensors called also depth cameras (e.g., [35, 36]),
- estimation of depth from video data by the use of video analysis on computers.

The depth sensors illuminate a scene with invisible infrared light and mostly exploit one of the following two technologies:

- by measurements of the time-of-flight [37] from the radiator to the object and back to the sensor,
- by analysis of structured light reflected from a scene illuminated with a specific pattern.

Currently, both technologies are under further development resulting in their improvements. Despite which technology is used, the usage of depth sensors is conceptually very attractive as they may produce the depth in real time with reasonable latency. Nevertheless, their practical employment still faces severe problems related to limited spatial and temporal resolutions of the acquired depth maps, limited distance ranges, synchronization of video and depth cameras, additional infrared illumination of the scene that may interfere with other equipment, mutual interference of several sensors working simultaneously at the same scene, and sensitivity to environmental factors including solar illumination. Currently, these sensors are only capable of acquiring low-resolution depth maps, which are usually enhanced by postprocessing methods based on interpolation and denoising filters. Also, the maximum and minimum depth value acquired by these sensors is limited. Furthermore, since depth sensors are physically independent of video cameras, they are positioned at slightly different positions, resulting in depth maps that do not exactly match the associated views. Already, substantial research work is done with the aim to overcome the abovementioned problems, see, e.g., [38–40]. Despite all the abovementioned problems, the technology of depth cameras is intensively developed for many potential applications including industrial computer vision, mobile robot navigation, control of autonomous cars, and many others.

Depth can be also estimated in the process of video analysis. The real views used for depth estimation should be corrected by compensation of the lens distortions, and possibly also by compensation of the differences in color characteristics of the cameras. Moreover, illumination differences also should be compensated.

The depth estimation may be described as follows. For the simplest case, consider two views. The pairs of characteristic points need to be found in the views. For each such pair, disparity d can be measured as the shift between the locations of the corresponding characteristic in the two views. Assume that the focal length of both cameras is f, and the distance between the optical centers of the cameras, i.e., the base distance is b. Assuming $f \ll z$ we get [41], we may calculate the depth of a point object

$$z = \frac{fb}{d}. \tag{2.1}$$

In order to use Formula 2.1, the values of focal length f and the base b need to be measured. It is done in the process of calibration of the multi-camera system, when some special calibration video is recorded, and the relevant camera parameters as well locations of the camera sensors are estimated using the data obtained from the calibration video [42].

Estimation of depth from a pair of views has been studied since many years (e.g.[43–45]. Some methods [46, 47] focus on the segmentation-aided depth estimation based on optimization performed on a graph. While achieving relatively high quality of estimated depth maps, these methods are designed for stereo pairs only. Moreover, main optimization process is performed on the pixel level, making the whole estimation very time-consuming. Exploitation of the outputs from more than two cameras provides the opportunity to produce more exact depth maps. For example, the method of [48] estimates depth maps for limited resolution in the real time, using the outputs from four cameras with parallel optical axes. The method of [49] proposes the estimation of the multiview depth based on the epipolar plane image. While providing the inter-view consistent depth of the high quality, this method is still limited to linear arrangements of cameras. Multiview depth estimation can be based on the Belief Propagation [50]. In the work described in [51], the inter-view consistency is ensured by depth maps cross-checking and multiview matching of views. The methods have been also proposed that provide the temporal consistency of the estimated depth maps [52, 53]. There exist a huge number of papers on various aspects of the depth estimation, and this paragraph provides sparse samples of the references rather than an entire review.

The depth estimation reference software [54] has been developed by MPEG, and currently, it is widely used a reference for multiview depth estimation.

Recently, it was shown that for highly occluded scenes, nonuniform distribution of cameras around a scene leads to better depth estimation [20]. Therefore, for such real scenes, it was proposed to acquire multiview video using camera pairs [55].

Obviously, depth maps can be represented as greyscale images. In practice, the name of depth map is used for the data sets, where the samples represent either depth or disparity. The depth or disparity samples have often 8-bit representation. If disparity representation is used, each sample value corresponds to the inverse of the distance from the given camera to a given scene point, or more exactly to the plane that contains this particular scene point and is perpendicular to the optical axis of the camera. It means that the range between the minimum and maximum depth distances is divided into 256 unequal intervals. Closer distances are represented more accurately while the further ones more sparsely. Therefore, for many applications, the depth sample representations with more than 8 bits are used.

Depth estimation allows to produce the multiview plus depth representation that may be used for the synthesis of virtual views, or, in other words, for depth-image-based rendering (DIBR) that is essential for free viewpoint television, augmented and virtual reality, lightfield displays, etc. The virtual view synthesis is also exploited in order to increase compression efficiency for multiview video [56].

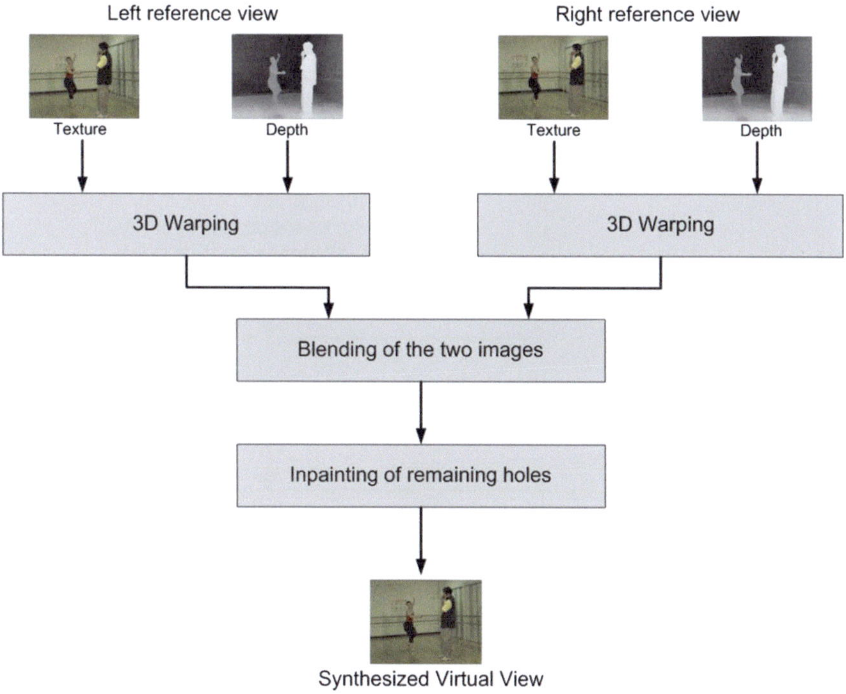

Fig. 2.3 Block diagram of the DIBR algorithm

Figure 2.3 presents a block diagram of the DIBR algorithm, based on two reference views and their associated depth maps. Any virtual view can be generated based on these two references. Usually, two nearest real views, labeled left and right reference views in Fig. 2.3, are selected from the multiview sequence and warped [57]. The warped images generated from the two views are then blended to form the new virtual position [58, 59]. Since some disoccluded regions and holes may still remain, in painting is applied to fill the missing data [57].

In order to reduce errors introduced by stereo matching algorithms, [60] proposes a depth map preprocessing algorithm based on temporal filtering, compensation for errors and spatial filtering. An illumination compensation technique is applied in [61] to reduce color discontinuities and improve visual quality of the synthesized views. The warped depth maps are processed by median and bilateral filters before inverse warping in [62] to improve the visual quality of the synthesized view. Furthermore, depth map pixels at edges are detected and are not warped in [63]. This technique reduces unreliable data in these regions from the warping operations.

Other DIBR techniques found in literature include the enhancement of virtual views through pixel classification, graph cuts, and depth-based inpainting [64]. The perceived depth quality and visual comfort in stereoscopic images are improved using stereoacuity before rendering the images in [65]. Furthermore, a just

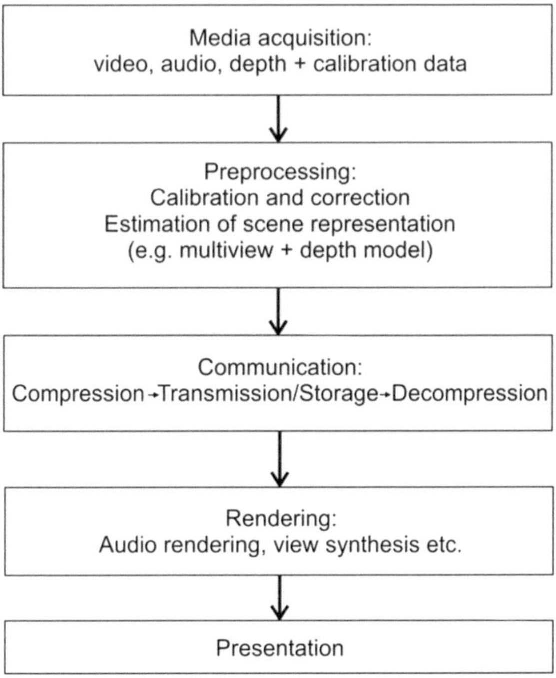

Fig. 2.4 The processing chain for spatial video associated by spatial audio [3] © IEEE 2017

noticeable depth difference (JNDD) model and saliency analysis are used in [66] to provide a better user perception of the rendered content. Recently, good-quality synthesis technique has been demonstrated for practical virtual navigation in a scene represented by multiview plus depth with real cameras sparsely located around a scene [55]. For research purposes, the view synthesis reference software [62] is available in the version adequate for the synthesis of the views from arbitrary locations.

The data processing pipeline for multiview plus depth representation of visual scene together with the corresponding audio data is depicted in Fig. 2.4.

2.3 Standardization—The Status and Current Activities

2.3.1 Standardization in Multimedia

Standardization is crucial for telecommunications where the transmitter and the receiver are often placed in the locations being very distant one from the other. In such cases, the interoperability of hardware and software delivered by different vendors is an issue of paramount importance. The means to ensure the interoperability is to observe standards agreed by all involved parties. In practice, such standardization agreements are obtained either in international institutions or by consortia of companies sharing substantial portions of the relevant markets.

The following international institutions play the primary role in multimedia standardization:

ISO—International Organization for Standardization,
IEC—International Electrotechnical Commission,
ITU—International Telecommunication Union.

In the area of multimedia, ISO and IEC work mostly jointly and they jointly issue international standards (IS). International standards are therefore numbered as, e.g., ISO/IEC IS 14496. Except the number, each standard has also its own generic name. The ISO/IEC standards are divided into parts, like Part 1 "Systems", Part 2: Video, Part 3 "Audio", etc. In fact, a part of a standard defines the minimum requirements for interoperability for a given technology, like video compression or audio compression. The parts of standards may also be recommendations of ITU. The standards (called recommendations) of ITU are grouped into Telecommunication Sector (ITU-T) and Radiocommunication Sector (ITU-R). Of course, some standards are independently developed and issued by only one institution, some are issued jointly by two or three of them. Moreover, some internationally recognized standards have been also defined by IEEE, i.e., the Institute of Electrical and Electronics Engineers and by SMPTE (Society of Motion Picture and Television Engineers).

Moreover, there also regional and national standardization organization. For example, the Chinese consortium for Audio Video Coding Standard plays an important role in the standardization of the compression of video and audio.

In many cases, the active role is played by an industrial consortium. For example, a group of big companies (Amazon, ARM, Cisco, Google, Intel, Microsoft, Mozilla, Netflix, NVidia) has recently created an Alliance for Open Media with the aim of producing a new standard for video compression called AV1.

For video and audio compression, the minimum interoperability requirements are related to the semantics and syntax of the bitstream, i.e., they define how to read the bitstream. It means that a standard defines the decoders, while having limited impact on the encoders (cf. Fig. 2.5).

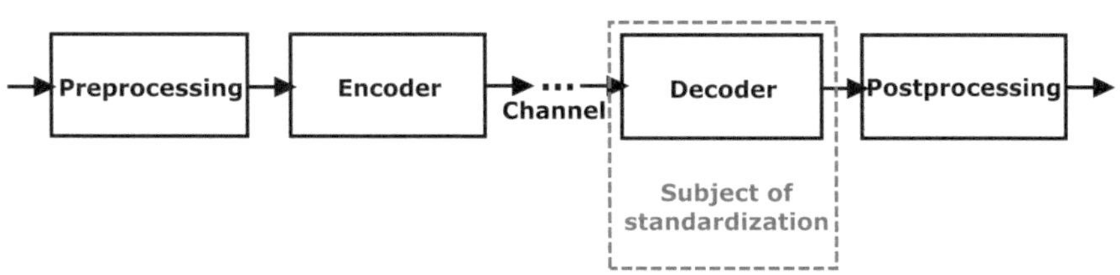

Fig. 2.5 Standardization of compression

2.3.2 Basic Technologies

In the recent years, significant efforts have been made in standardization of compression of multiview video, multiview plus depth video as well as other related aspects. These techniques mostly rely on the consecutive generations of monoscopic video coding. During the last 25 years, consecutive generations of monoscopic video coding technology have been accepted as the international standards, like MPEG-2 (MPEG-2) [67], Advanced Video Coding (AVC) [68], and High-Efficiency Video Coding (HEVC) [69]. Currently, the new generation of video compression technology is under development and is expected to be standardized around 2020–2021 as a part of the prospective MPEG-I (immersive) standard. These consecutive video coding generations have been developed thanks to huge research efforts that reach thousands of man-years recently.

Assuming the required quality level corresponding to the broadcast quality and a mature codec implementation, for demanding content, and for a given video format, the bitrate B of the compressed bitstream may be very roughly estimated using the formula [70–72, 22]

$$B \approx A \cdot V \quad \text{(Mbps)}, \tag{2.2}$$

where A is technology factor, where
$A = 4$ for MPEG-2,
$A = 2$ for AVC,
$A = 1$ for HEVC,
$A = 0.5$ for the prospective technology expected around year 2021 (Versatile Video Coding),
and V is video format factor, where
$V = 1$ for the Standard Definition (SD) format, (either 720×576, 25 fps or 720×480, 30 fps, chroma subsampling 4:2:0, i.e., one chroma sample from each chroma component C_R and C_B per 4 luma samples),
$V = 4$ for the High Definition (HD) format (1920×1080, 25/30 fps, chroma subsampling 4:2:0),
$V = 16$ for the Ultra High Definition (UHD) format (3840×2160, 50/60 fps, chroma subsampling 4:2:0).

The conceptually simplest way to implement the coding of multiview video is to encode each view as an independent video stream. Such type of compression is usually called simulcast coding. Simulcast coding exploits the commonly used relatively cheap video codecs may be efficiently applied. The total bitrate B_m of the bitstreams is

$$B_m = N \cdot B, \tag{2.3}$$

where N—the number of views,
B—the bitrate for a single view from Eq. 2.2.

2.3.3 Multiview Video Coding

The main idea of the multiview video coding is to exploit the similarities between neighboring views. One view, called the base view, is encoded like a monoscopic video using standard intraframe and temporal interframe predictions, therefore it is also called the independent view. The respective bitstream constitutes the base layer of the multiview video representation. The independent or the base view may be decoded from the base-layer bitstream using a standard monoscopic decoder. For encoding of the dependent views, i.e., the other views the inter-view prediction with disparity compensation may be used in addition to standard intraframe and interframe predictions. In inter-view prediction, a block in a dependent view is predicted using a block of samples from a frame from another view in the same time instant. The location of this reference block is pointed out by the disparity vector. This inter-view prediction is dual to the interframe prediction, where the motion vectors are replaced by the disparity vectors.

In multiview video coding, the pictures are predicted not only from temporal interframe references, but also from inter-view references. An example of a prediction structure is shown in Fig. 2.6.

Multiview video coding has been already standardized as extensions to the MPEG-2 standard [73], the AVC standard [74], and the HEVC standard [75]. The multiview extension of AVC is denoted as MVC (Multiview Video Coding) and

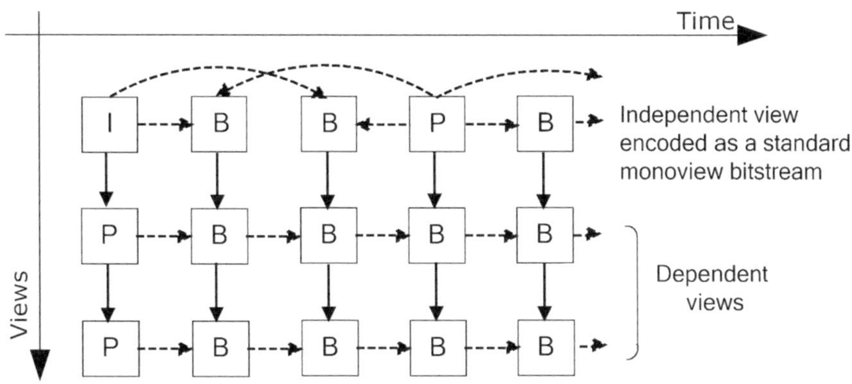

Fig. 2.6 Typical frame structure in multiview video coding using inter-view prediction with disparity compensation: solid line arrows denote interframe predictions while dashed line arrows correspond to temporal predictions. The letters I, P, and B denote I-frames (intraframe coded), P-frames (compressed using intra- and temporal interframe coding) and B-frames (compressed using two reference frames)

that of HEVC as MV-HEVC (Multiview HEVC). These multiview extensions have been standardized in such a way that low-level coding tools may are virtually the same as for monoscopic video coding. Therefore, some more advanced techniques for multiview coding are not included into the standards.

For the state-of-the-art multiview video coding technology is MV-HEVC [69].

The multiview coding provides the bitrate reduction of order 15–30%, sometimes reaching even 50% as compared to the simulcast coding. These high bitrate reductions are achievable for video that is obtained from cameras densely located on a line, and then rectified in order to virtually set all the optical axes parallel and on the same plane. For sparse and arbitrary camera locations, the gain with respect to the simulcast coding reduces significantly.

Recently [76], it was shown that the efficiency of the inter-view prediction is virtually the same for Multiview HEVC and for HEVC augmented by Intra Block Copy tool (originally designed for computer-generated content) using the same resolution of translation/displacement vectors. It is worth to add that the latter codec has simpler single-loop structure and is nearly compliant with standard HEVC Screen Content Codec. The result was obtained for rectified multiview video clips acquired using cameras with parallel optical axes, i.e., for the application scenario, for which Multiview HEVC was designed. This result put into question the need to develop multiview video codecs for future generations of video compression techniques.

2.3.4 3D Video Coding

Many 3D video coding tools have been already proposed including prediction based on: view synthesis, inter-view prediction by 3D mapping defined by depth, coding of disoccluded regions, advanced inpainting, special techniques for depth coding using platelets and wedgelets, etc. [77, 56, 78, 79, 33, 80]. Some of these tools have been already included into the standards of 3D video coding: 3D High Profile of AVC (AVC), [81] and 3D Main Profile of HEVC (HEVC), [75]. The latter defines the state-of-the-art technology for compression of 3D video with accompanying depth. This technology not only compresses the depth but also exploits the depth in order to improve coding performance of the multiview video.

For 3D-HEVC, the standardization requirement was to reuse the monoscopic decoding cores for implementations. MV-HEVC, 3D-HEVC, and the scalable extension of HEVC share nearly the same high-level syntax of the bitstreams, and the multi-loop structure of the encoders and decoders is the common architecture used in the implementations. Therefore, view encoding cannot depend on the corresponding depth. As compared to MV-HEVC, 3D-HEVC provides additional prediction types:

(1) Combined temporal and inter-view prediction of views that refers to pictures from another view and another time instant;

(2) View prediction that refers to a depth map corresponding to the previously encoded view;
(3) Prediction of depth maps using the respective view or a depth map corresponding to another view.

The compression gain of 3D-HEVC over MV-HEVC is expressed by 2–12% bitrate reduction. Nevertheless, the compression gains of both 3D-HEVC and MV-HEVC are smaller when cameras are not aligned on a line. For circular camera arrangements, in particular with the angles between the camera axes exceeding 10°, the gain over simulcast falls below 15%, often being around 5%. This observation stimulated research on the extensions of 3D-HEVC that uses true 3D mapping for more efficient inter-view prediction [82, 83]. Such extension of 3D-HEVC has been proposed in the context of transmission of the multiview plus depth representations of the dynamic scenes in the future free viewpoint television systems [22].

3D video coding is currently a research topic for several groups around the world, and also future standardization activities are expected. Recently, the MPEG-FTV, the body within MPEG, was exploring possible 3D-HEVC extensions for efficient coding of multiview video taken from arbitrary camera positions. Currently, this activity has been shifted to MPEG-I project. Hitherto practical deployment of 3D-HEVC is negligible but growing interests in the applications hitherto mentioned in this chapter will stimulate applications of this compression as well as, probably, standardization of its more efficient extensions. It is also expected that the coding tools of 3D-HEVC together with possible improvements will be included, probably with some delay, into the forthcoming video coding standard that is expected to be ready around 2020–2021 in its first version.

In general, depth maps are characterized by homogeneous regions separated by sharp edges at object boundaries. Despite the distinct characteristics, multiview video and depth maps represent the same scene. Therefore, video and depth map of a given view exhibit some correlation. The similarities between both streams can thus be exploited by the video coding methods. In such a scheme, a base view still needs to be encoded independently from other views and depth maps, allowing compatibility with legacy single-view displays. All the remaining views and depth maps will depend on this stream.

In the scope of the MVD coding, the ISO/IEC MPEG standardization process comes out with three solutions based on different coding technologies. MVC + D is proposed as a simple solution for sending views along with corresponding depth maps, using the multiview video coding (MVC) [84] algorithm. All changes are related to high-level syntax elements only providing a way to signal the presence of depth data [85]. Other two solutions incorporate specific tools for the independent compression of depth maps or for the joint compression of video and depth, based on advanced video coding (AVC) [68] and high-efficiency video coding (HEVC) [69] encoders. The first one is 3D-AVC algorithm, which is backward compatible with AVC and provides a fast and easy adoption of 3D video in the market. The other one is the current state-of-the-art solution for 3D video coding, known as 3D-HEVC [33, 69].

The different features of depth maps, associated to the fact that they are not displayed at the decoder, imply that compression of depth maps with the standard video encoder might not be optimal. In order to improve the coding efficiency of depth maps, and the quality of the synthesized views, it has been shown that preservation of depth map edges is very important. In this context, alternative methods based on different coding paradigms have been proposed outside of the scope of standardization groups. The platelet and wedgelet depth modeling, pattern-matching coding and linear-fitting modeling are some of the solutions suggested in literature [86–89].

Techniques to save even more bandwidth include the downsampling of the depth maps. These will require the upsampling of the decoded maps at the receiver side. In any case, preservation of the edges in the depth maps is very important for view synthesis. Thus, a joint video/depth edge-based upsampling method can be applied as in [90] to better define the edges in the depth map. This is possible because the edges in the depth map are also present in the video, which corresponds to the same scene and objects.

Compression efficiency can be improved by removing high-frequency components from both views depth maps. Each image may be divided into regions based on their depth values. Regions which are far away from the camera are low-pass filtered more coarsely than closer regions. This ensures that the removal of the detail does not severely degrade the quality of the image [91]. This method assumes that the viewer is more concerned with the foreground than with the background. Similarly, regions of image further from the camera can be quantized more than closer regions [70, 71]. Objects in the view and the depth map video streams move with similar direction and speed. This correlation can be exploited using a scalable video coding (SVC) architecture, where the base layer encodes the views and the enhancement layer carries the depth data. This idea is presented in [92] and [93] where an inter-view prediction scheme is coupled with an inter-layer motion prediction method. The inter-layer motion prediction is based on SVC. Currently, this approach is part of a 3D-HEVC standard where depth maps motion field can be predicted from corresponded motion field.

Although MVD requires the additional compression of depth information, it saves a high amount of bits by transmitting a reduced set of views. Furthermore, due to its characteristics, depth maps tend to result in a much smaller compressed bitstream when compared to the video. At the decoder side, a higher number of views can be generated using a synthesis algorithm. One of the most popular techniques is depth-image-based rendering (DIBR) in which the depth data and the view are used to generate the virtual image. This technique was selected by the motion picture experts group (MPEG) as the reference synthesis framework for free viewpoint video architectures, which relies on the multiview video-plus-depth format. In fact, the view synthesis reference software (VSRS) that was released by the ad hoc group on 3D audio and visual (3DAV) of MPEG is based on DIBR [94]. Although originally it was designed only for linear view arrangement, recently it was generalized to cope view general view arrangements as well [62].

2.3.5 New Standardization Projects

The international organizations work by their working groups of experts. For ISO and IEC there are two groups: JPEG (official name is ISO/IEC JTC1/SC29/WG1) and MPEG (official name is ISO/IEC JTC1/SC29/WG11). For ITU the relevant working group is VCEG. In order to create a new general video coding standard that will correspond to more modern compression technology, both ISO/IEC and ITU have created a joint group called Joint Video Exploration Team (JVET). This group is working towards a new video coding standard that will be related to the technology (recently called Versatile Video Coding) that halfes the bitrates of HEVC. Within ISO/IEC this standard will be a part of the forthcoming MPEG-I (from immersive) standard.

In 2017, the MPEG-I standardization project comprises also the extensive works on point cloud compression and lightfield video compression. The latter is also a work item for JPEG that has created already a working subgroup JPEG PLENO that is dealing with lighfield image compression. The lightfield images will be considered in the next two sections.

2.4 Lightfield Super-Multiview with Camera Array

In order to visualize a lightfield, it first needs to be captured. Regardless of the parametrization, the lightfield should ideally be captured on a sufficiently large plane, and with the smallest possible granularity both in the spatial and angular sense.

While it is possible to capture lightfield using a single sensor (as described in the next section), the physical baseline (distance between the leftmost and rightmost captured position) is limited by the physical size of the camera. This means that the viewing angle (Field of View) of the captured imagery can only be relatively small, unless the camera is capturing an object from close up. See Fig. 2.10. If we are about to capture larger scenes with a large field of view, we either have to use very big cameras (which do not exist in practice), or a camera array spanning the necessary baseline.

It is important to note that while the ultimate goal is to capture a (near) continuous lightfield, camera arrays can only capture a lightfield with a specific granularity due to the gap between adjacent cameras. That is, all these camera arrays are sampling the lightfield at regular intervals, which needs to be taken into account when working with the captured data.

The layout of camera arrays can be quite different depending on the scene and capture requirements. Some special cases include linear, converging linear, and arc setups. Camera arrays can also be 1D or 2D arrays. In a linear array, cameras are positioned next to each other, with equal distance between cameras, their optical axis is parallel, and perpendicular to the line on which cameras are arranged. In a converging linear array, the position of the cameras is similar, but they are rotated, so that their optical axis points towards a common point of convergence. In case of an arc/circular camera array, cameras are positioned on a circular path, all pointing to a point of convergence in the center. In case of a 1D array, the cameras are arranged in a single row (or column, but that's a quite unusual setting), while in case of a 2D array, cameras are arranged both horizontally and vertically. The advantage of regular camera arrays is that the rough position and orientation of cameras is known, which is later refined by a camera calibration process. Apart from this, an unstructured array of cameras that capture the same scene from different angles can be considered a camera array, and can be used for lightfield capture, however, the density of the captured data may vary over the field of view.

There are many examples of camera arrays in both research and industrial settings, used for a variety of purposes. A quite well known and early camera array is the Stanford Multi-Camera Array [95], consisting of 128 video cameras. These cameras can be arranged in various layouts, such as a linear array of parallel cameras having horizontal and vertical parallax, or a converging array of cameras. This large rig has been used for capturing lightfields for research purposes, for example for lightfield rendering, synthetic aperture imaging. Numerous other multi-camera rigs are known, such as the 100 camera array at Nagoya University [96], the 27-camera array at Holografika [97] or the recent horizontal and vertical parallax 16-camera system from Fraunhofer IIS [98]. These camera systems provide sufficient input to 3D lightfield displays, as the density (in terms of angular resolution) and width (in terms of baseline) of the captured lightfield allows for wide-angle visualization.

The main design constraint of camera arrays is the physical size of cameras and lenses, which pose an upper limit on how dense the arrangement of such cameras can be. For this reason, typically small camera modules are preferred, while some designers even use board level cameras to achieve an even narrower size per camera.

Using cameras with the possibility of triggering ensures that the frames captured by the individual units represent the same time instant, which is important to ensure consistency between images when capturing a moving scene.

Static scenes can, of course, be captured without strict synchronization. Going further, as a special case of a camera "array" one can use a moving camera to capture a static scene [99], or a static camera with a rotating object [100] to obtain a lightfield. These approaches work properly as long as the static scene indeed remains static during the capture session (for example, no people walking by, no changes in illumination due to different positions of the sun), and that camera positioning is precise enough to assume that no further camera adjustments are necessary.

Calibration of the individual cameras (resulting in intrinsic camera parameters) is just as important as with single camera capture, however, in the case of many-camera arrays, the relative position of cameras (resulting in extrinsic camera parameters) is just as important. Cameras pairs are typically calibrated by using stereo calibration techniques [42] (which can be performed for multiple camera pairs if they can see the same calibration patterns/features), and finding globally consistent extrinsic multi-camera parameters by using an optimization algorithm on the camera parameters [101].

Any kind of regular 2D cameras can be used to build a camera array. For video capture, typically machine vision cameras with trigger capabilities, or professional video cameras are used. DSLRs have also been used for capturing both static and animated lightfields. Camera arrays built of GoPro cameras have also been used, however, these cameras do not allow for real-time streaming of the captured video over a cable connection—in such cases the recoded lightfield needs to be downloaded after the capture session. In real-time lightfield capture settings however, the bandwidth required for transfer and store the resulting video data can be a concern.

Researchers not in the possession of a camera array wishing to do research on lightfields can do so using the many available public datasets. A good collection of these can be found in the MPEG-FTV Call for Evidence [102], which lists selected super-multiview and free viewpoint television content to be used for experimentation (Figs. 2.7, 2.8, 2.9).

Fig. 2.7 100 camera array of Nagoya University. (*Source* Masayuki Tanimoto)

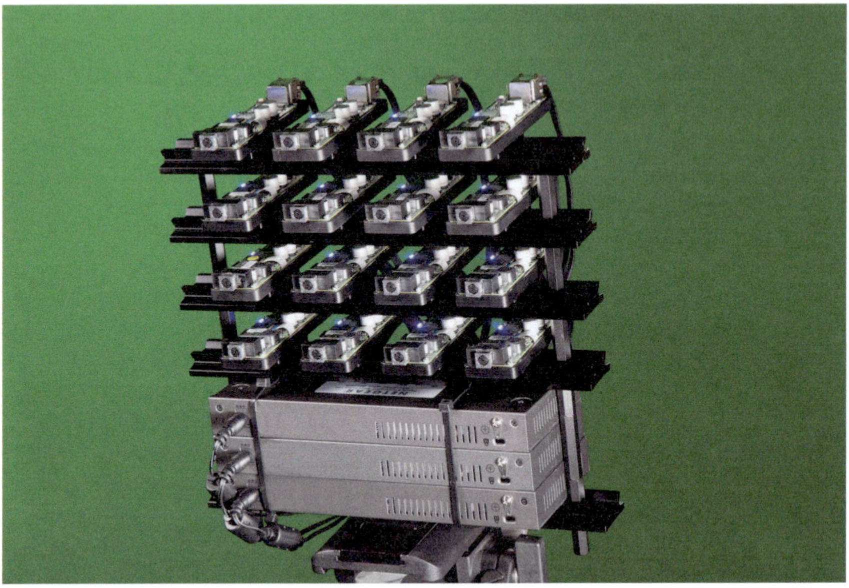

Fig. 2.8 16-camera full parallax camera array of Fraunhofer IIS. Copyright: Kurt Fuchs| Fraunhofer Institute for Integrated Circuits IIS

Fig. 2.9 36-camera matrix at Poznan University of Technology, Multimedia laboratory

2.5 Lightfield with Microlens Array

Lightfield with microlens array—also known as holoscopic [1] plenoptic [103] and integral imaging [104]—derives from the fundamentals of lightfield/radiance sampling [105], where not only the spatial information about the 3D scene is represented but also angular viewing direction, i.e., the "whole observable" scene.

The concepts behind this lightfield imaging technology were firstly proposed by G. M. Lippmann and referred to as integral photography in 1908 [106]. The conventional lightfield imaging system comprises a main lens and a regularly spaced array of microlenses, known as a "fly's eye" lens array [1] which is overlaid with the image sensor at the focal distance, *f*, as seen in Fig. 2.10. Therefore, different from a conventional camera that captures an image by integrating the intensities of all rays (from all directions) impinging each sensor element, in a lightfield camera each sensor element collects the light of a single ray (or of a thin bundle of rays as depicted in Fig. 2.10) that converges on the microlens from a given angular direction.

The traditional lightfield camera can be generalized to alternative camera setups, such as the one proposed in [103] and referred to as focused setup camera. In the focused camera, the main lens and the microlenses are focused in an image plane in front (or behind) of the microlens array plane. As a result, the main lens forms a relay system with each microlens. In practice, these differences in the optical geometry will only change the trade-off between providing maximal angular or spatial resolution in the captured lightfield image [107].

Among the advantages of employing a lightfield imaging system with microlens array is the ability to open new degrees of freedom in terms of content production and manipulation, supporting functionalities not straightforwardly available in conventional imaging systems, namely: postproduction refocusing, changing depth of field, and changing viewing perspective. Moreover, the interaction functionalities can also be enriched, for instance, by allowing the user to vary the plane of focus

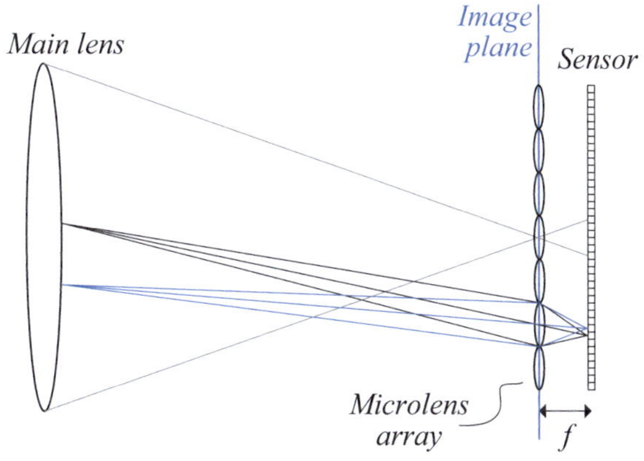

Fig. 2.10 Basic optical setup of the traditional lightfield camera comprising a main lens, a microlens array, and an image sensor

and depth of field interactively. In addition to this, it is still possible to derive from this type of content geometric information, such as depth/disparity and ray-space [19] representations.

Recently, lightfield imaging with microlens array has become a promising approach for 3D imaging and sensing, being applied in many different areas of research, e.g., 3D television, [1, 108] image recognition and medical imaging [104]. For this reason, novel initiatives on image and video coding standardization have also considered lightfield application scenarios. Notably, the JPEG working group started recently a new study activity—known as JPEG Pleno [109]—targeting richer image capturing, visualization, and manipulation. In addition, the MPEG group started the third phase of free viewpoint television (FTV), in August 2013, targeting SMV, free navigation and full parallax imaging applications [110].

However, introducing lightfield image and video applications with its appealing functionalities will require to identify the requirements and challenges in this type of systems, as well as to understand the users' needs in terms of lightfield content interaction. Regarding the challenges, to provide a lightfield representation with convenient spatial resolution and viewing angles, a huge amount of data is required and thus efficient coding is of utmost importance. In addition to this, as the imaging technology moves toward richer representations, novel data representations are essential to support the new applications and functionalities that arise [109]. In this sense, a scalable coding architecture is desirable to support a very flexible scaling of the lightfield content with a diverse range of consumption environments and devices. Moreover, this makes it possible to accommodate in a single compressed bitstream a variety of sub-bitstreams appropriate for users with different preferences and various application scenarios: from the user who wants to have a simple 2D version of the lightfield content without actively interacting with it; to the user who wants full immersive and interactive 3D lightfield visualization. Additionally, providing supplementary data—such as disparity/depth, ray-space, and 3D model—to be incorporated into the scalable bitstream is also important to support lightfield applications that are adaptable to various display interfaces, e.g., stereo, multiview, super-multiview, and also lightfield displays. Hence, it would facilitate the support for displays with different sizes, and with larger number of viewpoints and angular resolutions. Another important requirement is to provide backward compatibility with the current state-of-the-art in image and video coding technologies so as to support interoperability with the widely used 2D and 3D representation formats [109].

Towards the goal of identifying more powerful lightfield representation and coding solutions, several image and video coding schemes have been recently proposed in the literature for the lightfield with microlens array case and try to take advantage of its characteristic planar intensity distribution to achieve more efficient compression. Notably, as a result of the used optical system, the lightfield raw image corresponds to a 2D array of microimages (MIs), also known as elemental images, where both light intensity and direction information are recorded, as illustrated in Fig. 2.11a. Due to the small baseline between adjacent microlenses

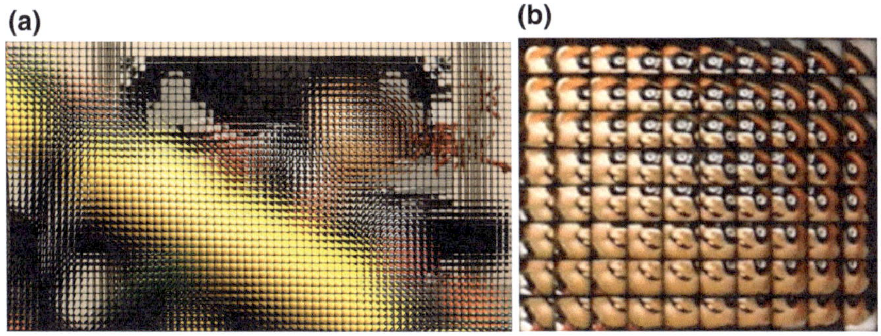

Fig. 2.11 Lightfield image captured with a focused setup camera using a 250 μm pitch microlens array **a** Full image with resolution of 1920 × 1088; **b** Enlargement of 280 × 224 pixels showing the array of microimages

used in the lightfield acquisition process, a significant cross-correlation exists between neighboring MIs (see Fig. 2.11b).

In terms of the possible different ways to organize the lightfield data for coding and transmission, the following three main approaches can be identified.

2.5.1 Lightfield Raw Data-Based Approach

This category corresponds to cases in which encoding and transmission of the lightfield image are done in its entirety, represented as a 2D grid of MIs. For this, a special lightfield prediction scheme is introduced in a state-of-the-art 2D codec to exploit the nonlocal spatial redundancy between different MIs for improving the coding efficiency. Figure 2.12 illustrates a basic coding diagram based on the high-efficiency video coding standard (HEVC) [111] for introducing a lightfield prediction scheme.

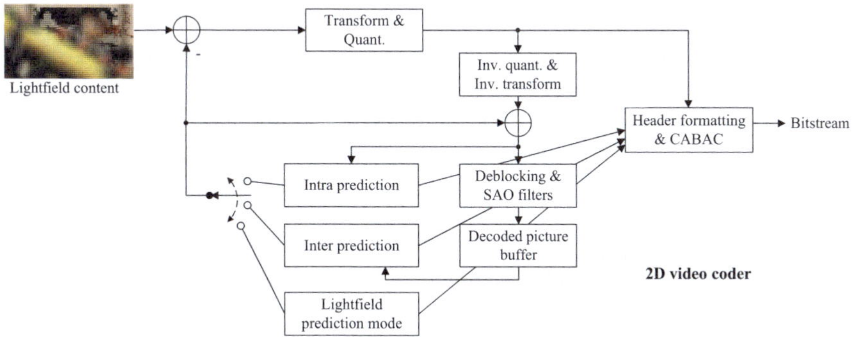

Fig. 2.12 Basic diagram for a lightfield raw data-based coding approach based on HEVC

Following this approach, a scheme for displacement intra prediction, referred to as self-similarity (SS) estimation and compensation, was proposed in [112] to improve the performance of the H.264/AVC standard for lightfield image coding. Later, in [113, 114], the authors proposed to introduce the SS prediction into the HEVC standard for image and video coding so as to take advantage of the flexible partition patterns used in this type of video codecs. In [115], the authors investigate alternative nonlocal spatial prediction, and also propose to include a prediction framework based on locally linear embedding into HEVC for lightfield image coding. More recently, in [116], a displacement intra prediction with multiple hypothesis method is proposed for both lightfield image and video content. Please refer to the Chap. 6 for more details on this multiple hypothesis lightfield coding method.

The advantage of these coding schemes is that they explore the particular correlation of the lightfield content without requiring any explicit knowledge about the used optical system (e.g., microlens' size, focal length, and distance of the microlenses to the image sensor). Although these parameters may be provided by camera makers, many of them are highly dependent on the manufacturing process, being different from camera to camera. For instance, the fabrication process results in microlenses that may vary slightly in shape, size, and relative position, needing a very careful and complex calibration process in the lightfield camera. For this reason, using compression and rendering tools that are less dependent on these calibration processes would be advantageous for supporting a vaster selection of devices without increasing the complexity.

On the other hand, although these coding schemes achieve significant compression gains when compared to the existing state-of-the-art alternatives, transmitting the entire lightfield data without a scalable bitstream may represent a serious problem since the user needs to wait until the entire content of each picture arrives before it can be visualized, independently of the used type of display and level of interaction the user may want to do with it.

2.5.2 *Multiview-Based Approach*

Some coding schemes propose to decompose the lightfield data into several viewpoint sequences to be represented as a multiview video [117–119] which is then coded with a standard multiview video coder, as illustrated in Fig. 2.13. A viewpoint image (a.k.a. sub-image) represents an orthographic projection of the complete captured scene in a particular direction, and can be constructed by simply extracting one pixel with the same relative position from each MI. In [117–119], a coding approach based on the multiview video coding (MVC) [74] extension of H.264 standard is proposed to jointly exploit temporal motion and disparity between adjacent viewpoint images. Therefore, the sequence of each viewpoint is encoded using MVC by defining different scanning orders and coding configurations.

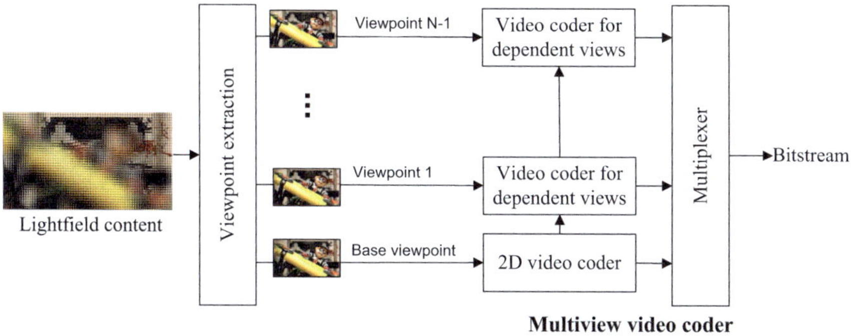

Fig. 2.13 Basic architecture for a lightfield coding scheme based on a multiview video codec

Although scalability and backward compatibility are guaranteed by using a standard multiview video codec, a drawback of these coding schemes is that they usually consider computer-generated sequences with a small number of viewpoint images (up to 9), while this number is typically much higher for natural lightfield content (usually, more than 50). Consequently, these coding schemes become more complex and with a larger amount of header information, when applied to natural content.

Since rendering viewpoint images usually produce very low-resolution images with aliasing [120], an alternative to the multiview representation based on these viewpoint images is presented in [121], as shown in Fig. 2.14. In this case, the lightfield content is decimated into 2D views with larger resolution than viewpoint images by using the rendering algorithms proposed in [103]. Hence, a scalable coding solution is proposed to support backward compatibility with 2D representation (base layer) and also with the current stereo and multiview representation (in one or more enhancement layers). Finally, the top enhancement layer supports the

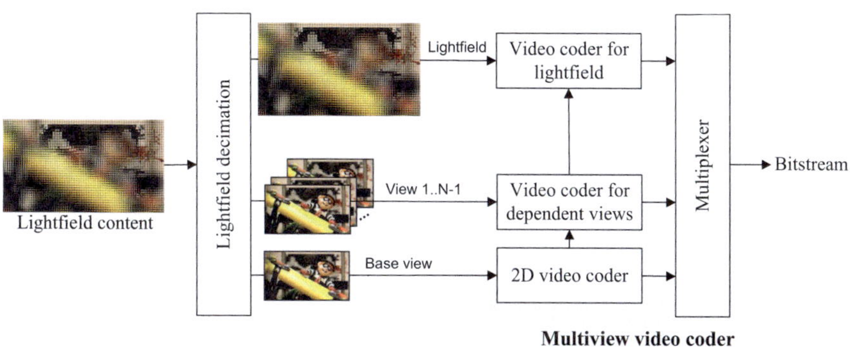

Fig. 2.14 Basic scalable lightfield coding architecture for backward compatibility with 2D, stereo, and multiview representation

entire lightfield content. For more details about this scalable coding approach, please refer to the Chap. 6.

This scalable coding architecture is able to support a diverse range of consumption environments and devices. On the other hand, the end-user still needs to receive the entire lightfield bitstream to have a viewing experience with the novel and appealing interaction functionalities supported by this type of content (such as changing focus and depth of field).

2.5.3 *Subsampled Grid of MIs Plus Disparity Approach*

Other coding schemes propose to represent the lightfield data by a subsampled set of MIs with their associated disparity information [122–125]. As first proposed in [122], the grid of MIs is subsampled to remove the redundancy between neighboring MIs and to achieve compression. Thus, only the remainder subsampled set of MIs and associated disparity data are encoded and transmitted, as depicted in Fig. 2.15a. At the decoder side, the lightfield data is reconstructed by simply applying a disparity shift (in [123, 125]) or by using a depth-image-based rendering

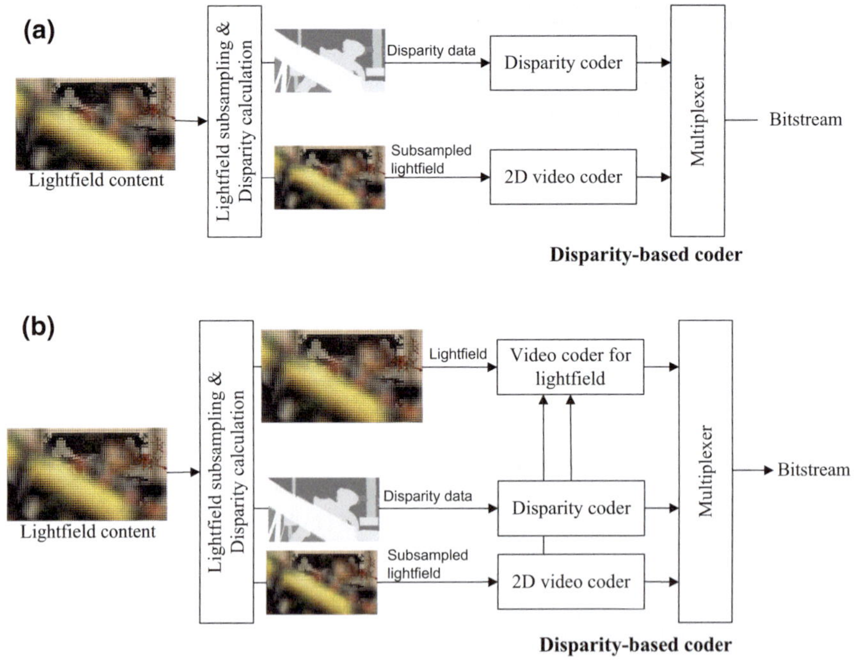

Fig. 2.15 Basic lightfield coding architectures for subsampled grid of MIs plus disparity approach

(DIBR) algorithm modified to support the multiple MIs as input views in [124], and followed by an inpainting algorithm to fill in the missing areas.

However, in real-world images, the disparity/depth information is estimated from the acquired lightfield raw data, which introduces inaccuracies. Hence, the quality of the reconstructed MIs—and, consequently, the quality of rendered views—is severely affected by these inaccuracies at the encoder side. Additionally, due to occlusion problems and quantization errors when (lossy) encoding this disparity/depth maps, some synthesized MIs might present too many missing areas to be filled, thus introducing even further inaccuracies. The reconstruction artifacts are even more challenging for MI synthesis because of the small angle-of-view (which is intrinsically limited by the pitch of the microlenses).

For this reason, instead of uniformly selecting the MIs, the selection is performed adaptively in [123, 124], so as to obtain better view reconstruction. For this, extra MIs are selected by identifying possible hole-causing regions, increasing considerably the bits consumption. In [125], the entire lightfield image is also encoded and transmitted in an enhancement layer, as shown in Fig. 2.15b, so as to provide better rendering views. More details about this coding approach can be seen in the Chap. 6.

The main advantage of incorporating the disparity information into the bitstream is that it facilitates the support of a larger variety of displays and larger levels of user's interaction. However, a common characteristic of the aforementioned approaches is that the quality of rendered views is negatively affected by the inaccuracies in the synthesis of the missing MIs.

2.6 Free Navigation and Free Viewpoint Television

Free viewpoint television (FTV) is an interactive video service that provides the ability for a viewer to navigate freely around a scene [19]. Such service is also simply called virtual navigation or free virtual navigation. A viewer watches the scene from an arbitrary direction and from virtual viewpoints on an arbitrary navigation trajectory. At each virtual viewpoint, the corresponding view has to be synthesized and made available at the receiver. At the same time instant, possibly many viewers share the same FTV service, and each viewer navigates independently. View synthesis may use either the distributed model where views are synthesized independently in each receiver, or the centralized model where views requested by all viewers are synthesized in the servers of the service provider. The distributed model requires high transmission bandwidth in the server-to-viewer downlinks and significant processing power of viewer terminals. On the other hand, the centralized model suffers from delays in the bidirectional server-to-terminal communications, similarly to networked gaming. Therefore, both models are considered for future applications.

An FTV system requires efficient techniques for multi-camera system calibration and video correction, depth estimation, and view synthesis as pointed out in

Fig. 2.16 Tripods with wireless camera modules designed and produced at Poznań University of Technology © IEEE 2016

previous sections. In a practical FTV system, the number of cameras should be limited, and therefore the distances between cameras are large. The cameras are located around a scene, in a roughly circular camera setup (see Fig. 2.16).

Recently, the generic structure of FTV systems has been proposed as shown in Fig. 2.17. Throughout this paper, we are going to use this structure that consists of the following functional blocks:

- The video and audio acquisition system,
- The representation server that produces a visual representation of the spatial dynamic scene,
- The rendering servers that serve the requests for the synthesis of video and audio at particular virtual locations around a scene,

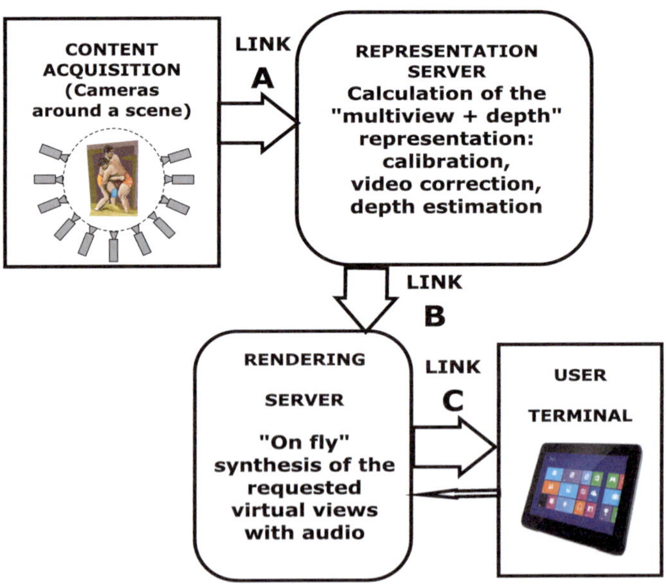

Fig. 2.17 The general structure of an FTV system—from [20, 21] © IEEE 2016

- The user terminal.

The video and audio acquisition system has to provide data necessary to compute the spatial representation of a scene. Except video and audio, the data include also some depth information obtained either from pure multiview video analysis or also from depth sensors. The depth acquisition using the depth sensors is conceptually very attractive, but its practical application still faces severe problems related to limited resolutions of the acquired depth maps, limited distance ranges, additional infrared illumination of the scene, synchronization of the video and depth cameras, and sensitivity to the environmental factors including solar illumination. In particular, in this paper, we focus on the multiview recording of real events where additional infrared illumination might be unacceptable. Therefore, the considerations in this paper base on the assumption that the depth information is obtained by the video analysis only, and the special depth sensors are not used.

The video and audio data together with the system calibration data are transmitted via Link A that belongs to the contribution environment, thus needs the high–fidelity compression. As the video data in Link A are yet neither calibrated nor corrected, for video, standard single-view compression techniques may be used, including both intraframe techniques like M-JPEG 2000 or HEVC All Intra, or interframe studio profiles of AVC or HEVC. Note that simple FTV systems will probably rarely use nonlinear edition as the FTV material does not need any choice of the camera during the production process. The FTV video material does not need camera changes and zooming, as that is done individually by a viewer. If the nonlinear edition is not needed, there is also no need for the random frame access and no need for small error accumulation in the multiple encoding–decoding cycles. Therefore, the requirement to use the intraframe coding may be released, and the standard interframe compression techniques may be used for video. In that way, the requested bitrate may be significantly reduced but still the total bitrate will be determined by simulcasting the video streams from multiple cameras plus the audio streams from many microphones.

Acknowledgements This book chapter was partially supported by COST Action IC1105—3D-ConTourNet.

The book chapter was partially supported by National Science Centre, Poland according to the decision DEC-2012/05/B/ST7/01279.

References

1. Aggoun, A., Tsekleves, E., Swash, M.R., Zarpalas, D., Dimou, A., Daras, P., et al.: Immersive 3D Holoscopic Video System. IEEE Multimed. **20**, 28–37 (2013)
2. Isgro, F., Trucco, E., Kauff, P., Schreer, O.: Three-dimensional image processing in the future of immersive media. IEEE Trans Circuits Syst. Video Techn. **14**, 288–303 (2004)
3. Domański, M., Stankiewicz, O., Wegner, K., Grajek, T.: Immersive visual media—MPEG-I: 360 video, virtual navigation and beyond. In: International Conference on Systems, Signal and Image Processing, Poznań, May 2017

4. Benesty, J., Chen, J., Huang, Y.: Microphone array signal processing. Springer-Verlag, Berlin (2008)
5. Ziegler, M., Zilly, F., Schaefer, P., Keinert, J., Schöberl, M., Foessel, S.,Dense lightfield reconstruction from multi aperture cameras. In: 2014 IEEE International Conference on Image Processing (ICIP), Paris 2014, pp. 1937–1941
6. Herre, J., Hilpert, J.: A. Kuntz,J. Plogsties, MPEG-H 3D Audio—The new standard for coding of immersive spatial audio. IEEE J Select Topics Signal Proces **9**, 770–779 (2015)
7. Blauert, J. (ed.): Technology of binaural listening. Springer-Verlag, Berlin/Heidelberg (2013)
8. ISO/IEC IS 23003-1: 2007, "MPEG audio technologies—Part 1: MPEG Surround"
9. "Spatial Audio Object Coding (SAOC)", ISO/IEC IS 23003-2: 2016, 2nd Ed
10. "3D audio", ISO/IEC International Standard 23008-3 (2015)
11. "Unified Speech and Audio Coding (USAC)", ISO/IEC IS 23003-2: 2016 (2nd Ed.)
12. Ishida, T., Shibata, Y.: Proposal of tele-immersion system by the fusion of virtual space and real space. In: 2010 13th International Conference on Network-Based Information Systems (NBiS), Takayama, Gifu, Japan (2010)
13. EBU Technical Report TR 039, "Opportunities and challenges for public service media in vr, ar and mr", Geneva, April 2017
14. https://en.wikipedia.org/wiki/360-degree_video, as Octiber 28th, 2017
15. "Omnidirectional Media Format", ISO/IEC DIS 23090-2, Doc. ISO/IEC JTC1/SC29/WG11 N16824 April 2017, Hobart, Australia
16. Requirements for Omnidirectional Media Format. ISO/IEC JTC1/SC29/WG11 Doc. N 16773, April 2017, Hobart, Australia
17. Call for Proposals for Point Cloud Coding V2. ISO/IEC/JTC1/SC29/WG11, Doc. N16763, April 2017, Hobart, Australia
18. Lafruit, G., Domański, M., Wegner, K., Grajek, T., Senoh, T., Jung, J., Kovács, P., Goorts, P., Jorissen, L., Munteanu, A., Ceulemans, B., Carballeira, P., García, S., Tanimoto, M.: "New visual coding exploration in MPEG: Super-MultiView and Free Navigation in Free viewpoint TV", in IST Electronic Imaging, pp. 1–9. Stereoscopic Displays and Applications XXVII, San Francisco (2016)
19. Tanimoto, M., Panahpour, M., Fujii, T., Yendo, T.: FTV for 3-D spatial communication. Proc. IEEE **100**(4), 905–917 (2012)
20. Domański, M., Bartkowiak, M., Dziembowski, A., Grajek, T., Grzelka, A., Łuczak, A., Mieloch, D., Samelak, J., Stankiewicz, O., Stankowski, J.: Krzysztof Wegner. New results in free-viewpoint television systems for horizontal virtual navigation. In: 2016 IEEE International Conference on Multimedia and Expo (ICME), Seattle, WA, 2016, pp. 1–6
21. Domański, M., Dziembowski, A., Grzelka, A., Mieloch, D.: Optimization of camera positions for free-navigation applications, Int Con Signals Elect Syst. ICSES 2016, Kraków, Poland, September 5–7 2016
22. Domański, M.: Approximate video bitrate estimation for television services. ISO/IEC JTC1/SC29/WG11 Doc. MPEG M3671, Warsaw, June 2015
23. Domański, M., Dziembowski, A., Grajek, T., Grzelka, A., Kowalski, L., Kurc, M., Łuczak, A., Mieloch, D., Ratajczak, R., Samelak, J., Stankiewicz, O., Stankowski, J.: Krzysztof Wegner. Methods of high efficiency compression for transmission of spatial representation of motion scenes. In: IEEE International Conference on Multimedia and Expo Workshops, Torino (2015)
24. Miller, G., Starck, J., Hilton, A.: Projective surface refinement for free-viewpoint video. 3rd European Conf, pp. 153–162. CVMP, Visual Media Production (2006)
25. Smolic, A., et al.: 3D video objects for interactive applications. European Signal Proc. Conf, EUSIPCO (2005)
26. Tanimoto, M.: Overview of free viewpoint television. Signal Proc. Image Communic. **21**, 454–461 (2006)

27. Wei, K.-Ch., Huang, Y.-L., Chien, S.-Y.: Point-based model construction for free-viewpoint tv. In: IEEE International Conference on Consumer Electronics ICCE 2013, Berlin, pp. 220–221
28. Müller, K., Merkle, P., Wiegand, T.: 3D Video Representation Using Depth Maps. Proc. IEEE **99**(4), 643–656 (2011)
29. "3D world largest 200-inch autostereoscopic display at Grand Front Osaka", published: 28 April 2013, https://wn.com/3d_world_largest_200-inch_autostereoscopic_displayat_grand_front_osaka
30. Holografik.:, "HoloVizio C80 3D cinema system", Budapest, http://www.holografika.com/Products/NEW-HoloVizio-C80.html, retrieved on April 21, 2017
31. NICT News, Special Issue on Stereoscopic Images, no. 419, November 2011
32. Adelson, E.H., Bergen, J.R., Landy, M., Movshon, J.A. (eds.): The plenoptic function and the elements of early vision. In Computational Models of Visual Processing, pp. 3–20. MIT Press, Cambridge, U.K. (1991)
33. Müller, K., Schwarz, H., Marpe, D., Bartnik, C., Bosse, S., Brust, H., Hinz, T., Lakshman, H., Merkle, P., Rhee F.H., Gerhard, T., Winken, M., Wiegand, T.: 3D High-Efficiency Video Coding for Multi-View Video and Depth Data. IEEE Trans. Image Proces. **22**(9), 3366–3378 (2013)
34. Domański, M., Grajek, T., Klimaszewski, K., Kurc, M., Stankiewicz, O., Stankowski, J., Wegner, K.: Poznan multiview video test sequences and camera parameters ISO/IEC JTC1/SC29/WG11 Doc. MPEG M17050, Xian, China, October (2009)
35. Stamos., Allen, P.K.: Integration of range and image sensing for photo-realistic 3D modeling. In: Proceedings 2000 ICRA. Millennium Conference. IEEE International Conference on Robotics and Automation. Symposia Proceedings, San Francisco, CA, 2000, pp. 1435–1440 vol. 2
36. Sandberg, D., Forssen P.E., Ogniewski, J.: Model-based video coding using colour and depth cameras. In: 2011 International Conference on Digital Image Computing: Techniques and Applications, Noosa, QLD, 2011, pp. 158–163
37. Gokturk, S., Yalcin, H., Bamji, C.: A time-of-flight depth sensor—system description, issues and solutions. In: Proceedings of the IEEE Conference on Computer Vision and Pattern Recognition Workshop, Jun. 2004
38. Kang, Y.S., Ho, Y.S.: High-quality multi-view depth generation using multiple color and depth cameras. IEEE Int Conf Multi Expo **2010**, 1405–1410 (2010)
39. Sen, X., Li, Y., Qiong, L., Zixiang, X., A gradient-based approach for interference cancelation in systems with multiple Kinect cameras. In: 2013 IEEE International Symposium on Circuits and Systems, pp. 13–16 (2013)
40. Wang, Q.: Computational models for multiview dense depth maps of dynamic scene. In: 2015 IEEE Computer Society Conference on Computer Vision and Pattern Recognition (2015)
41. Hartley, R., Zisserman, A.: Multiple view geometry in computer vision, 2nd edn. Cambridge Univ Press, (2015)
42. Zhang, Z.: A Flexible New Technique for Camera Calibration. IEEE Trans. Pattern Anal. Mach. Intell. **22**(11), 1330–1334 (2000)
43. Atzpadin, N., Kauff, P., Schreer, O.: Stereo analysis by hybrid recursive matching for real-time immersive video conferencing. Circ. Syst. Video Technol. IEEE Trans. **14**(3), 321–334 (2004)
44. Lee, S., Ho, Y.: View-consistent multiview depth estimation for three-dimensional video generation. In: 3DTV-Conference: The True Vision—Capture, Transmission and Display of 3D Video (3DTV-CON), pp. 1-4, June 2010
45. Min, D., Yea, S., Vetro, A.: Temporally consistent stereo matching using coherence function. 3DTV-Conference: The True Vision—Capture, Transmission and Display of 3D Video (3DTV-CON), pp. 1-4, June 2010
46. Bleyer, M., Gelautz, M.: Graph-based surface reconstruction from stereo pairs using image segmentation. Proc. SPIE Int. Soc. Optical Eng. **5665**, 288–299 (2005)

47. Hong, L., Chen, G.: Segment-based stereo matching using graph cuts. In: 2004 IEEE Computer Society Conference on Computer Vision and Pattern Recognition, pp. 74–81 (2004)
48. Zilly, F., Riechert, C., Muller, M., Eisert, P., Sikora, T., Kauff, P.: Real-time generation of multi-view video plus depth content using mixed narrow and wide baseline. J. Vis. Commun. Image R. **25**(4), 632–648 (2014)
49. Jorissen, L., Goorts, P., Rogmans, S., Lafruit, G., Bekaert, P.: Multi-camera epipolar plane image feature detection for robust view synthesis. In: 3DTV-Conference: The True Vision—Capture, Transmission and Display of 3D Video (3DTV-CON) (2015)
50. Sun, J., Zheng, N.N., Shum, H.Y.: Stereo Matching Using Belief Propagation. IEEE Trans. Pattern Analy. Machine Intell. **25**(7), 787–800 (2003)
51. Montserrat, T., Civit, J., Escoda, O., Landabaso, J.: Depth estimation based on multiview matching with depth/color segmentation and memory efficient belief propagation. In: IEEE International Conference on Image Processing, pp. 2329–2332 (2009)
52. Stankiewicz, O., Domański, M.: Krzysztof Wegner. Estimation of Temporally-Consistent Depth Maps from Video with Reduced Noise. In: 3DTV Conference: The True Vision-Capture, Transmission and Display of 3D Video, 3DTV-Con 2015, Lisbon, Portugal, 8-10 July 2015
53. Mieloch, D., Dziembowski, A., Grzelka, A., Stankiewicz, O., Domański, M.: Graph-based multiview depth estimation using segmentation. IEEE Int. Conf. Multimedia Expo ICME 2017, Hong Kong, 10–14 July 2017
54. Stankiewicz, O., Wegner, K., Tanimoto, M., Domański, M.: Enhanced Depth Estimation Reference Software (DERS) for Free-viewpoint Television. ISO/IEC JTC1/SC29/WG11 Doc. MPEG M31518, Geneva, 2013
55. Dziembowski, A., Grzelka, A., Mieloch, D., Stankiewicz, O.: Krzysztof Wegner. In: Domański, M (ed). Multiview Synthesis—improved view synthesis for virtual navigation, 32nd Picture Coding Symposium, PCS 2016, Nuremberg, Germany, December 4–7, 2016
56. Domański, M., Stankiewicz, O., Wegner, K., Kurc, M., Konieczny, J., Siast, J., Stankowski, J., Ratajczak, R., Grajek, T.: High Efficiency 3D Video Coding Using New Tools Based on View Synthesis. IEEE Trans. Image Process. **22**(9), 3517–3527 (2013)
57. Tanimoto, M., Tehrani, M., Fujii, T., Yendo, T.: Free-viewpoint TV—A Review of the Ultimate 3DTV and its Related Technologies. IEEE Signal Proces. Mag. pp. 67–76, January 2011
58. Do, L., Zinger, S., Morvan, Y., With, P.: Quality Improving Techniques in DIBR for Free-viewpoint Video. In: Proceedings of 3DTV Conference: The True Vision—Capture, Transmission and Display of 3D Video, May 2009
59. Mori, Y., Fukushima, N., Fujii, N., Tanimoto, M.: View Generation with 3D Warping using Depth Information for FTV. In: Proceedings of 3DTV Conference: The True Vision—Capture, Transmission and Display of 3D Video, May 2008
60. Oh, K., Yea, S., Vetro, A., Ho, Y.: Virtual View Synthesis Method and Self-Evaluation Metrics for Free Viewpoint Television and 3D Video. Int. J. Imaging Syst. Technol. **20**(4), 378–390 (2010)
61. Yang, X., Lui, J., Sun, J., Li, X., Liu, W., Gao, Y.: DIBR based View Synthesis for Free-viewpoint Television. In: Proceedings of 3DTV Conference: The True Vision—Capture, Transmission and Display of 3D Video, May 2011
62. Wegner, K., Stankiewicz, O., Tanimoto, M., Domanski, M.: Enhanced View Synthesis Reference Software (VSRS) for Free-viewpoint Television. ISO/IEC JTC1/SC29/WG11 MPEG2013/M31520 October 2013, Geneva, Switzerland
63. Zarb, T., Debono, C.: Depth-based Image Processing for 3D Video Rendering Applications. In: Proceedings of the 21st International Conference on Systems, Signals and Image Processing, pp. 215–218, May 2014
64. Tran, A., Harada, K.: View Synthesis with Depth Information based on Graph Cuts for FTV. In: Proceedings of the 19th Korea-Japan Joint Workshop on Frontiers of Computer Vision, pp. 289–294, February 2013

65. Xu, J., Yan, F., Cao, X.: Stereoacuity-guided Depth Image based Rendering. In: Proceedings of the IEEE International Conference on Multimedia and Expo, July 2014
66. Lei, J., Zhang, C., Fang, Y., Gu, Z., Ling, N., Hou, C.: Depth Sensation Enhancement for Multiple Virtual View Rendering. IEEE Trans. Multimedia **17**(4), 457–469 (2015)
67. "Generic coding of moving pictures and associated audio information: Video", ISO/IEC Int. Standard 13818-2: 2013 and ITU-T Rec. H.262 (V3.1),2012
68. "Advanced video coding", ISO/IEC International Standard 14496-10, 8th Ed., September 2014, and ITU-T Rec. H.264 (V12), 12th Ed., April 2017
69. "High Efficiency Video Coding", ISO/IEC IS 23008-2, 3rd Ed., October 2017, and ITU-T Rec. H.265, 4th Ed., December 2016
70. Domański, M., Grajek, T., Karwowski, D., Klimaszewski, K., Konieczny, J., Kurc, M., Łuczak, A., Ratajczak, R., Siast, J., Stankiewicz, O., Stankowski, J., Wegner, K.: New coding technology for 3D video with depth maps as proposed for standardization within MPEG. In: 19th International Conference on Systems, Signals and Image Processing, IWSSIP 2012, Vienna, Austria, 11–13 April 2012, pp. 401–404
71. Domański, M., Grajek, T., Karwowski, D., Konieczny, J., Kurc, M., Łuczak, A., Ratajczak, R., Siast, J., Stankowski, J.: Krzysztof Wegner. Coding of multiple video + depth using HEVC technology and reduced representations of side views and depth maps. 29th Picture Coding Symposium, PCS 2012, Kraków, May 2012, pp. 5–8
72. Domański, M., Dziembowski, A., Mieloch, D., Łuczak, A., Stankiewicz O., Wegner, K.: A practical approach to acquisition and processing of free viewpoint video. In: 2015 Picture Coding Symposium (PCS), Cairns, QLD, 2015, pp. 10–14
73. Haskell, B.G., Puri, A., Netravali, A.N.: Digital video: an introduction to MPEG-2. Chapman & Hall, New York (1996)
74. Vetro, A., Wiegand, T., Sullivan, G.J.: Overview of the stereo and multiview video coding extensions of the H.264/MPEG-4 AVC standard. Proc. IEEE **99**, 626–642 (2011)
75. Tech, G., Chen, Y., Ohm, K.M.J.-R., Vetro, A., Wang, Y.-K.: Overview of the multiview and 3D extensions of high efficiency video coding. IEEE Trans. Circ. Syst0 Video Technol. **26**(1), 35–49 (2016)
76. Samelak, J., Stankiewicz, O., Domański, M.: Do we need multiview profiles for future video coding generations?, Doc. ISO/IEC JTC1/SC29/WG11 M41499 October 2017, Macau, China
77. Chen, Y., Zhao, X., Zhang, L., Kang, J.-W.: Multiview and 3D Video Compression Using Neighboring Block Based Disparity Vector. IEEE Trans Multimedia **18**(4), 576–589 (2016)
78. Gao, Y., Cheung, G., Maugey, T., Frossard, P., Liang, J.: Encoder-driven inpainting strategy in multiview video compression. IEEE Trans. Image Process. **25**, 134–149 (2016)
79. Merkle, P., Bartnik, C., Müller, K., Marpe, D., Wiegand, T.: 3D video: Depth coding based on inter-component prediction of block partitions. 29th Picture Coding Symposium, PCS 2012, Kraków, May 2012, pp. 149–152
80. Shao, F., Lin, W., Jiang, G., Yu, M.: Low-Complexity Depth Coding by Depth Sensitivity Aware Rate-Distortion Optimization. IEEE Trans. Broadcast. **62**(1), 94–102 (2016)
81. Hannuksela, M.M., Rusanovskyy D., Su, W., Chen, L., Li, R., Aflaki, P., Lan. D., Joachimiak, M., Li, H., Gabbouj, M.: Multiview-Video-Plus-Depth Coding Based on the Advanced Video Coding Standard. IEEE Trans Image Proces **22**(9), 3449–3458 (2013)
82. Stankowski, J., Kowalski, L., Samelak, J., Domański, M., Grajek, T.: Krzysztof Wegner. 3D-HEVC Extension for Circular Camera Arrangements. In: 3DTV Conference: The True Vision-Capture, Transmission and Display of 3D Video, 3DTV-Con 2015, Lisbon, Portugal, 8-10 July 2015
83. Samelak, J., Stankowski, J., Domański, M.: Adaptation of the 3D-HEVC coding tools to arbitrary locations of cameras. Int. Conf. Signals Elect. Syst. Kraków (2016)
84. Chen, Y., Wang, Y., Ugur, K., Hannuksela, M., Lainema, J., Gabbouj, M.: The emerging MVC standard for 3D video services. EURASIP J. Adv. Signal Process. **2009**, 1–13 (2008)
85. Chen, Y., Hannuksela, M., Suzuki, T., Hattori, S.: Overview of the MVC + D 3D video coding standard. J. Visual Commun. Image Rep. (2013)

86. Wegner, K., Stankiewicz, O., Domański, M.: Fast View Synthesis using platelet-based depth representation. In: 21th International Conference on Systems, Signals and Image Processing, IWSSIP 2014, Dubrovnik, Croatia, May 2014
87. Merkle, P., Muller, K., Marpe, D., Wiegand, T.: Depth intra coding for 3d video based on geometric primitives. IEEE Trans Circuits Syst Video Technol (2015)
88. Graziosi, D., Rodrigues, N., Pagliari, C., Faria, S., Silva, E., Carvalho, M.: Compressing depth maps using multiscale recurrent pattern image coding. Electron. Lett. **46**(5), 340–341 (2010)
89. Lucas, L., Wegner, K., Rodrigues, N., Pagliari, C., Silva, E., Faria, S.: Intra Predictive Depth Map Coding using Flexible Block Partitioning. IEEE Trans. Image Process. **24**(11), 4055–4068 (2015)
90. Deng, H., Yu, L., Qui, J., Zhang, J: A Joint Texture/Depth Edge-Directed Up-Sampling Algorithm for Depth Map Coding. In: Proceedings of the IEEE International Conference on Multimedia and Expo, July 2012
91. Aflaki, P., Hannuksela, M., Homayouni, M., Gabbouj, M.: Joint depth and texture filtering targeting MVD compression. In: Proceedings of the 2014 IEEE Visual Communications and Image Processing Conference, pp. 410–413 (2014)
92. Zhang, J., Hannuksela, M., Li, H.: Joint Multiview Video Plus Depth Coding. In: Proceedings of the 2010 IEEE 17th International Conference on Image Processing, September 2010
93. Tao, S., Chen, Y., Hannuksela, M., Wang, Y., Gabbouj, M., Li, H.: Joint Texture and Depth Map Video Coding Based on the Scalable Extension of H.264/AVC. In: Proceedings of the IEEE International Symposium on Circuits and Systems, pp. 2353–2356, May 2009
94. "Report on Experimental Framework for 3D Video Coding," ISO/IEC JTC1/SC29/WG11, N11631, October 2010
95. Wilburn, B., Joshi, N., Vaish, V., Talvala, E.-V., Antunez, E., Barth, A., Adams, A., Horowitz, M., Levoy, M.: High performance imaging using large camera arrays. ACM Trans. Graphics **24**(3), 765–776 (2005)
96. Tanimoto, M., Fujii, T., Senoh, T., Aoki, T., Sugihara, Y.: Test Sequences with Different Camera Arrangements for Call for Proposals on Multiview Video Coding. ISO/IEC JTC1/SC29/WG11/M12338, Poznan (2005)
97. Balogh, T., Kovács, P.T.: Real-time 3D light field transmission. In: Proceedings on Real-Time Image and Video Processing, Proc. SPIE 7724, Brussels (2010)
98. Zilly, F., Schoberl, M., Ziegler, M., Keinert, J., Foessel, S.: Light-Field Acquisition System That Facilitates Camera and Depth-of-Field Compositing in Post-Production. SMPTE Motion Imaging Journal **124**(1), 16–21 (2015)
99. Kim, C., Zimmer, H., Pritch, Y., Sorkine-Hornung, A., Gross, M.: Scene reconstruction from high spatio-angular resolution light fields. ACM Trans. Graph **32**(4), art. 73 Jul 2013
100. Jones, A., McDowall, I., Yamada, H., Bolas, M., Debevec, P.: Rendering for an interactive 360° light field display. ACM Trans. Graphics **26**(3) art. 40 Jul 2007
101. Bo, L., Heng, L., Koser, K., Pollefeys, M.: A multiple-camera system calibration toolbox using a feature descriptor-based calibration pattern. In: Intelligent Robots and Systems (IROS), 2013 IEEE/RSJ International Conference on, pp. 1301–1307, 3-7 Nov. 2013. https://doi.org/10.1109/iros.2013.6696517
102. Call for Evidence on Free-Viewpoint Television: Super-Multiview and Free Navigation—update. ISO/IEC JTC1/SC29/WG11 Doc. N15733, October 2015, Geneva, Switzerland
103. Georgiev, T., Lumsdaine, A.: Focused plenoptic camera and rendering. J. Electron. Imaging *19*, 021106–021106 (2010). https://doi.org/10.1117/1.3442712
104. Xiao, X., Javidi, B., Martinez-Corral, M., Stern, A.: Advances in three-dimensional integral imaging: sensing, display, and applications [Invited]. Appl. Opt. **52**, 546–560 (2013)
105. Levoy, M., Hanrahan, P.: Light field rendering. In: Proc. 23rd Annu. Conf. Comput. Graph. Interact. Tech. - SIGGRAPH '96. New Orleans, LA, US, pp 31–42 (1996)
106. Lippmann, G.: Épreuves Réversibles Donnant la Sensation du Relief. J. Phys. Théorique. Appliquée **7**, 821–825 (1908)

107. Ng, R.: Fourier slice photography. ACM, New York, NY, USA, pp 735–744 (2005)
108. Arai, J.: Integral three-dimensional television (FTV Seminar). ISO/IEC JTC1/SC29/WG11 Doc. MPEG M34199, Sapporo, Japan (2014)
109. Ebrahimi, T.: JPEG PLENO Abstract and executive summary. ISO/IEC JTC 1/SC 29/WG1 Doc. JPEG N6922, Sydney, Australia (2015)
110. Tehrani, M.P., Shimizu, S., Lafruit, G. et al.: Use cases and requirements on free-viewpoint velevision (FTV). ISO/IEC JTC1/SC29/WG11 MPEG Doc. MPEG N14104, Geneva, Switzerland (2013)
111. Sullivan, G.J., Ohm, J.-R., Han, W.-J., Wiegand, T.: Overview of the high efficiency video coding (HEVC) standard. IEEE Trans. Circuits Syst. Video Technol. **22**, 1649–1668 (2012)
112. Conti, C., Lino, J., Nunes, P., et al.: Spatial prediction based on self-similarity compensation for 3D holoscopic image and video coding. Proc - Int. Conf. Image Process ICIP (2011). https://doi.org/10.1109/ICIP.2011.6116721
113. Conti, C., Nunes, P., Soares, L.D.: New HEVC prediction modes for 3D holoscopic video coding. In: 2012 19th IEEE Int. Conf. Image Process. Orlando, FL, US, pp 1325–1328 (2012)
114. Conti, C., Soares, L.D., Nunes, P.: HEVC-based 3D holoscopic video coding using self-similarity compensated prediction. Signal Process. Image Commun. (2016).https://doi.org/10.1016/j.image.2016.01.008
115. Lucas, L.F.R., Conti, C., Nunes, P., et al.: Locally linear embedding-based prediction for 3D holoscopic image coding using HEVC. In: 2014 Proc. 22nd Eur. Signal Process. Conf. Lisbon, Portugal, pp 11–15 (2014)
116. Li, Y., Sjostrom, M., Olsson, R., Jennehag, U.: Coding of focused plenoptic contents by displacement intra prediction. IEEE Trans. Circuits Syst. Video Technol. **26**, 1308–1319 (2016). https://doi.org/10.1109/TCSVT.2015.2450333
117. Adedoyin, S., Fernando, W.A.C., Aggoun, A., Kondoz, K.M.: Motion and disparity estimation with self adapted evolutionary strategy in 3D video coding. IEEE Trans. Consum. Electron. **53**, 1768–1775 (2007). https://doi.org/10.1109/TCE.2007.4429282
118. Dick, J., Almeida, H., Soares, L.D., Nunes, P.: 3D Holoscopic video coding using MVC. In: 2011 IEEE EUROCON - Int. Conf. Comput. as a Tool. Lisbon, Portugal, pp 1–4 (2011)
119. Shi, S., Gioia, P., Madec, G.: Efficient compression method for integral images using multi-view video coding. In: 2011 18th IEEE Int. Conf. Image Process. Brussels, Belgium, pp 137–140 (2011)
120. Bishop, T.E., Favaro, P.: Plenoptic depth estimation from multiple aliased views. In: 2009 IEEE 12th Int. Conf. Comput. Vis. Work. ICCV Work. Kyoto, Japan, pp 1622–1629 (2009)
121. Conti, C., Nunes, P., Soares, L.D.: Inter-layer prediction scheme for scalable 3-D holoscopic video coding. IEEE Signal Process. Lett. 20:819¬–822 (2013). https://doi.org/10.1109/LSP.2013.2267234
122. Piao, Y., Yan, X.: Sub-sampling elemental images for integral imaging compression. In: 2010 Int. Conf. Audio, Lang. Image Process. Shanghai, China, pp 1164–1168 (2010)
123. Choudhury, C., Chaudhuri, S.: Disparity based compression technique for focused plenoptic images. In: Proc. 2014 Indian Conf. Comput. Vis. Graph. Image Process. - ICVGIP '14. Bangalore, India, pp 1–6 (2014)
124. Graziosi, D.B., Alpaslan, Z.Y., El-Ghoroury, H.S.: Depth assisted compression of full parallax light fields. In: Proc. SPIE 9391, Stereosc. Displays Appl. XXVI. San Francisco, CA, US (2015)
125. Li, Y., Sjöström, M., Olsson, R.: Coding of plenoptic images by using a sparse set and disparities. In: 2015 IEEE Int. Conf. Multimed. Expo. IEEE, pp 1–6 (2015)

Chapter 3
3D Content Acquisition and Coding

Dragan Kukolj, Libor Bolecek, Ladislav Polak, Tomas Kratochvil, Ondrej Zach, Jan Kufa, Martin Slanina, Tomasz Grajek, Jarosłw Samelak, Marek Domański and Dragorad A. Milovanovic

Abstract This chapter starts by addressing the impact of the inaccurate camera system alignment on the spatial reconstruction accuracy and stereo perception. An experimental study is described, using a stereoscopic camera setup and its deterministic relations derived by trigonometry, spatial model, and basic stereoscopic formulas. The significance of errors that can occur for possible cameras system setup is analyzed in order to find the appropriate settings and physical constraints of the camera system, which minimize the error. Then, the chapter presents an overview of the compression tools used in current stereoscopic and multiview video encoders. It includes the stereoscopic frame-compatible formats using spatial multiplex in the side-by-side and top-and-bottom fashion; the video plus depth representation, the layered coding approach, and multiview encoding. Furthermore, an extension of multiview video compression for the arbitrary camera arrangements is presented. The current status of MPEG exploration experiments of next-generation video codec technologies is also outlined in the last section. First, the UltraHD compression performance beyond HEVC is presented and second, the

D. Kukolj (✉)
Faculty of Technical Sciences, University of Novi Sad, Novi Sad, Serbia
e-mail: dragan.kukolj@rt-rk.com

L. Bolecek · L. Polak · T. Kratochvil · O. Zach · J. Kufa · M. Slanina
Department of Radio Electronics, SIX Research Center, Brno University of Technology (BUT), Brno, Czech Republic
e-mail: bolecek@feec.vutbr.cz

L. Polak
e-mail: polakl@feec.vutbr.cz

T. Kratochvil
e-mail: kratot@feec.vutbr.cz

O. Zach
e-mail: ondrej.zach@phd.feec.vutbr.cz

J. Kufa
e-mail: xkufaj00@stud.feec.vutbr.cz

M. Slanina
e-mail: slaninam@feec.vutbr.cz

P. A. Assunção and A. Gotchev (eds.), *3D Visual Content Creation, Coding and Delivery*, Signals and Communication Technology,
https://doi.org/10.1007/978-3-319-77842-6_3

recent developments in HDR/WCG format conversion and coding are presented. Finally, the testing procedures and 3D projection formats for 360° video are addressed.

3.1 Introduction

The amount of acquired digital data is always increasing with novel 3D video formats. Consequently, more demanding video coding schemes are required. The evolution of video coding technologies is continuously pushing performance boundaries and capabilities of existing and new codecs. In this technological context, the following chapter presents an accuracy analysis of spatial reconstruction in stereo system design. Moreover, a comparison of 3D content coding techniques available today with preliminary simulation results, and benchmarking of the next-generation video codec is given.

The next section—3.2, prepared by Bolecek, Polak and Kratochvil, explores the impact of the inaccurate camera system alignment on the spatial reconstruction accuracy and stereo perception. A practical experiment with a stereoscopic camera setup was carried out, where setup's deterministic relations are derived by trigonometry, spatial coordinates, and basic stereoscopic formulas. Authors tried to reveal in the experiment how significant errors can occur for possible cameras system setup and to find the appropriate settings and placement of the camera system in order to minimize the error. In Sect. 3.3, written by Polak, Zach, Kufa, Slanina and Kratochvil, an overview of the compression tools for stereoscopic and multiview video available today is given. It includes a survey of novel coding techniques including the stereoscopic frame-compatible formats using spatial multiplex in the side-by-side and top-and-bottom fashion; the video plus depth representation, the layered coding approach and multiview encoding. Further, an extension of multiview video compression for the arbitrary camera arrangements is presented. It generates the derivation of the disparity vectors from depth data for sequences captured using cameras located on an arc. Extensive experiments on widely recognized multiview test sequences are described. Then, as the most important part of Sect. 3.3, there is an overview of currently available coding tools.

T. Grajek · J. Samelak · M. Domański
Chair of Multimedia Telecommunications and Microelectronics,
Poznań University of Technology, Poznań, Poland
e-mail: tgrajek@multimedia.edu.pl

M. Domański
e-mail: domanski@et.put.poznan.pl

D. A. Milovanovic
University of Belgrade, Beograd, Serbia
e-mail: dragoam@gmail.com

In Sect. 3.4, written by Grajek, Samelak and Domanski, efficient modifications of 3D-HEVC codec toward arbitrary view setup are presented, with particular emphasis on the arc view arrangement. The current status of MPEG exploration experiments of next-generation video codec technologies is outlined in the last section, which is prepared by Milovanovic and Kukolj. In its first part, the UltraHD compression performances beyond HEVC are presented. In the second part, the recent developments in HDR/WCG format conversion and coding are highlighted. In the last part of this section, the testing procedure and 3D projection formats for 360° video are evaluated.

3.2 Effect of an Incorrect Camera Alignment on the Accuracy of the Spatial Reconstruction and Stereo Perception

The importance of obtaining spatial models of the objects from two-dimensional photographs and scenes of the real world has been rising. For instance, digital photogrammetry and computer vision deal with creating of these models, which can be used in many areas of human activity. Photogrammetry allows reconstructing the objects without physical contact with them and analyzes their characteristics [1–3].

Among other factors, the 3D reconstruction of the object depends on the type of the used camera system. In general, two systems can be considered. In the first one, called as normal or stereo, mutual positions of the cameras differ only in their horizontal position. The angles of mutual rotation between the cameras equal to zero. In the second case, marked as generalized system, cameras have arbitrary positions and angles of their mutual rotation are nonzero. In the remaining part of this section, we will deal solely with a stereo system. The accuracy of 3D reconstruction can be influenced by many factors. Inaccuracy in the camera model, in exterior calibration and inaccuracy occurring at image processing are the best-known factors [2]. In [3], influences of the stereo base with various sizes and focal distance on the depth resolution are investigated. In this work, an accurate determination of the mentioned parameters was considered.

Methods to guarantee the precise alignment were proposed in [4, 5]. Results showed that the change of the stereo base has a higher impact on the overall accuracy than changes in the focal distance. Another important factor influencing achievable accuracy is the image discretization, to be more precise, the finite size of the pixel [6, 7]. Another fundamental aspect is an accurate determination of the corresponding points [8]. The depth of the object has a crucial impact on the accuracy of spatial coordinates, which decreases quadratically with an increasing depth. Hence, Kamencay et al. in [9] proposed a system with variable stereo base, which, independently on the depth, has a constant error. Furthermore, the accuracy in some specific issues is examined in [10, 76].

In this section, the influence of various errors (occurring in the camera alignment) on the performance of 3D reconstruction is explored. For this purpose, a geometric-based description and studies presented in [7, 11, 12] are used. The problem of misalignments was investigated in [13]. Practical experiments reveal that formulas in [12] are simplified and can be used only in special scenarios (e.g., the point lies on the horizontal axis of the image). Presented results extend the original contribution of the work in [14] which contains examination of error in all three spatial coordinates. Analysis of potential errors in all three spatial coordinates depending on position of the point in the space and system parameters is presented in detail. Furthermore, two possible views on the aspects influence errors coming from camera alignment error are specified. To be more precise, two aspects are considered: parameters of the sensing system and properties of the reconstructed spatial point.

Remaining parts of this section are organized as follows. Equations for estimating of errors caused by wrong camera alignment are deduced in Sect. 3.2.1. Here, consequences of errors in camera alignment are mathematically analyzed. Section 3.2.2 contains description of the influence of the camera system parameters and spatial position of the object, while an experimental results discussion is given in Sect. 3.2.3.

3.2.1 The Influence of Inaccurate Camera Alignment

Geometry of the camera has a direct impact on the accuracy of the 3D reconstruction. There is considered a stereoscopic (normal) scenario. In this case, positions of the corresponding points differ only in horizontal position. Of course, it is true at correct camera alignment. However, this assumption is not valid if the cameras are not in normal (so-called perfect) position. Here, the corresponding points cannot be found because they are being searched only in the same row.

In this study, the errors in camera alignment, represented by error angles α, β, and γ, are explored. The considered scenario is clearly shown in Fig. 3.1, where possible changes of parameters of the sensing system and changes in position of captured objects are outlined.

The error in camera alignment influences the coordinate system. This causes changes in image coordinates of the spatial points. In the first step, basic (general) formulas to calculate errors in spatial coordinates (marked as ΔX, ΔY, ΔZ), respecting incorrectly determined image coordinates (x_2', y_2'), are deduced. The relation for the error in depth (ΔZ) was presented in [12]. Remaining equations to calculate ΔX and ΔY are given in this study. In the second step, equations for calculation of x_2' and y_2' are obtained. For this purpose, trigonometric functions are used, which are coming from considered geometric situations.

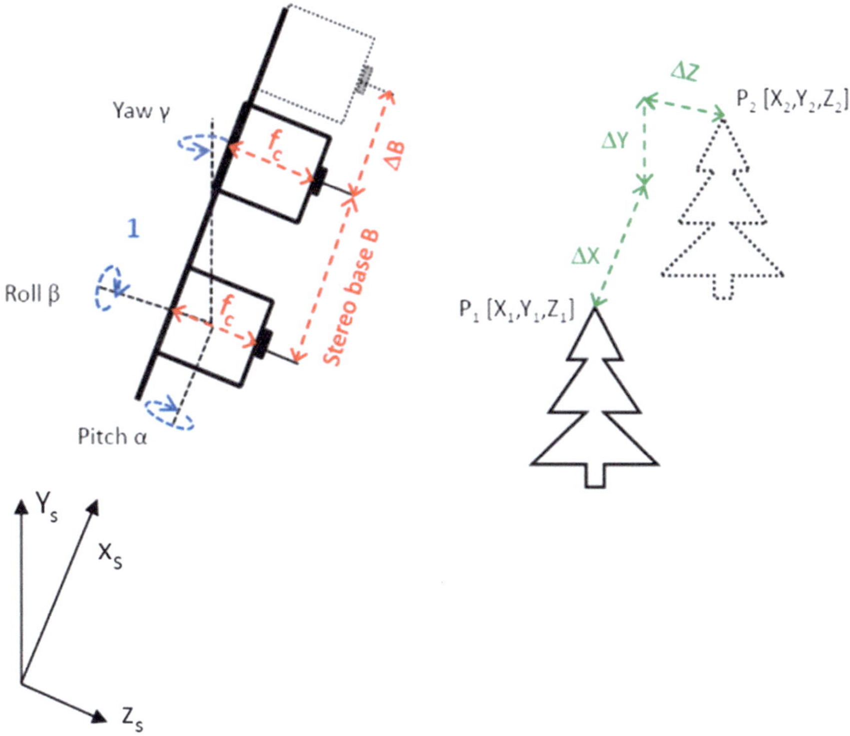

Fig. 3.1 Vision system with two cameras with possible fault angles α,β,γ

Theory Background

First, formulas to calculate spatial coordinate (*Z*) for two scenarios of camera alignment (with and without error) are presented. Both equations were obtained from the basics of stereophotogrammetry theory [15, 12]:

$$Z_{\text{true}} = f_c\left(\frac{B}{x_2' - x_1} - 1\right) \cong f_c\left(\frac{B}{x_2' - x_1}\right), \tag{3.1}$$

$$Z_{\text{obs}} = f_c\left(\frac{B}{x_2 - x_1} - 1\right) \cong f_c\left(\frac{B}{x_2 - x_1}\right), \tag{3.2}$$

where Z_{true} and Z_{obs} are the real (true) and observed absolute depth from image plane to the object, respectively, f_c is the focal length of both cameras, *B* marks horizontal distance between the cameras (stereo base), x_1 and x_2 are the true (correct) positions of the measured point in the first image (captured by the first camera) and second image (captured by the second camera), respectively, and x_2'

is the observed error position of the particular pixel in the second image (captured by the second camera at incorrect rotation).

The error of spatial coordinate ΔZ can be calculated as a difference between the observed and real depth of the point [14]:

$$\Delta Z = Z_{\text{true}} - Z_{\text{obs}}. \tag{3.3}$$

After substituting (3.1) and (3.2) into (3.3), the following mathematical operations are executed:

$$\begin{aligned}
\Delta Z &= f_c\left(\frac{B}{x_2' - x_1}\right) - f_c\left(\frac{B}{x_2 - x_1}\right), \\
\Delta Z &= f_c B \frac{x_2 - x_2'}{(x_2' - x_1)(x_2 - x_1)}, \\
\Delta Z &= Z_{\text{true}}\left(\frac{x_2 - x_2'}{x_2 - x_1}\right).
\end{aligned} \tag{3.4}$$

From (3.4), it is clearly visible that occurring error in the spatial coordinate depends on the incorrect horizontal position in the second image (x_2').

In the next steps, general formulas for ΔX and ΔY [14] are obtained:

$$X_{\text{true}} = B\left(\frac{x_2}{x_2 - x_1} - 1\right) \cong B\left(\frac{x_2}{x_2 - x_1}\right), \tag{3.5}$$

$$X_{\text{obs}} = B\left(\frac{x_2'}{x_2' - x_2} - 1\right) \cong B\left(\frac{x_2'}{x_2' - x_2}\right), \tag{3.6}$$

where X_{true} and X_{obs} are the real (true) and observed absolute spatial horizontal coordinates, respectively. The general form of ΔX can be expressed as [14]

$$\Delta X = \frac{B\left[x_2\left(x_2' - x_2\right) - x_2'(x_2 - x_1)\right]}{(x_2 - x_1)\left(x_2' - x_2\right)}. \tag{3.7}$$

Finally, the equation for ΔY was derived

$$Y_{\text{true}} = B\left(\frac{y}{x_2 - x_1} - 1\right) \cong B\left(\frac{y}{x_2 - x_1}\right), \tag{3.8}$$

$$Y_{\text{obs}} = B\left(\frac{y'}{x_2' - x_2} - 1\right) \cong B\left(\frac{y'}{x_2' - x_2}\right), \tag{3.9}$$

where Y_{true} and Y_{obs} are the true (real) absolute spatial vertical coordinates, and Y_{obs} is the observed absolute spatial vertical coordinate.

Based on previous steps, the final general formula of ΔY will be

$$\Delta Y = \frac{B\left[y\left(x_2' - x_2\right) - y'\left(x_2 - x_1\right)\right]}{\left(x_2 - x_1\right)\left(x_2' - x_2\right)}. \tag{3.10}$$

The vertical image coordinate (y_{im}) can also be changed. Such change introduces a problem which was not considered before [12] and moreover, this parameter is not included in Eqs. (3.1) and (3.2). However, the change of vertical position can cause that the corresponding point will be not found.

Errors in the Rotation of the Camera

In this section, errors in image coordinates caused by incorrect camera alignment are derived, based on the geometric model of considered scenarios. At the beginning, attention is focused on the error in roll with rotation angle α between two cameras. Let assume that the first camera is perfectly calibrated. In this case, its optical axis is the same with axis z. Optical axes of the second camera are parallel to the optical axes of the first camera, but the second camera has incorrect calibration. Hence, the error is in the angle α. Taking into account the error in image coordinates and previously derived expressions (3.4), (3.7) and (3.10), errors in all spatial coordinates, caused by error angle α, can be calculated as

$$\Delta Z_\alpha = Z_{\text{true}} \frac{X_2(\cos\alpha - 1) - Y\sin\alpha}{B}, \tag{3.11}$$

$$\Delta X_\alpha = \frac{BX}{X + B\cos\alpha + X\cos\alpha + Y\sin\alpha} - X, \tag{3.12}$$

$$\Delta Y_\alpha = \frac{XY + B^2\sin\alpha + Y^2\sin\alpha - BX\sin\alpha - XY\cos\alpha}{X + B\cos\alpha + X\cos\alpha + Y\sin\alpha}. \tag{3.13}$$

Now, we consider that the calibration of both cameras is perfect. However, at the configuration of the second camera, we assume an error for a certain rotation angle β about a line which is parallel to the bar. It is important to mention that the epipolar line is no longer parallel with the bar [14]. After basic mathematical operations, errors in all spatial coordinates, caused by error angle β, can be expressed as

$$\Delta X_\beta = Y + \frac{BXZ}{XZ(1 - \cos\beta) - Bf_c - Xf_c + BZ\cos\beta + BY\sin\beta - XY\sin\beta}, \tag{3.14}$$

$$\Delta Y_\beta = Y + \frac{B\left(Yf_c\left(\frac{\cos 2\beta}{2} + 0.5\right) - Zf_c\cos\beta\sin\beta\right)}{\left(\frac{Xf_c}{Z\cos\beta - f_c + Y\sin\beta + \frac{f_c(B-X)}{Z}}\right)(Z\cos\beta - f_c + Y\sin\beta)}, \tag{3.15}$$

$$\Delta Z_{\beta} = \frac{XZ^2}{B(Z\cos\beta - f_c + Y\sin\beta)} - \frac{XZ}{B}. \tag{3.16}$$

Finally, we assume that the first camera is again perfectly calibrated, but its optical axis represents the z-axis of the ordinate system with the center in the focus. However, in the calibration of the second camera is an error at a certain rotation angle γ about the y-axis [14]. In such case, errors in all spatial coordinates will be

$$\Delta X_{\gamma} = Y - \frac{(B - X)\left(\frac{Z^2 \sin 2\gamma}{2} + X^2 \sin\gamma - Xf_c + XZ\left(\cos\gamma - \left(\frac{\cos 2\gamma}{2} + 0.5\right)\right)\right)}{Bf_c - Xf_c - BZ\cos\gamma - BX\sin\gamma + XZ\left(\frac{\cos 2\gamma}{2} + 0.5\right)}, \tag{3.17}$$

$$\Delta Y_{\gamma} = Y - \frac{BYZ}{Z\left(\frac{Xf_c\left(\left(\frac{\cos\gamma}{2}\right) + 0.5\right) - \frac{Zf_c\cos 2\gamma}{2}}{Z\cos\gamma - f_c + X\sin\gamma} + \frac{f_c(B - X)}{Z}\right)}, \tag{3.18}$$

$$\Delta Z_{\gamma} = \frac{Xf_c\left(\frac{\cos\gamma}{2} + 0.5\right) - Zf_c\cos\gamma\sin\gamma}{Bf_c(Z\cos\gamma - f_c + X\sin\gamma)} - \frac{XZ}{B}. \tag{3.19}$$

3.2.2 Influence of the Camera System Parameters and Spatial Position of the Object

In the following experiments, the impact of camera system parameters as well as influence of object position on errors caused by incorrect camera alignment is explored. Dependencies of the errors in spatial coordinates on the parameters of the sensing camera system and on the position of the point in the scene are studied first. After that, the dependency of the accuracy on the positions of the object in the scene is explored.

The study of errors in spatial coordinates is important, because spatial coordinates are the final output of 3D reconstruction which is independent of the content of image and used 3D display. These coordinates are calculated from image coordinates of the corresponding points. It is evident that possible errors in spatial coordinates depend on the occurring errors in image coordinates. The spatial perception is important in the evaluation of the video quality and it is directly affected by errors in image coordinates. Nevertheless, the spatial coordinates are used for their generality and independence.

The stereo camera system, used in this study, can be characterized by two parameters: horizontal distance between cameras, called as stereo base (B), and focal length (f). The influence of the focal length on the error in determining the spatial coordinates (X, Y, Z) was observed within the range from 6 to 100 mm. The viewing angle decrease is less than 28° for a focal length higher than 180 mm.

The using of teleobjective for stereo is typical for images acquired by satellites. Altogether, five values of error angles are considered: 0.2, 0.5, 1, 2, and 3. Obtained results are shown in Fig. 3.2. In all graphs, relative errors X_{rel} are plotted because more aptly inform about error severity. It is calculated as $\delta X = (\delta X_{\text{abs}}/X)\,100$, where δX_{abs} is the absolute error, and X is a given variable. It is visible that relative errors of all spatial coordinates are independent on the focal length for the error angle α. Such outcome is caused by that the focal length does not appear in the equations for error calculation in the image coordinates for the error angle α. In the case of the error angle pitch β, the situation is opposite, because the errors for all coordinates are increasing with the increasing focal length. Once again, the worst relative error occurs for the vertical spatial coordinate Y. Here, occurring errors depending mainly on the size of the focal length and on the size of error angle. Finally, errors for X and Z coordinate are growing with the increasing value of angle yaw γ. The worst X_{rel} can be found for the spatial coordinate Z. It is an interesting fact that errors in the spatial coordinates X and Y for the pitch are practically independent on the size of error angle.

From the viewpoint of obtaining 3D video, there is a critical error in the image vertical coordinate. The formation of the stereoscopic 3D effect is not possible due to the change of the vertical position in one image disallows, because the 3D effect is based on the condition that both eyes see the same object in various positions (horizontal positions only) of their view. Human eyes always see the object in the same vertical position whereas; the change of horizontal positions defines the distance from the observer. Consequently, possible errors in the horizontal coordinates are not so critical. Errors of the corresponding point in horizontal position cause only wrongly perceived depth while the loss of vertical conformity is resulted in the loss of the 3D effect (because of loss of correspondences). In the image processing, corresponding points are searched only in the same row of the stereo images. It is the reason that why correspondence for the error in the vertical position was not found.

The influence of the stereo base length B on the error in the determination of spatial coordinates (X, Y, Z) was examined in the range from 35 mm to 1.2 m. The especially interesting baseline length is approximately 70 cm. The length of the stereo base directly influences the perceived spatial effect. Longer stereo baseline causes more significant stereo perception in longer distance. All obtained results are plotted in Fig. 3.3. It is visible that errors in spatial coordinates are practically independent on the stereo base for the roll. The error is decreasing with the increasing stereo base for the pitch. Moreover, this decrease in the small length of the stereo base causes the most harp form in the obtained curve. Stereo base longer than 50 cm is optional from the view of the accuracy of 3D reconstruction.

Dependences of the accuracy on the positions of the object in the scene are shown in Figs. 3.4, 3.5 and 3.6. The highest errors occur for the yaw, but errors for the yaw are independent on the vertical positions of the object. Observed dependencies are increasing with the increasing object distance from the origin of the coordinate system. Therefore, the sensing of the object in the proximity of cameras is advantageous for the minimization of the danger of the errors. This requirement is unfortunately in conflict with the requirement for relative long stereo base.

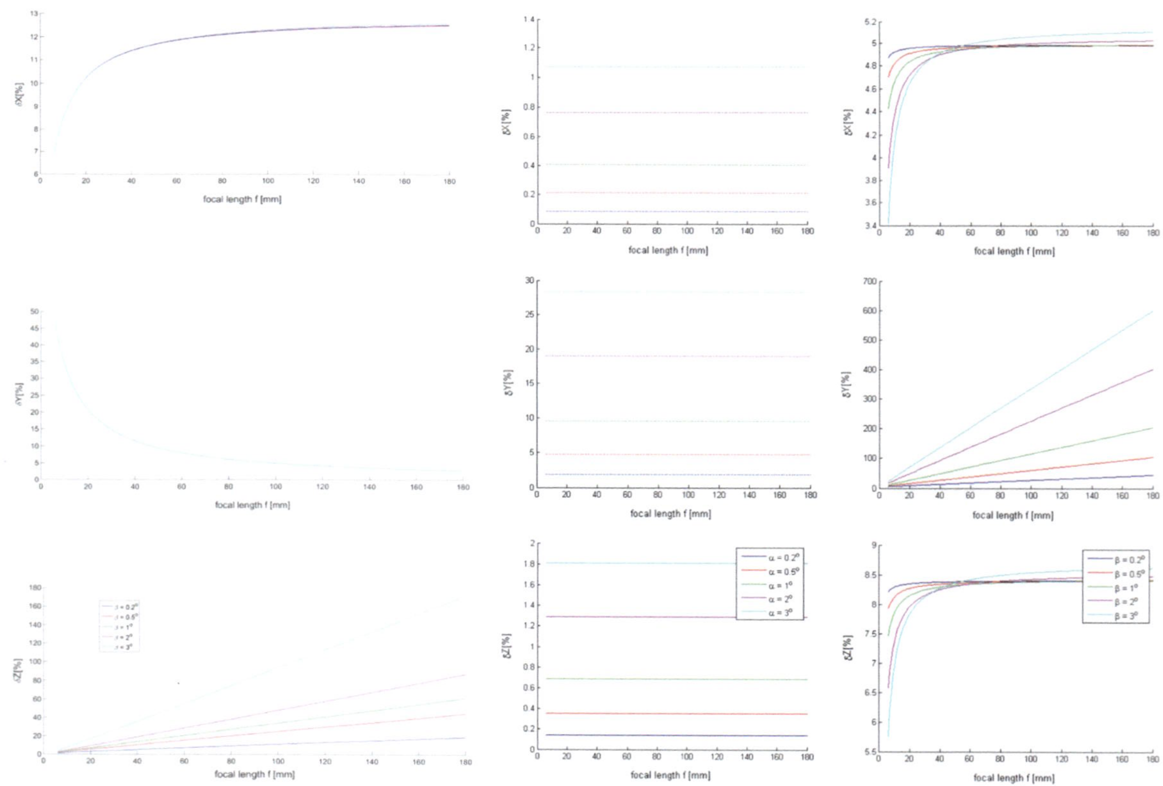

Fig. 3.2 The dependency of the relative errors ΔX, ΔY, ΔZ of the coordinate X, Y, Z on the focal length (f) for all types of alignment errors: first row—roll (α), center row—pitch (β), and third row—yaw (γ). The parameter of dependency is the size of the angles

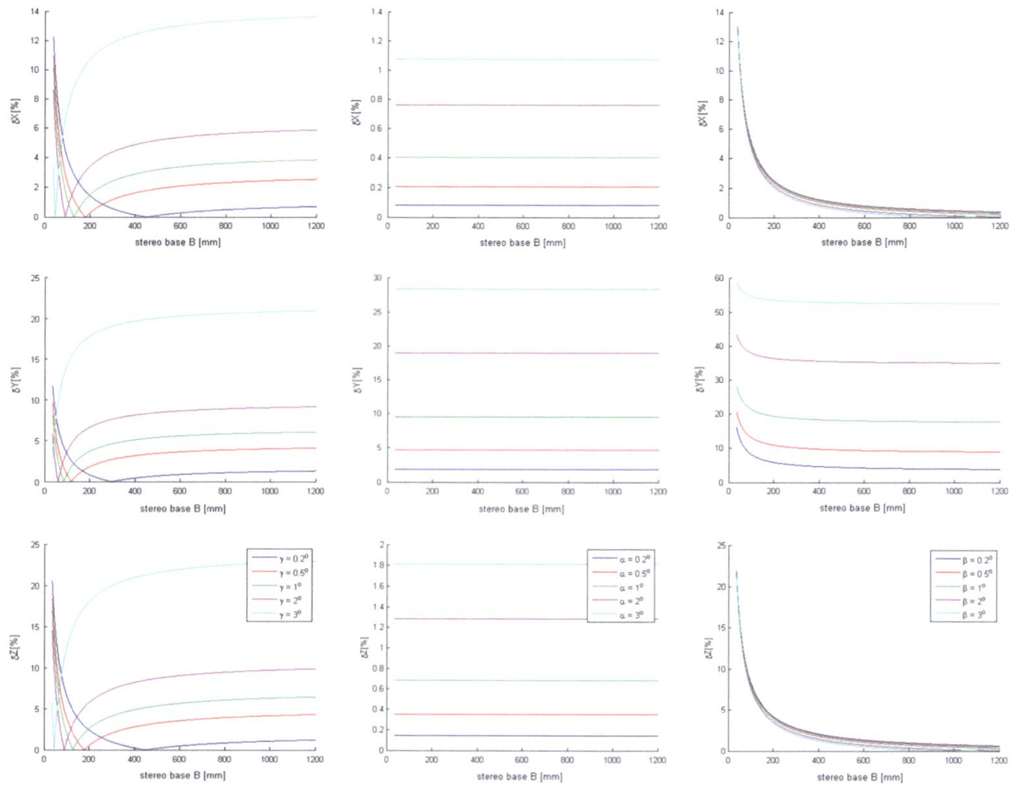

Fig. 3.3 The dependency of the relative errors ΔX, ΔY, ΔZ of the coordinate X, Y, Z on the stereo base for all types of alignment errors: first row—roll (α), center row—pitch (β), and third row—yaw (γ). The parameter of dependency is the size of the angles

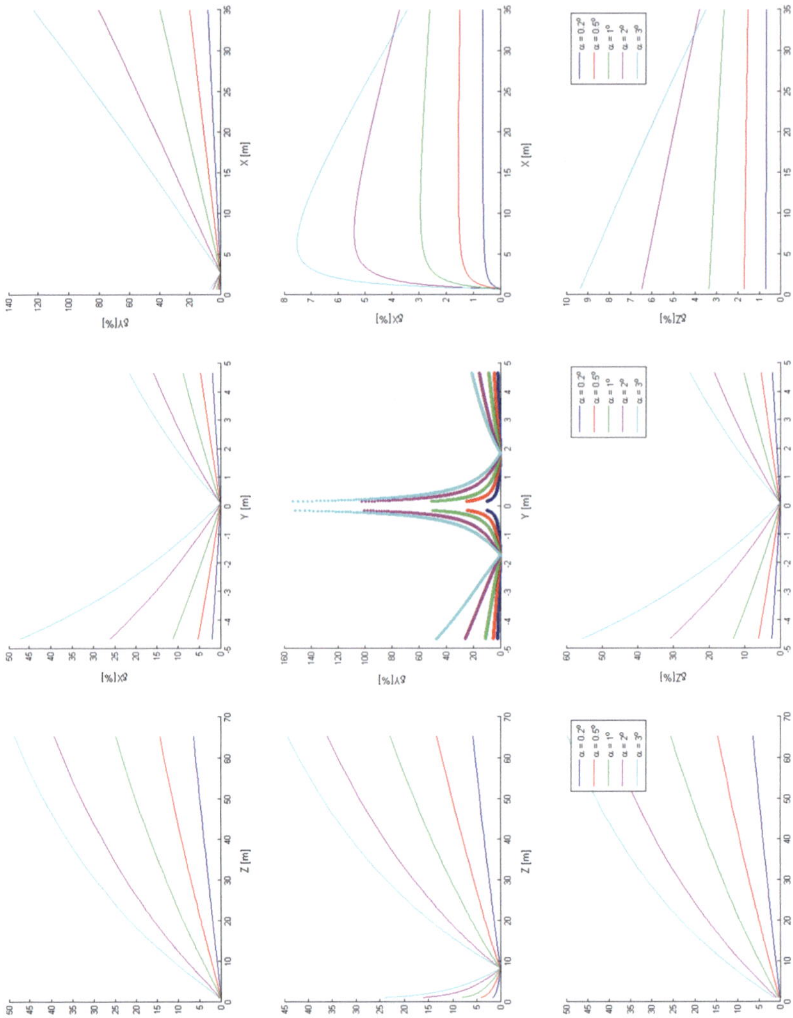

Fig. 3.4 The dependency of the relative errors Δ*X*, Δ*Y*, Δ*Z* of the coordinate *X*, *Y*, *Z* on the spatial coordinates—error roll: first row—*X*, center row—*Y*, and third row—*Z*. The parameter of dependency is the size of the angles

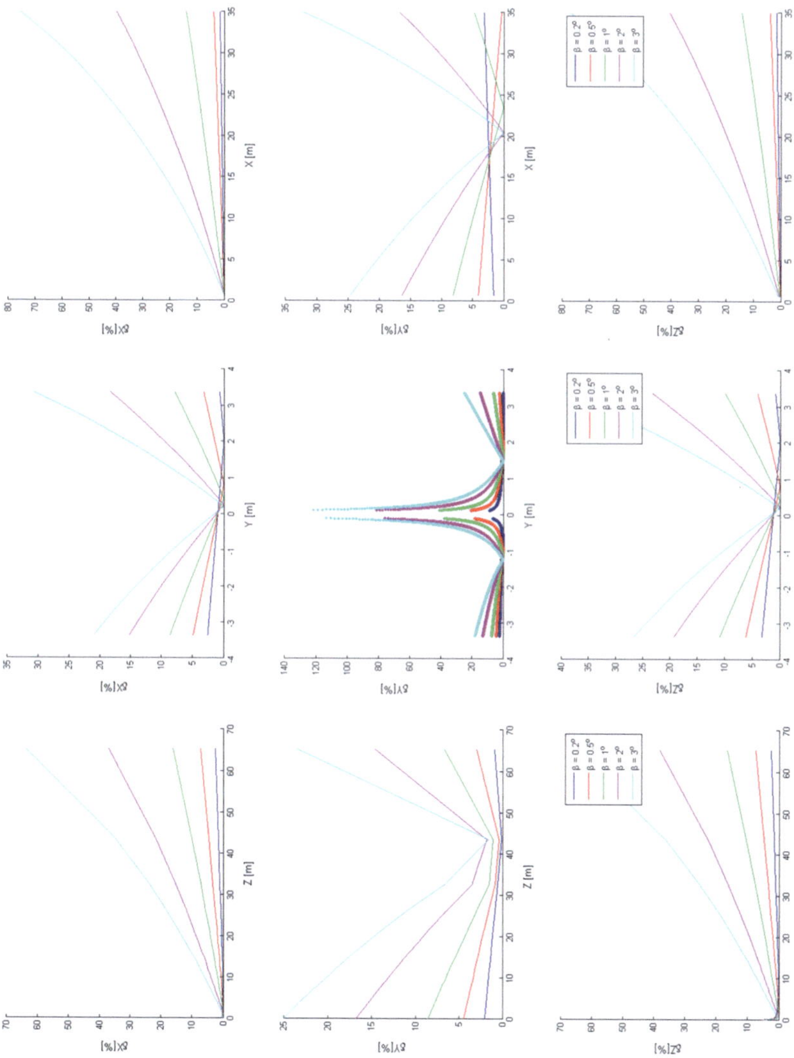

Fig. 3.5 The dependency of the relative errors ΔX, ΔY, ΔZ of the coordinate X, Y, Z on the spatial coordinates—error pitch: first row—X, center row—Y, and third row—Z. The parameter of dependency is the size of the angles

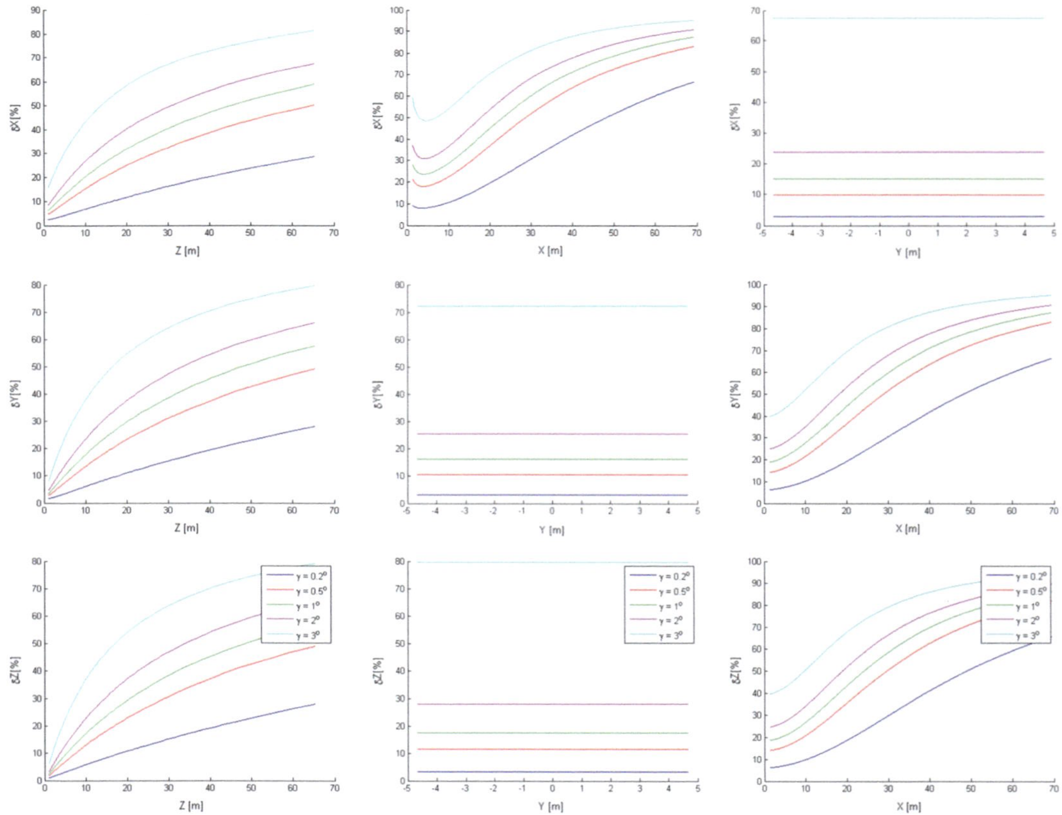

Fig. 3.6 The dependency of the relative errors ΔX, ΔY, ΔZ of the coordinate X, Y, Z on the spatial coordinates—error yaw: first row—X, center row—Y, and third row—Z. The parameter of dependency is the size of the angles

3.2.3 Remarks

In this section, the influence of an inaccurate camera system alignment on the 3D reconstruction accuracy and stereo perception was studied. Attention was devoted to the camera system parameters, namely, stereo base and focal length of cameras (marked as *B* and *f*). This study extends previous work in this field [14, 12].

Relative error of all spatial coordinates for all considered errors in alignment was investigated. Results revealed that the usage of the stereo base longer than 50 cm is a good compromise for the minimization of errors in 3D reconstruction. The roll error is independent on the parameter of the sensing system whereas error pitch is decreasing for the focal length and with the increasing stereo base. The influence of the position of the spatial point in the spatial coordinate system was explored too. Regarding the obtained results, it is observed that:

- The yaw (γ) between the cameras has the highest influence on the accuracy of the spatial coordinate *X*.
- The size of the angle between the yaw (γ) of the cameras has high influence on the error of the coordinate *X*.
- The accuracy of the spatial coordinates *Y* and *Z* is very sensitive on the pitch (β) and on the yaw (γ) between the cameras, respectively.

3.3 Compression Tools for Stereoscopic and Multiview Video

The recent decade in the field of Digital Video Broadcasting (DVB) and multimedia video systems can be characterized as "the decade of significant revolution". One of the important milestones in this revolution is related to display technology and new television (TV) services, including Three-Dimensional TV display technology (3DTV). The idea of a 3DTV (stereoscopic) display was presented for the first time by Sir Charles Wheatstone in 1838 [16]. With the advancement in modern stereoscopic display technologies both in the cinema and at home, 3D TV has been receiving a lot of interest apparently after James Cameron's blockbuster 3D movie Avatar, released in 2009.

Since then, the interest for stereoscopic multimedia content has been increasing in many fields including TV broadcasting, visualization, medicine and security. In all these fields, ensuring an appropriate visual quality of 3D video is among the most important issues. The techniques of capture, post-production, coding, transmission, decoding, and display are the key factors which influence the overall quality of experience (QoE) of 3D video [17–23].

In this section, an overview of the available compression tools for stereoscopic and multiview video is provided. This is an extremely important component from the video image quality point of view. First of all, in Sect. 3.3.1, the basic stereoscopic frame-compatible formats are briefly described. Section 3.3.2 presents an overview of the dominant compression tools, applicable to 3D video. A comparison of performance of the considered compression tools and their flexibility is provided in Sect. 3.3.3. Finally, Sect. 3.3.4 emphasizes the main findings of this survey.

3.3.1 Stereoscopic Frame-Compatible Formats

Frame-compatible formats define the arrangement of the left and right images in a spatial multiplex, which results in an image which can be treated like a normal High Definition TV (HDTV) image by the demodulator and compression decoder in the receiver [24, 25]. For the final 3D image, the left and the right images are packed together in the samples of a single video frame. In such format, half of the coded samples represent the left view and the other halves represent the right view. Consequently, one full coded frame consists of two coded frames with half the spatial resolution of a full single-view frame. For packing of the left and the right images, there exist several basic methods, such as side-by-side (SbS), top-and-bottom (TaB), and quincunx ("checkerboard"), as shown in Fig. 3.7.

"*The SbS format is defined as the arrangement of the frame-compatible spatial multiplex such that the horizontally anamorphic left-eye picture is placed in spatial multiplex to occupy the first half of each line, and the right-eye picture is placed in the spatial multiplex to occupy the second half of each line*" [24]. It means that for, e.g., 1280 × 800 pixels in a full frame the left view and the right view multiplexed in the frame consist of 640 × 800 pixels each. The main advantage of the SbS format is in its simplicity, low bandwidth requirements, and use of passive 3D glasses. However, its main disadvantage is decreased QoE due to reduced horizontal resolution of the frames corresponding to each of the views [26].

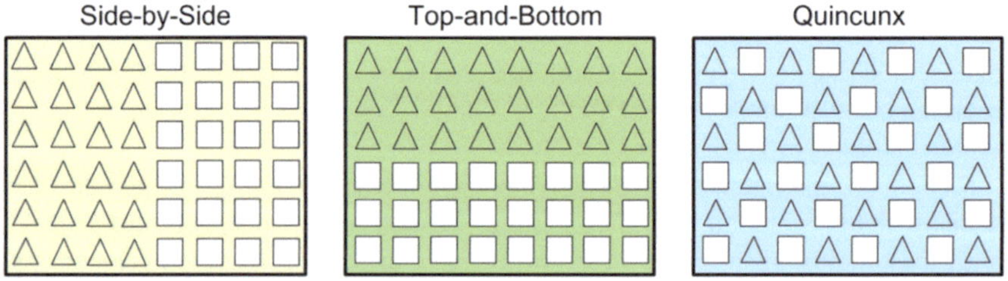

Fig. 3.7 Common 3D frame-compatible formats: side-by-side (SbS), top-and-bottom (TaB), and quincunx ("checkerboard"). Symbols "*triangle*" and "*square*" correspond to the samples belonging to the left view and right view images, respectively (based on [32])

"The (TaB) format is defined as the arrangement of the frame-compatible spatial multiplex such that the vertically anamorphic left-eye picture is placed in the spatial multiplex to occupy the first (top) half of a single HDTV video frame, and the right-eye picture is placed in the spatial multiplex to occupy the second (bottom) half of a single HDTV video frame" [24]. Consequently, the TaB format is almost the same as the SbS; however, frames for each eye are scaled down vertically. It means that for, e.g., the frame resolution of 1280 × 800 pixels, the left view and the right view are both scaled down to the spatial resolution of 1280 × 400 pixels.

A quincunx is a geometric pattern containing five coplanar points, four of them forming either a rectangle or a square with the fifth point being at its center. The "checkerboard" (quincunx) sampling [23] can be applied to each view, with the two views interleaved in alternating samples in both the horizontal and the vertical dimension (see Fig. 3.7). The pixels are then packed in SbS or TaB for transmission. The stereoscopic TV separates the respective interleaved images and displays them sequentially (half resolution images). The main drawback of this format is in the increased unwanted artifacts, at least without enhancement metadata. Moreover, it requires active shutter glasses to be used with the display. If the 3DTV does not include an appropriate decoder, then enhancement of metadata is ignored. This is the main reason why this format is not preferred for 3DTV broadcasting [27].

If the presentation device cannot process and display the SbS format, then 3D content can be converted to sequential form in a straightforward manner. In principle, for the left view and the right view, each frame of the stereoscopic content is split into two temporally successive subframes. In order to create a sequential stereoscopic content, the left and the right parts of a typical SbS frame are stretched from 640 × 800 to 1280 × 800 pixels—involving upscaling and interpolation of pixel values. Such conversion results in doubling the frame rate.

Based on the features of SbS and TaB formats, it can be concluded that both formats are appropriate for TV sets with active 3D glasses. On the other hand, using passive 3D glasses is better for the TaB format because passive 3DTV systems, in general, reduce the vertical frame resolution by a half.

Using content conforming to the SbS to be displayed by a stereoscopic display with passive glasses is currently the worst case scenario since half the resolution is lost in horizontal direction due to frame packing, and then half the resolution is lost in the vertical direction due to differentiation of odd and even lines for different light polarization in order for the two views to be correctly separated by the glasses.

The main advantage of the above described stereoscopic formats is their compatibility with the widely deployed delivery infrastructure (encoding, signal processing, transmission, reception, signal processing, and decoding) and broadcasting systems. In advance, the deployed and emerging compression tools can be also applied to such video formats [28]. On the other hand, the main disadvantage of such processing and representation of stereoscopic content is the loss of temporal or spatial resolution. Furthermore, in order to correctly distinguish between the left view and the right view and to perform deinterleaving, an "out-of-band" signaling is necessary. More details can be found in [23].

Frame Sequential Stereoscopic Format

Sequential stereoscopic format, sometimes marked as "page-flip", is the simplest 3D format available. A frame sequential signal in full spatial resolution picture is carried at typically 120 frames per second, corresponding to a 60 Hz progressive scanning format for each of the two views. The frames are arranged into a sequence and received as left/right/left/... frames. Such solution does not need additional signal processing for the decoding of the received signal in the 3DTV receiver. This is the reason why this 3D format is the simplest in terms of display pre-processing when used as an input to a 3DTV receiver with active shutter glasses stereoscopic display approach. As already mentioned, the successive frames in a sequence are intended for different eyes. To be more precise, particular left view frames are shown only to the left eye, whereas the right view frames are shown only to the right eye at the same time. This causes that 3D content, sent at 120 frames per second (fps), for instance, is actually displayed at 60 fps for each eye. The main advantage of the frame sequential format, apart from its simplicity, is an expectation of high QoE (no spatial resolution losses in horizontal and vertical dimensions). However, among its disadvantages are the requirements for active shutter stereoscopic glasses and demand for a higher bandwidth [29].

Video Plus Depth (V + D)

Video plus depth (one stream with associated depth map, marked as "V + D") is another class of 3D formats. It contains a video signal and a per-pixel depth map. Additional depth data (either monochromatic or luminance video signal) for each pixel can be generated from either calibrated stereo or multiview video by its depth estimation. The depth range that can be described by the depth map is restricted to the minimum (z_{near}) and maximum (z_{far}). This range is typically described by 8-bit values. Consequently, the nearest and farthest points are represented by the values 255 (z_{near}) and 0 (z_{far}), respectively, such as a gray scale image. Such depth images in a sequence can be converted into a YUV 4:0:0 format video signal and processed by a desired video compression tool [30, 28].

The main advantage of the V + D format can be found in the advanced possibilities of stereoscopic signal processing (e.g., to modify the level of depth perception related to display size or viewing distance). However, its main drawback is nontrivial processing needed to obtain a depth map from a pair of aligned video sequences captured by a stereo camera. Moreover, the accessibility of stereo signals by this format is limited. More details about video plus depth can be found in [28, 31, 32].

3.3.2 Compression Tools for Stereoscopic Video

The stereoscopic content is subject to several processing steps with variability in their setup. The technique of capturing and depth map presentation as well as applied display technology directly influences the overall picture quality of 3D video content [20, 33]. Thanks to increasing data volume in multimedia content, appropriate video codecs whose performance meets with video image quality, device compatibility, and hardware and software complexity are necessary [34, 18, 20, 22].

In this section, the performance of the currently available compression tools for stereoscopic and multiview videos is evaluated and compared by 2D objective video quality metrics. We consider video coding algorithms available in ITU-T H.26x Recommendations, namely, H.264 AVC, H.265 HEVC, and their multiview (MV) extensions, namely, MVC and MV-HEVC. Furthermore, we consider the possibilities of coding media for the two views independently, using symmetrical or asymmetrical approach.

Symmetrical and Asymmetrical 3D Video Coding

As mentioned above, methods to compress 3D video content and related depth map play a key role in guaranteeing a certain perceptual quality level in 3DTV services. Generally, compression techniques can be symmetrical or asymmetrical. The symmetrical 3D video compression is based on applying any 2D video coding tool on the left and on the right images independently. However, synchronization for playback between the two views should be maintained. The relatively low computational complexity of decoding is the main advantage of such solution since no inter-sequence prediction is employed in the compressed bitstream. On the other hand, higher (double) bandwidth to transmit two 2D streams is its main disadvantage [35].

In asymmetrical 3D video compression, different compression tools may be used for the left view and the right view. As a result, target bitrates in both views can be different. Although the view with lower bitrate has lower decompressed 2D video quality, the perceived binocular visual quality is still similar to the symmetrical 3D video coding [35]. Hence, asymmetrical 3D video compression is mainly based on the advantage of the human visual system (the perceived quality of the joined left and right views with different quality is close to the higher quality view) [36]. It can be concluded that in symmetrical compression, the required time for compression/decompression is almost the same whereas for asymmetrical compression the time for decompression of each view may differ.

Advanced Video Coding (AVC)—H.264

Advanced video coding (AVC) [37] is a result of joint collaboration of International Telecommunication Union (ITU) and Motion Picture Experts Group (MPEG). Since its standardization in 2003, it has become the gold standard of video encoding. AVC is used for distribution of the content in DVB systems and on the Internet, as well as a storage format on Blu-Ray discs. AVC has a wide variety of profiles and levels to reflect the demands of different delivery scenarios, from small devices such as smartphones with limited resources up to high-quality scenarios with 4 K resolution. In this study, AVC is mainly used as a benchmark to other video coding algorithms. In order to encode used video sequences according to AVC standard, an open-source implementation was used—x264 [38]. Compared to the reference JM implementation [39], x264 offers good performance in terms of rate-distortion efficiency, computational speed, a broad variety of settings, and cross-platform compatibility.

Multiview Video Coding (MVC)—H.264 Annex H

Multiview Video Coding (MVC) is an extension of AVC standardized for the first time in 2010. In addition to AVC, new features and coding tools are available to process more than one view. New syntax was added to take advantage of new inter-view prediction mode. In 3D video capture setup, both views belong to the same scene. Therefore, there is a correlation between the different views. In such a case, one of the views can be used as reference for encoding the other view, which can lead to increase of coding efficiency compared to encoding the views independently. An example of *inter-view* prediction is depicted in Fig. 3.8. The *inter-view* prediction enables to lower the bitrate while preserving the quality for all the views used [40]. However, the additional syntax may lead to increase of the final bitrate due to increased overhead volume.

As a tool for encoding the video sequences conforming to the MVC standard, the FRIM encoder [41] was used. FRIM serves as a free MVC encoder and is available on MS Windows platform. The main advantage of FRIM compared to the reference JMVC encoder [42] is the possibility of direct setting of target bitrate and higher computational speed.

High Efficiency Video Coding (HEVC)—H.265

High Efficiency Video Coding (HEVC) [43] is a direct successor to H.264/AVC. It is built on the same basics as AVC; however, the coding tools are improved to require just half of the AVC bandwidth with preserving the same perceivable video quality. The new features include improved basic coding structure called Coding Tree Unit (CTU) with size up to 64 × 64 pixels, new directional modes for intra-picture prediction, Content-Adaptive Binary Arithmetic Coding (CABAC) as the only entropy coding, Sample Adaptive Offset (SAO) filter, and other improvements. However, the increase of the video quality is at the cost of higher both encoder and decoder complexity, which leads to increase of computational demands. The main advantage of HEVC compared to AVC is high-quality scenarios up to 4 K (3840 × 2160) and 8 K (7680 × 4320) resolution. HEVC does

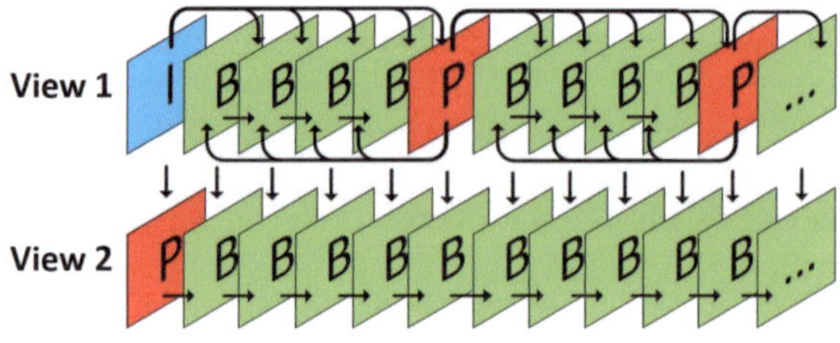

Fig. 3.8 Principles of inter-view prediction for two views (based on [40])

not have implicit support for 3D or multiview video content encoding, therefore, each view has to be treated separately.

To encode the video content in accordance with HEVC standard, its reference implementation—the HM reference model [44]—was employed. The reference implementation offers wide variety of settings, and its source code is easily modifiable, which makes it widely used in research.

Multiview High Efficiency Video Coding (MV-HEVC)

Similarly as MVC, Multiview High Efficiency Video Coding (MV-HEVC) is based on the single-view variant of the codec. The coding tools of MV-HEVC are the same as in HEVC, and the main changes are made in syntax only, to provide the support of encoding more than one view. The inter-view prediction mode is also available in MV-HEVC. In addition, the syntax changes also allow for using both multiview and scalable video encoding [45].

As a software tool to encode the 3D video sequences according to MV-HEVC, the reference implementation of the standard MVHM v15.1 [46] was used. The encoding core is pure HM reference model; however, additional settings enable to opt for encoding the input video sequences as 3D or multiview video content.

3.3.3 Performance Analysis of Compression Tools

In this section, the quality of the encoded/decoded stereoscopic videos by the compression tools listed in previous sections is evaluated and discussed. All the selected algorithms use raw video data only as their inputs. Consequently, no depth map is necessary. The results presented in this section are a continuation of previous work [22].

Video Sequences

To demonstrate the encoding/decoding [47] performance of the considered compression tools, several short 3D video sequences were used. As source sequences, we used eight video sequences available in databases [48] (Fig. 3.9a, c, d, e, f, g, h and [49] (Fig. 3.9b). All the sequences had Full HD resolution of 1920×1080 pixels in each frame (1080p) of one view stream, 25 frames per second. The total length of each sequence was 10 s. The selected sequences cover a wide variety of scene characteristics which are represented by the Temporal Information (TI) and Spatial Information (SI) parameters (see Fig. 3.10). These parameters were calculated according to ITU-T P.910 [50].

For a symmetrical scenario, sequences (a) to (h) depicted in Fig. 3.9 were used; in case of asymmetrical video coding, we used sequences (e) to (h).

Fig. 3.9 Preview of the used 3D test video sequences

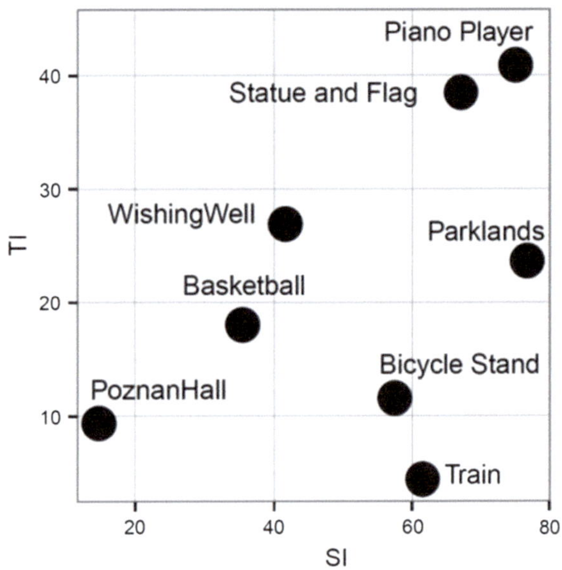

Fig. 3.10 SI and TI values of the used 3D test video sequences

Encoding Parameters Settings

For the symmetrical 3D video coding, four quality levels were considered, defined by a concrete bitrate. The bitrate values varied from 0.5 to 4 Mbps per view. Consequently, the total bitrate for the complete stereoscopic sequence is double. Additional parameters of the encoders were selected to follow the values recommended in the documentation of the encoders. Video encoder settings used in this study is clearly presented in Table 3.1. Symbol "/" denotes a default setting.

In the case of asymmetrical 3D video coding, a different approach was used. We consider a scenario, where the left view is always encoded using H.264/AVC at a static bitrate 10 Mbps (typical value for Satellite Digital Video Broadcasting

Table 3.1 Encoding parameters settings [22]

Parameters					
Codec	Version of the encoder	Profile	Level	Preset	Motion estimation range
AVC	x264-r25697	High	5.1	Very slow	/
MVC	x64 1.25	High	4.0	1	/
HEVC	HM 15	Main	5.1	/	64
MV-HEVC	HTM 15.1	Main	None	/	64

(DVB-S) of Full HD content). The additional right view is encoded compliant to H.265/HEVC with bitrate values in the range from 250 kbps to 1 Mbps. The left view is considered as a backward-compatible video stream, and the right view can be broadcasted as an extending service for compatible devices.

2D Objective Quality Metrics

To evaluate the quality of the test video sequences, compressed by the above briefly introduced video codecs, established 2D objective metrics were used, namely: Peak Signal-to-Noise Ratio (PSNR) [51], Structural Similarity Index (SSIM) [52], and Video Quality Metric (VQM) [53]. The PSNR, typically expressed in logarithmic (dB) scale, is the ratio between the maximum power of the signal (reference) and the power of distorting noise in an image [51]. The SSIM index is a full reference objective metric which measures the similarity (luminance, contrast, and structure) between two images (reference and compressed). It is based on the human visual system (HVS) [52]. Finally, VQM is an advanced video quality metric which measures the perceptual effects (impairments) of the video. Output values from this metric have a high correlation with scores from subjective quality metrics [53].

A performance comparison of the used video codecs, evaluated by 2D objective quality metrics, is clearly shown in Figs. 3.11 and 3.12. Each row is related to one test video sequence, containing an evaluation of PSNR versus bitrate, SSIM versus bitrate, and VQM versus bitrate, respectively. Higher PSNR and SSIM values reflect higher video quality while low VQM values denote high video quality. All the described 2D video quality metrics were applied to 3D videos [54]; hence, such metrics were computed for the left and for the right view separately. Results for the left view only are plotted in Figs. 3.11 and 3.12, because results for both views showed minimal differences.

Symmetrical 3D Video Coding—Performance Analysis

Compression efficiency of the video codecs with similar features (AVC and MVC; HEVC (H.265) and MV-HEVC) are comparable for *Basketball* video sequence. Small differences are visible only in SSIM values (see Fig. 3.11b). Overall, HEVC and MV-HEVC outperform AVC and MVC.

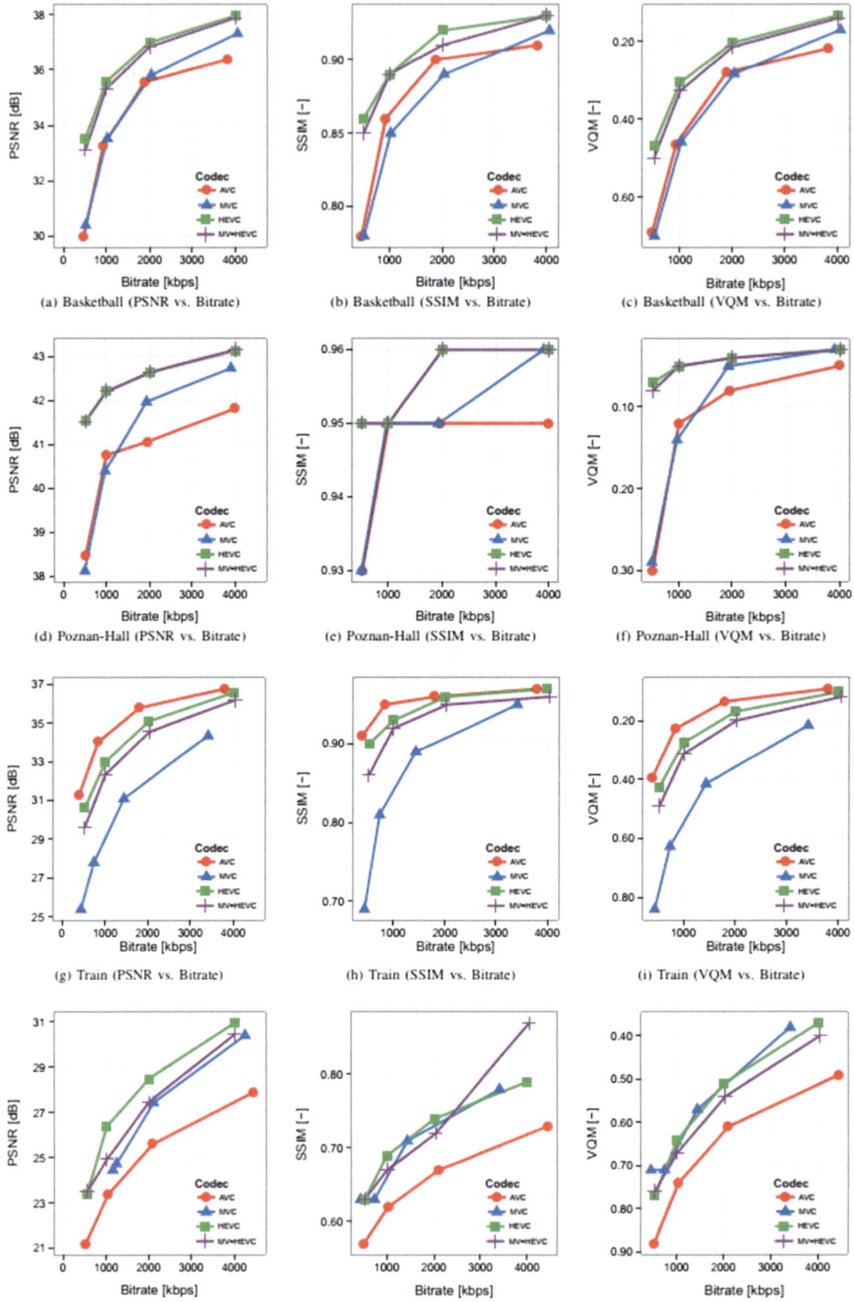

(a) Basketball (PSNR vs. Bitrate) (b) Basketball (SSIM vs. Bitrate) (c) Basketball (VQM vs. Bitrate)

(d) Poznan-Hall (PSNR vs. Bitrate) (e) Poznan-Hall (SSIM vs. Bitrate) (f) Poznan-Hall (VQM vs. Bitrate)

(g) Train (PSNR vs. Bitrate) (h) Train (SSIM vs. Bitrate) (i) Train (VQM vs. Bitrate)

(j) Wishing-well (PSNR vs. Bitrate) (k) Wishing-well (SSIM vs. Bitrate) (l) Wishing-well (VQM vs. Bitrate)

Fig. 3.11 Mean PSNR, SSIM, and VQM values versus video bitrates for sequences Basket, Poznan Hall, Train, and Wishing Well

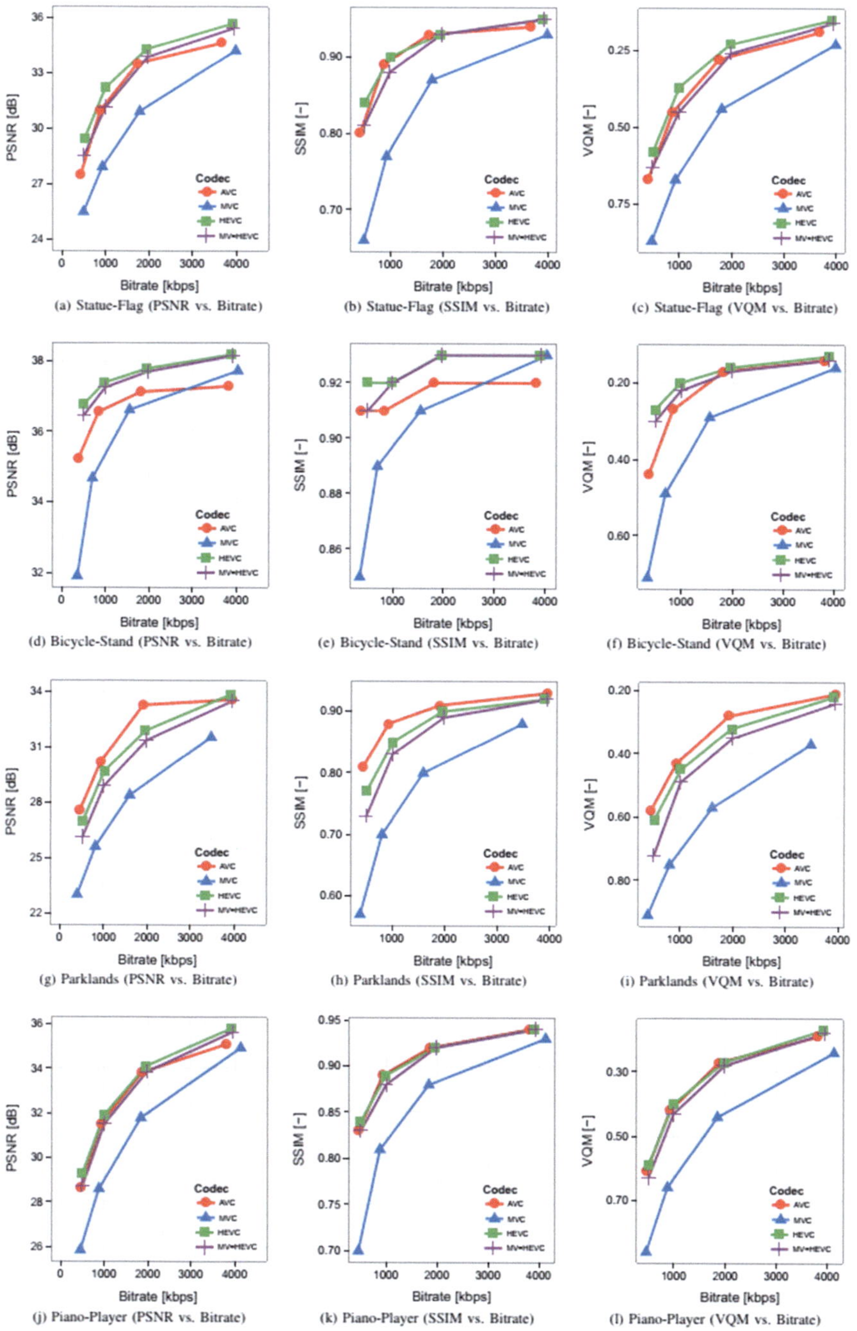

(a) Statue-Flag (PSNR vs. Bitrate) (b) Statue-Flag (SSIM vs. Bitrate) (c) Statue-Flag (VQM vs. Bitrate)

(d) Bicycle-Stand (PSNR vs. Bitrate) (e) Bicycle-Stand (SSIM vs. Bitrate) (f) Bicycle-Stand (VQM vs. Bitrate)

(g) Parklands (PSNR vs. Bitrate) (h) Parklands (SSIM vs. Bitrate) (i) Parklands (VQM vs. Bitrate)

(j) Piano-Player (PSNR vs. Bitrate) (k) Piano-Player (SSIM vs. Bitrate) (l) Piano-Player (VQM vs. Bitrate)

Fig. 3.12 Mean PSNR, SSIM, and VQM values versus video bitrates for sequences Statue-Flag, Bicycle-Stand, Parklands and Piano-Player

Higher differences between the performance of new video codecs (HEVC and MV-HEVC) and their predecessors (AVC and MVC) are visible for the *Poznan Hall* video. All objective values for the HEVC and MV-HEVC are practically the same while differences for AVC and MVC are gradually increasing from 2 Mbps approximately. The higher performance of MVC, in comparison with AVC, can be explained by a dominant global motion in the scene.

The *Train* video sequence is characterized by high SI and low TI values. Probably, such features can play a key role in that the classical compression tools (HEVC and AVC) have higher performance than their multiview versions. Moreover, AVC codec outperforms both advanced HEVC and MV-HEVC codecs.

Different performance between the AVC and HEVC codecs is visible for the *Wishing Well* video sequence. The MVC definitely outperforms the AVC codec. On the other hand, HEVC is slightly better than its MV extension. Interestingly, their performance at lower bitrates is comparable.

The same performance analysis has been carried out for video sequences *Statue-Flag*, *Bicycle-Stand*, *Parklands,* and *Piano-Player,* and the corresponding results are shown in Fig. 3.12. The *Statue-Flag* is characterized by high SI and TI values (see Fig. 3.10). All considered objective metrics show high performance of HEVC codec whereas the difference between MV-HEVC and AVC codecs is negligible. Interestingly, the worst values from objective metrics are assigned to the MVC codec. For content *Bicycle-Stand*, both HEVC and MV-HEVC outperform their predecessors (AVC and MVC). Only VQM metric for the bitrate values higher than 2 Mbps shows comparable performance for AVC, HEVC, and MV-HEVC.

Analysis of objective metrics for the video sequence *Parklands*, which has the highest SI value, shows the same performance of all considered video codecs as previously in the case of video sequence *Train*. Such a phenomenon can be explained by similar high values of Spatial Index. Interestingly, almost the same performance of AVC, HEVC, and MV-HEVC can be observed in the case of video sequence *Piano-Player*. This video sequence has the highest TI value. All the mentioned video codecs outperform multiview version of AVC (MVC) for all considered bitrates.

In general, the PSNR results show comparable performance for codecs with similar features (AVC and MVC; HEVC and MV-HEVC). Furthermore, results from 2D objective quality metrics proved theoretical assumption that HEVC and MV-HEVC perform better than AVC and MVC compression algorithms. However, results also showed that this possible gain of multiview coding is highly dependent on the content of the considered video sequence.

Asymmetrical 3D Video Coding—Performance Analysis

Asymmetrical approach offers the possibility of encoding each view separately using different coding algorithms [55]. In this scenario, we encoded the left view at a constant bitrate using AVC, the right view was encoded using HEVC (see subsection 3.3.3). The values of computed video quality metrics for left view can be

Table 3.2 Objectively measured quality of the left views in asymmetric scenario

Sequence name	PSNR [dB]	SSIM [-]	VQM [-]
Statue-Flag	34.48	0.935	0.381
Bicycle-Stand	37.56	0.925	0.198
Parklands	32.97	0.914	0.334
Piano-Player	35.01	0.934	0.387

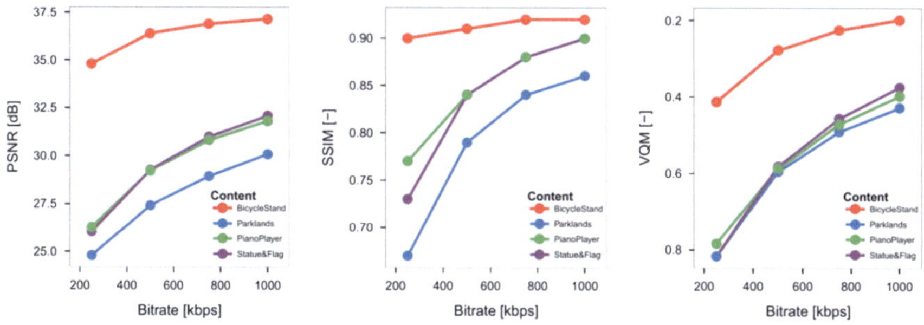

Fig. 3.13 Mean PSNR, SSIM, and VQM values versus video bitrates for asymmetrical 3D Video Coding

found in Table 3.2. When taking a look at the values, it can be seen that the scores give evidence of a *good* quality sequence.

The objectively measured quality for the HEVC encoded right views is depicted in Fig. 3.13. Once again, the x-axis shows the bitrate values, vertical axis depicts the objective quality scores. For example, video sequence *Bicycle-Stand* always achieves higher scores than other sequences. This is in accordance with the fact that this sequence has comparatively low temporal activity (see Fig. 3.10), with respect to other sequences. In general, due to the fact we used lower bitrate range for this scenario, we achieved lower objective ratings. However, as stated in [35], the 3D video quality experience as perceived by the real viewer might be similar as when using both views encoded to the same quality.

3.3.4 Remarks

This section presented the different frame formats of stereoscopic video content to be used in 3DTV broadcasting services, in online streaming, stored on magnetic or optical media, and more. It is clear that the frame formats have a direct impact on the perceived visual quality, mainly due to their tendency to influence the spatial resolution of the transmitted content.

Furthermore, the different mainstream video coding algorithms and standards available for either asymmetrical or symmetrical compression of stereoscopic content were described. A simple experimental evaluation shows that an absolute comparison of the different codecs is hardly possible since a very significant content dependence is observed. Thus, the benefit of using a stereoscopic or multiview extension of a video compression standard does not necessarily lead to a tangible increase in the perceptual quality.

3.4 Multiview Video Compression for Arbitrary Camera Locations

As mentioned earlier in this book, 3D-HEVC [56] is the state-of-the-art compression technology for 3D video in "multiview video and depth" format (MVD) [57]. The technology has been developed by Collaborative Team on 3D Video Coding, Extensions Development (JCT-3V), formed by ISO/IEC and ITU-T members. This technology is incorporated into the HEVC standard (High Efficiency Video Coding) as Annex I of ISO/IEC MPEG-H Part 2 and ITU-T Recommendation H.265.

The 3D-HEVC is built on the top of the MV-HEVC [56] codec. The MV-HEVC codec utilizes inter-view prediction between the views. In other words, the side views are used as another source for inter-prediction, just like previous frames of the same view are used for motion compensated inter-prediction. It should be stressed here that no information about scene structure (e.g., depth maps) is used during encoding with MV-HEVC.

The goal of the 3D-HEVC is to exploit the information about 3D scene structure (in form of depth maps) to increase coding efficiency of 3D video. However, during the development of the 3D-HEVC technology, explicit 1D parallel (linear) views arrangement has been assumed. Thus, inter-view prediction of samples, motion vectors, etc., are done as purely horizontal shifts defined by the disparity values. Such a prediction is unable to effectively remove the inter-view redundancy from the views with optical axes in the significantly different directions. Moreover, many modern Super Multiview (SMV) displays require circular (arc) view arrangements for better experience of a user. Therefore, efficient compression technology for nonlinear (e.g., arc) camera arrangements is of great interest.

This section present extension of 3D-HEVC toward arbitrary view setup, in particular the arc view arrangement.

3.4.1 Adaptation of 3D-HEVC to Nonlinear Camera Arrangements

One of the proposed techniques for efficient compression of multiview video acquired with nonlinear camera setup was developed by Zhejiang University [58]. The proposal was based on the HTM 13.0 [59], which is a reference

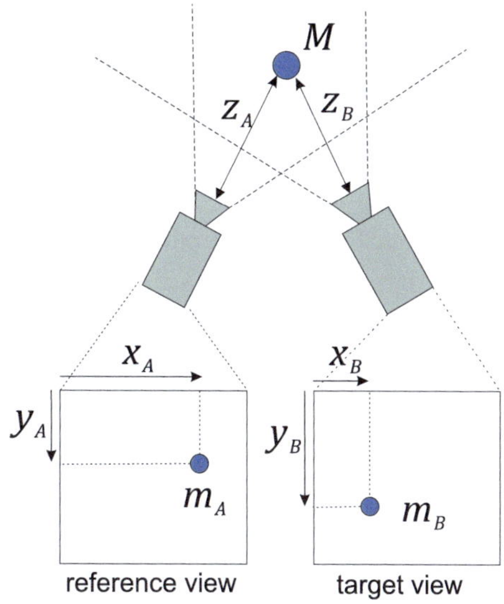

Fig. 3.14 Mapping points in 3D space © 2016 IEEE [74]

implementation of the 3D-HEVC codec. It introduces 2D disparity vectors in order to improve the efficiency of the Disparity Compensated Prediction. Additionally, several coding tools have been adjusted to nonlinear camera arrangements, e.g., View Synthesis Prediction, View Synthesis Optimization, Neighboring Block Disparity Vector, and Depth Rate-Distortion Optimization. This proposal provides a significant improvement in compression efficiency of sequences acquired using nonlinear camera setups. Nevertheless, it is not compatible with 3D-HEVC because of modifications in the syntax of the bitstream.

Another solution for adapting 3D-HEVC to nonlinear camera arrangements was developed by Poznan University of Technology [60]. The proposal is implemented on top of the HTM 13.0 and is also based on 2D disparity vectors. In order to derive the 2D vectors, a true mapping of samples in 3D space is performed as presented in Fig. 3.14.

In Fig. 3.14, the reference view is already encoded, and the target view is currently processed by the encoder. Both views contain projection of point M, but the positions of these projections differ. In case of nonlinear camera arrangements, the difference may appear both in horizontal and vertical direction. Thus, the 2D disparity vector can be defined as follows:

$$dv = m_A - m_B = \begin{bmatrix} x_A \\ y_A \end{bmatrix} - \begin{bmatrix} x_B \\ y_B \end{bmatrix} = \begin{bmatrix} x_A - x_B \\ y_A - y_B \end{bmatrix}. \tag{3.20}$$

When encoding the target view, the position of m_B can be calculated using position of m_A in the reference view, depth information and projection matrices for both views (3.21).

$$\begin{bmatrix} z_B \cdot x_B \\ z_B \cdot y_B \\ z_B \\ 1 \end{bmatrix} = \mathbf{P_B} \cdot \mathbf{P_A^{-1}} \cdot \begin{bmatrix} z_A \cdot x_A \\ z_A \cdot y_A \\ z_A \\ 1 \end{bmatrix}. \tag{3.21}$$

Values z_A and z_B represent the distances between the cameras and point M and can be calculated from depth value v using following formula:

$$z = \left(\frac{v}{2^{\text{Depth Map Bit Depth}}} \cdot \left(\frac{1}{Z_{\text{near}}} - \frac{1}{Z_{\text{far}}} \right) + \frac{1}{Z_{\text{far}}} \right)^{-1}, \tag{3.22}$$

where Z_{near} and Z_{far} define the range of depth maps, and Depth Map Bit Depth is a number of bits representing a single depth sample (usually 8 or 16 bits per sample is used).

Projection matrices for the reference and target view ($\boldsymbol{P_A}$ and $\boldsymbol{P_B}$, respectively) can be derived from multiplication of intrinsic and extrinsic camera parameters using formula (3.23).

$$\boldsymbol{P} = \begin{bmatrix} f_x & c & o_x & 0 \\ 0 & f_y & o_y & 0 \\ 0 & 0 & 1 & 0 \\ 0 & 0 & 0 & 1 \end{bmatrix} \cdot \begin{bmatrix} \boldsymbol{R} & -\boldsymbol{R} \cdot \boldsymbol{T} \\ \boldsymbol{O}^T & 1 \end{bmatrix}. \tag{3.23}$$

In 3D-HEVC, according to the assumption of 1D parallel views assumption, only some of the parameters used in (3.23) are transmitted in the bitstream (Table 3.3). With 2D disparity vectors, the remaining camera parameters are also

Table 3.3 Camera parameters needed in the case of linear and arbitrary camera locations © 2016 IEEE [74]

Parameter name	Parameters needed for arbitrary locations	Parameters needed for linear locations
Horizontal focal length	f_x	f_x
Vertical focal length	f_y	–
Horizontal optical center	o_x	o_x
Vertical optical center	o_y	–
Skew factor	c	–
Nearest distance to camera	Z_{near}	Z_{near}
Farthest distance to camera	Z_{far}	Z_{far}
Translation	$\mathbf{T} = [t_x\ t_y\ t_z]$	t_x
Rotation	**R**	–

required. The transmission of the additional parameters is the only modification of the bitstream introduced by the solution proposed by Poznan University of Technology.

Using a true mapping of points in 3D space along with the 2D disparity vectors resulted in adjusting following coding tools to nonlinear camera arrangement:

- Disparity Compensated Prediction (DCP),
- Neighboring Block Disparity Vector (NBDV),
- Depth-oriented NBDV (DoNBDV),
- View Synthesis Prediction (VSP),
- Inter-view Motion Prediction (IvMP), and
- Illumination Compensation (IC).

In the improved 3D-HEVC extension, the bitstream is modified only by adding a dedicated extension to the Video Parameter Set, which contains calculated projection matrices for each of the encoded views. This additional data is negligibly small compared to the amount of data produced by 3D-HEVC encoder.

3.4.2 Methodology of Evaluation

The evaluation of the improved 3D-HEVC codec was performed by comparing its efficiency with the unmodified 3D-HEVC codec, MV-HEVC, and HEVC simulcast. The HEVC encoder was used as a reference in order to present the efficiency of multiview encoders. All the encoders were compiled using HTM 13.0 reference software in order to obtain a reliable comparison. During the experiments, the encoders were configured as follows:

- Number of encoded frames: 50,
- Quantization parameter for views: {25, 30, 35, 40},
- Quantization parameters for depth (3D-HEVC only): {34, 39, 42, 45},
- GOP size: 8, and
- View synthesis optimization (3D-HEVC only): off.

The remaining configuration parameters were set according to JCT-3V Common Test Conditions (CTC) [61]. The evaluation was performed using three views of 10 multiview test sequences. The views in those sequences are arranged either on a line (Poznan Street, Poznan Hall 2 [62], Dancer [63], Balloons, Kendo [64], and Newspaper [65]), or an arc (Poznan Blocks [66], Big Buck Bunny Flowers [67], Ballet, and Breakdancers [68]). The quality of the views was assessed using averaged bitrate reduction for luma PSNR (Peak Signal-to-Noise Ratio), calculated with the Bjøntegaard formula [69]. It is an objective measure, commonly used for evaluation of video quality.

3.4.3 Results

Figure 3.15 presents the results of evaluation, divided into two groups to stress out the difference in compression efficiency in case of encoding views arranged on a line ("linear" sequences) and on an arc ("circular" sequences).

The achieved results show that the inter-view prediction can highly improve the compression efficiency of multiview video. For the linear camera arrangement, the multiview encoders produce roughly 30–50% lower bitrate than HEVC simulcast. On the other hand, the gain for compression of views arranged on an arc is much lower and oscillates in the range of 5–30%.

The improved 3D-HEVC encoder provides a significant improvement in compression of circularly arranged views, compared to the unmodified 3D-HEVC. This observation proves that the 2D disparity vectors allow predicting the content between two views more accurately than simple horizontal shifts. Additionally, this improvement introduces negligibly small raise in bitrate for compression of 1D parallel views. As mentioned before, this is caused by the necessity of transmitting full set of camera parameters, which was not the case in the original 3D-HEVC. Nevertheless, a major advantage of the improved encoder is that it is not restricted to any camera arrangement.

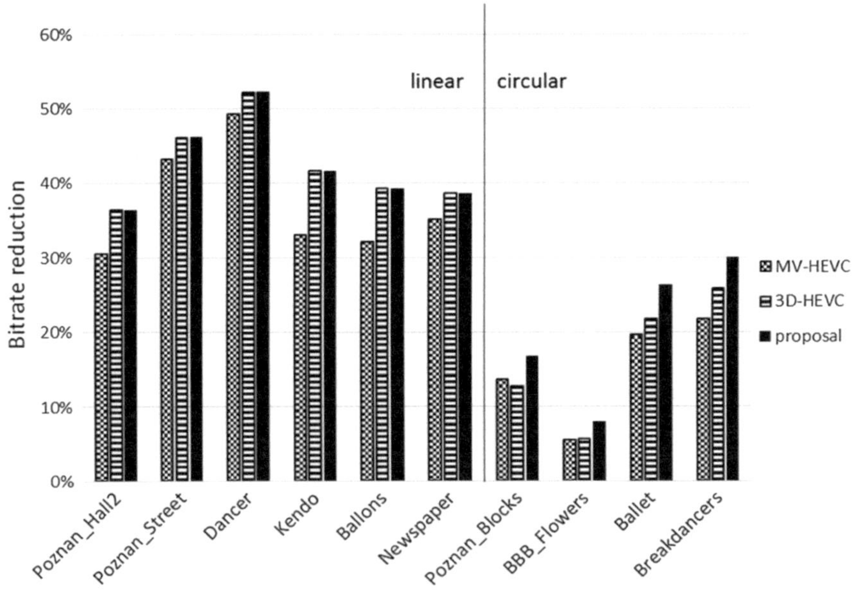

Fig. 3.15 Bitrate reduction against HEVC simulcast © 2016 IEEE [74]

3.4.4 Remarks

The number of applications that utilize multiview video acquired with 1D parallel camera setup is very limited. A more desired solution is to distribute the cameras freely, because it allows for obtaining much more information about the scene. Unfortunately, the state-of-the-art techniques for compression of multiview video do not support efficient coding of such content. The modifications of 3D-HEVC codec, developed by Zhejiang University and Poznan University of Technology, provide a significant improvement in compression of multiview video acquired with camera setup other than linear. This shows that there is still room for improvements in multiview video compression. As the multi-camera systems produce enormous amount of data, it is very important to develop efficient and universal technique for compression of such content.

3.5 Recent Developments in Video Compression with Capabilities Beyond HEVC

Digital video codecs continuously evolve beyond HEVC performance and with new capabilities to compress formats such as UltraHD, HDR, and omnidirectional 360° video. The current status of MPEG exploration experiments of next-generation video codec (NGVC) technologies is outlined in this section. In the first part, the high-quality UltraHD format compression performances beyond HEVC are presented. In the second part, the recent developments in extended dynamic/color-volume HDR/WCG format conversion and coding are highlighted. In the third part, NGVC testing procedure and 3D projection formats for 360° video are evaluated.

Currently, the MPEG investigates benefits of next-generation video coding technology (NGVC) which could improve the *compression performance* or give *new functionality*, as compared to HEVC. Further, test cases and evaluation methodologies for assessment of such benefits are investigated. As a first step, MPEG addressed in April 2017 the *Call for Evidence* (CfE) to interested parties which are in possession of technology providing better compression capability than the existing standard, either for conventional video material or for other domains such as HDR/WCG or 360° (VR) video. Based on the outcome of the call, and promising evidence that potential technology exists, MPEG and VCEG will produce a formal *Call for Proposals* (CfP) later in the year, with the intent to enter a more rigid standardization phase for the next generation of video compression standards beyond HEVC. A preliminary target date for completion of a new standard on the subject is late 2020. The planned phases for the 5-year standardization and roadmap are shown in Fig. 3.16 [MP20 Workshop on *MPEG Standardization roadmap*, Oct. 2016].

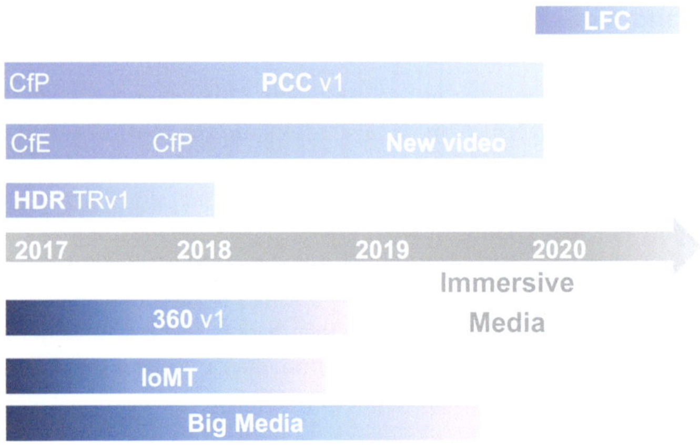

Fig. 3.16 Standardization roadmap for 3D media coding and systems and tools. [Doc. N16775 *MPEG Strategic standardization roadmap*, April 2017, Doc. N16804 *MPEG Time line*, April 2017]

New functionalities in ***immersive video*** technology (MPEG-I) enable to surround the user with a large field of view video (up to 360°) through VR headset or large 3D video walls. The user is presented different viewpoints to surroundings, corresponding to body/head movements in a limited volume around a central position [Doc. W16719 MP20 Workshop on *Global media technology standards for an immersive age*, Jan. 2017]. ***Free viewpoint television*** (MPEG-I FTV) technology synthesizes surround from a lower number of input camera views (for cost reasons) and real-time render of a large number of multiview images on super multiview 3D displays in application free navigation (FN) scenario [Doc. N15095 *Draft Call for Evidence on FTV*, Feb. 2015]. ***Omnidirectional*** **(360°)** ***video*** (MPEG-I/JVET) provides immersive experiences based on interactivity between the user and the content. However, the market fragmentation due to lack of appropriate standards on storage and delivery format for such content is becoming one of the strong concerns by the industry. MPEG plans to standardize optimized video coding technologies, delivery mechanism, the application format, and other relevant technologies [Doc. N16542 Summary of the *Survey on Virtual Reality*, Oct. 2016]. Based on survey feedback results, MPEG aligned its standardization roadmap with the expected deployment timelines. A standard addressing video and audio coding for six degrees-of-freedom (6DOF) where users can freely move around is on 5-year roadmap. ***Light Field Coding*** (MPEG-I/JPEG Pleno) cameras have been introduced as a new technology that has the potential to dramatically increase the sense of immersion that can be achieved by images and video captured by cameras, and arrays of cameras focused on a particular scene [70]. A key issue with the technology describing light information at all positions and from all viewing directions of the scene is the incredibly large amount of data necessary to achieve the desired sense of immersion, without unwelcome impacts on the user (i.e., motion sickness, viewer fatigue, and eye strain) [Doc. N16532 *Call for light field test material including plenoptic cameras and camera arrays*, Oct. 2016]. ***3D Point clouds*** (MPEG-I PCC) have recently emerged as representations of the real world enabling more immersive forms of interaction and communication to better understand and navigate it. They are typically captured using various setups of multiple cameras, depth sensors, LiDAR scanners, etc., but can also be generated synthetically. Point clouds can have attributes such as colors, material properties, and/or other attributes. The standard targets both efficient geometry and attributes compression, scalable/progressive coding, and coding of sequences of point clouds captured over time. In addition, the compressed data format should support random access to subsets of the point cloud [Doc. N16763 *Call for Proposals for Point cloud compression*, Apr. 2017]. Next, MPEG identifies the need for ensuring the interoperability among *multimedia-centric* ***Internet of Things*** (MPEG IoMT). A standard specifies interaction commands, the protocols and format of the aggregated and synchronized data which facilitate the large-scale deployment of complex media systems that can exchange data interoperably between media things and media wearables [Doc. N16535 *Call for Proposals on Internet of Media Things and Wearables*, Oct. 2016—Jan. 2018 FDIS approval]. ***Big Media*** (MPEG Exploration Part 21) has current focus on standardization gap analysis and on the development of a conceptual model for multimedia-related functionalities in context of Big Data. A huge amount of data comes from audiovisual sources or has a multimedia nature. However, audiovisual data are currently not incorporated in the Big Data standardization paradigm. MPEG has started analyzing the need for Big Media standards [Doc. N16565 *Liaison Statement from SC29/WG11 to JTC1/WG9 on Big Media*].

3.5.1 UltraHD Video Compression Performance Beyond HEVC

A next-generation of video compression (NGVC) technology would be needed by the beginning of the next decade that has sufficiently higher compression capability than the HEVC standard and give new functionality which supports professional high quality, high resolution (UltraHD), and extended dynamic/color-volume (HDR/WCG) video, as well as omnidirectional (360°) video. For some of the markets, the new standard will strive to reduce the bitrate for storage and transport of video by 50%. Special attention is given to support developing markets like augmented and virtual reality (AR/VR), unicast streaming, automotive applications, and multimedia-centric Internet of Things (IoMT). For these markets, special attention will be given to seamlessly enable the required functionality by close integration with transport and storage to provide efficient personalized interactive services, as well as appropriate 360° projections to enable VR applications.

Therefore, though the further improvement of compression performance is expected to play the major role in this development, adaptation capabilities for usage in various network environments, a variety of capturing/content generation, and display devices are considered important as well [Doc. N16359 *Requirements for a future video coding standard v3*, June 2016]. Professional video format UltraHD (*Ultrahigh Definition*), formally specified in Recommendation ITU-R BT.2020 and SMPTE ST2036-1, defines higher spatial resolutions with 4 K-UHD (3.840 × 2.160) and 8 K-UHD (7.680 × 4.320) with aspect ratio of 16 × 9 and progressive image sample structures and higher frame rates (HFR) up to 120 Hz as well as higher sample bit depths (DCR) up to 12 bits and a wider color gamut (WCG) encoding and digital representation [ITU-R BT.2020-2 *Parameter values for ultrahigh definition television systems for production and international programme exchange*, Oct. 2015]. There are the three different UltraHD format elements: the programme *production* standard, the broadcast *delivery* standard, and the *display* standard. The UltraHD production image standard was agreed in the ITU-R BT.2020 in 2012 in all elements except one—high-dynamic-range (HDR) system. At the February 2016 meeting of ITU-R WP6C, the recommendation BT. 2100 was approved and published [ITU-R BT. 2100 *Image parameter values for high-dynamic-range television for use in production and international programme exchange*, July 2016].

The MPEG adopted an open standardization approach in the development of the video codec specifications. All inputs and contributions to an MPEG meeting are made by documents, which are registered in a publicly accessible document repository. A set of deliverables, which turn to become normative or remain to be supplemental in their final form, are also publicly accessible. These comprise the *specification text* itself, the *reference software*, a *conformance specification*, and the *test model*. Furthermore, a *verification report* is produced which documents and demonstrates the achieved performance. Three steps of MPEG standardization development are based on formal subjective assessment of the video quality:

- *Call for Evidence* (**CoE**) purpose is to explore *in house* whether the coding efficiency and the functionality of the current version of HEVC standard can be further improved.
- *Call for Proposals* (**CfP**) on video coding technology is open to external parties with primary goal to define compression technology. To evaluate the proposed technologies, formal subjective tests are performed. Results of these tests are made public.
- *Verification tests* (**VT**) for video coding technology include test conditions, evaluation methodology, and timeline to assess the improvement of the coding performance.

Verification Tests

HEVC (*High Efficiency Video Coding*) is the joint video coding standardization project of the ITU-T Q.6/SG16 (*Video Coding Experts Group*) and ISO/IEC JTC1/SC29/WG11 MPEG (*Moving Picture Experts Group*). The JCT-VC (*Joint Collaborative Team on Video Coding*) was established to work on HEVC project and publish specifications:

- Edition 1.0 ITU-T Recom. H.265 V1 (2013-04-13) *First base specification.*
- Edition 2.0 ITU-T Recom. H.265 V2 (2014-10-29) *RExt, SHVC, MV extensions.*
- Edition 3.0 ITU-T Recom. H.265 V3 (2015-04-29) *3D-HEVC extension.*
- Edition 4.0 ITU-T Recom. H.265 V4 (2016-12-22) *SCC extension.*
- Edition 5.0 ITU-T Recom. H.265 V5 (2017-12-??) *HDR/WCG extension.*

A major goal for the development of the HEVC standard was to achieve a substantial improvement in compression capability relative to its predecessor, the AVC (*Advanced Video Coding*) standard. A subjective evaluation was conducted comparing the HEVC Main profile to the AVC High profile. The verification test compared visual quality for 20 video sequences with resolutions ranging from UltraHD to HD Ready that were encoded at various bitrates or quality levels [Doc. Q1011 *Report on HEVC compression performance verification testing*, March 2014]. Analysis of the subjective test results shows that HEVC test points at half or less than half the bitrate of the AVC reference were found to achieve comparable quality in 86% of the cases. Estimation of the bitrate savings from these results confirmed that the HEVC Main profile achieves the same subjective quality as AVC High profile while requiring on average approximately 59% fewer bits. The average bitrate savings for test sequences with 4 K-UltraHD resolution is estimated at approximately 64%. The tests were conducted according to the verification test environment and methodology as described in the HEVC verification test plan [Doc. P1011 *HEVC Verification test plan*, Jan. 2015]. The verification test was conducted using the HM12.1 (Reference HEVC codec) and JM18.5 (reference AVC codec). Each picture resolution was represented by five test sequences, giving a total of 20 test sequences. For each test sequence, four test points encoded using the HM12.1 and JM18.5 reference software were chosen. The JM18.5 reference points were chosen such that the quality levels span the entire range of the MOS (*Mean Opinion Score*) scale and PSNR (*Peak Signal-to-Noise Ratio*) range. The test results were collected, processed, and presented in the in Table 3.4 for the 4 K-UltraHD resolution test sequences.

JVET Coding Experiments

As previously planned and announced at its 109th meeting, MPEG (together with ITU-T SG 16 VCEG) hosted a brainstorming workshop at the October 2014 to

Table 3.4 HEVC video compression performance results versus AVC in terms of bitrate BR savings for five 4 K-UltraHD test sequences in objective (PSNR) and subjective (MOS) evaluation [Doc. JCTVC-Q1011, *HEVC verification test report*, April 2014] [75]

UltraHD test sequence	BR savings (PSNR) (%)	BR savings (MOS) (%)
BT709 Birthday	62	75% average in range [66–83]
Book	59	66% average in range [50–76]
Homeless sleeping	76	*No available*
Manege	33	56% average in range [39–68]
Traffic	41	58% average in range [44–70]

explore use cases, requirements, and potential timelines for the development of future video coding (FVC) standards [Doc. M34630 *Brainstorming panel discussion session on future video coding*, Oct. 2014]. MPEG has recognized a need to further study future application requirements, and the availability of technology developments to fulfill these requirements. Toward establishing a roadmap for future video coding standardization, MPEG has established two ad hoc groups to conduct this study:

1. AHG on *Future Video Coding technology*:

   [Doc. AZ01 *Report of AHG1 on Coding efficiency improvements*, June 2015]

- to discuss and identify challenges in video coding technology beyond HEVC,
- to identify suitable test cases and materials, and
- to solicit contributions on potential improvements in video compression.

2. AHG on *Industry needs for Future Video Coding*

   [Doc. F0102 *Industry recommendation for FVC standard development,* April 2017]

- collect information on industry needs for future video coding,
- identify new use cases for existing and emerging markets for video coding technology, and
- solicit presentations on emerging markets.

At the October 2015 meeting of ITU-T SG16/Q6 VCEG and ISO/IEC JTC1 SC29WG11 MPEG, both organizations have founded a joint team for exploration of future video coding technology. This meeting saw a large number of contributions addressing the next generation of video coding (NGVC). In order to make this study more efficient, it was agreed to establish **JVET** (*Joint Video Exploration Team*) with MPEG. The reference software for the JVET group was named JEM (*Joint Exploration Model*). The software effort was originally started by VCEG under the name HM KTA [4th-generation *Key Technology Areas* study Jan-Oct. 2015]. The JEM software is based on the HM (*HEVC Model*) software that has been jointly developed as reference (example) software for the HEVC standard.

Initial simulation software HM KTA had 10% improvement over HEVC HM 16.6 Main 10 [Doc. M36782 *Report of AHG on Future video coding standardization challenges*, June 2015]. New test model JEM 1.0 (*Joint Exploration Model*) based on well-understood and straightforward techniques had Y = 21% U = 17% V = 15% average improvement in color components of four test sequences with *Class A-D* resolutions. Next version JEM 2.0 (Feb. 2016) had Y = 15.4% U = 23.5% V = 20.1% (*AllIntra* codec configuration) and Y = 20.8% U = 29.9% V = 23.8% (*RandomAccess* codec configuration) average improvement in color components of UltraHD test sequences based on enhanced techniques (Table 3.5).

In HEVC hybrid video coding scheme shown in Fig. 3.17, the pictures of the input UltraHD video sequence are fed to the encoder. A prediction signal generated from information available on both, the encoder and the decoder side of the system

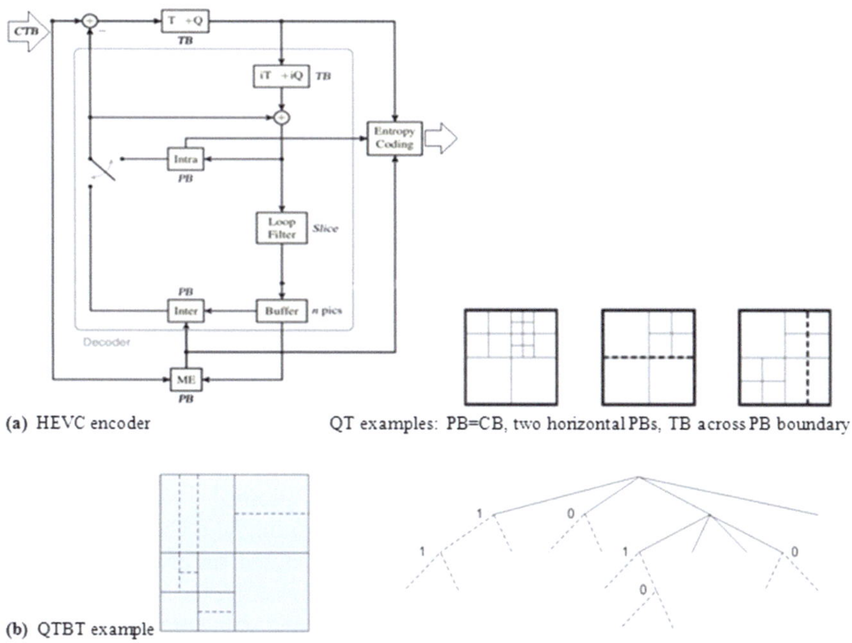

Fig. 3.17 **a** HEVC encoder scheme (Motion Estimation, Quantization, Transform) and example quadtree (QT) partitionings into prediction blocks PBs (*dashed lines*) and transform blocks TBs (*solid lines*). **b** JEM 2.0 algorithm improvements based on quadtree plus binary tree segmentation (QTBT) [Doc. B1001 *Algorithm description of JEM2*, Feb. 2016]

Table 3.5 Enhanced techniques implemented in simulation software JEM 2.0 (*Joint Exploration Model*) based on tradeoffs compression versus complexity

Technique	Experimental coding tools
Segmentation	Quadtree plus binary tree segmentation (QTBT), larger coding tree units and larger transform blocks (CTU size is set to be 256 × 256 by default)
Intra-prediction	Intra-mode coding with 67 prediction modes, four-tap intra-interpolation filter, boundary prediction filters, cross-component prediction, position-dependent prediction combination, adaptive reference sample smoothing
Inter-prediction	ATMVP improvement: sub-PU based motion vector prediction, adaptive motion vector resolution, higher precision motion vector storage, overlapped block motion compensation, local illumination compensation, affine motion compensation prediction, pattern-matched motion vector derivation, bidirectional optical flow
Transform	Adaptive multiple core transforms (AMT), secondary transforms (applied between forward core transform and quantization—at encoder, and between de-quantization and inverse core transform—at decoder side), signal-dependent KLT transform
Entropy coding	CABAC improvement: context model selection for transform coefficient levels, multiple adaption rate probability estimation with context-dependent updating speed, adaptive initialization for context models

is subtracted from the input signal. The residual, representing the resulting prediction error, is transformed, quantized, and encoded into the bitstream. The prediction parameters needed to reproduce the prediction signal at the decoder side are encoded into the bitstream as well. Under the assumption of error-free transmission, the encoder and decoder sides are synchronized since the encoder includes the complete prediction structure of the decoder. The HEVC specification text distinguishes between *blocks* and *units*. While the former address a specific area in a sample array (e.g., luma, *Y*), the latter comprise the collocated blocks of all encoded color components (*Y*, Cb, Cr, or monochrome) as well as all syntax elements and prediction data that is attached to the blocks (e.g., motion vectors). The base entities are the CTB (*Coding Tree Block*) and the corresponding CTU (*Coding Tree Unit*). The CTU contains the CTBs of the encoded color components and forms a complete entity in the bitstream syntax. A CTB is the root of a quadtree partitioning into CB (*Coding Blocks*). A Coding Block is partitioned into one or more PB (*Prediction Blocks*) and forms the root of a quadtree partitioning into TB (*Transform Blocks*). A corresponding set of units is specified which comprise the block and the respective syntax structure, each. Accordingly, a CU (*Coding Unit*) contains the PU (*Prediction Units*) and the tree-structured set of TU (*Transform Units*). While a PU contains the joint prediction information for all color components, a TU contains a separate residual coding syntax structure for each color component. The location and size of the CBs, PBs, and TBs of the luma component apply to the corresponding CU, PU, and TU. Accordingly, the locations and sizes for the chroma blocks are derived from the corresponding luma blocks.

The current version of JEM 6.0 test model is build up on top of the HEVC test model. The basic encoding and decoding flowchart of HEVC is kept unchanged in the JEM; however, the design elements of most important modules, including the modules of block structure, intra- and inter-prediction, residue transform, loop filter, and entropy coding, are somewhat modified and additional coding tools are added [Doc. N16887 *Algorithm description of Joint Exploration Test Model 6 (JEM6)*, Apr. 2017]. The description of encoding strategies used in experiments for the study of the new technology in the JEM is also provided. Exploration Experiments (EEs) are performed in order to get better understanding of technologies considered for inclusion to the next version of JEM, analyze and verify their performance, complexity, and interaction with existing JEM tools [Doc. N16889 *Description of Exploration Experiments on coding tools*, Apr. 2017]. The characteristics of test sequences that were offered for usage in evaluation of new compression technology are described in work plan. The objective is to understand the characteristics of 4 K-UltraHD test sequences and the coding performance difference between HM and JEM. New preselected HDR sequences are to be encoded with HM using the JVET HDR and WCG test conditions at the specified base QP values. One other purpose is to study test sequence and rate points for CfE/CfP test conditions [Doc. N16512 *Work plan for assessment of test material*, Oct. 2016].

Recently, JVET development enters a more rigorous evaluation phase, by issuing a preliminary CfE (*Call for Evidence*) in January 2017, potentially followed by a formal CfP (*Call for Proposals*). This preliminary CfE will be updated in early

April 2017, and responses to the CfE will be evaluated in July 2017. The possible subsequent CfP would follow the evaluation of the responses to the CfE. Assessment will be made based on objective criteria (such as rate savings judged by PSNR quality) as well as subjective quality evaluation (experts viewing) of conventional video material, or other domains such as HDR/WCG or 360° (VR) video. As test cases, the call defines rate points and materials in all of these latter categories, anchors with HEVC encodings are also provided. The scope of technology consideration includes a broad variety of video source content, e.g., camera-view content, screen content, VR/360 video and high-dynamic-range video for such applications as broadcast (with live or pre-authored content), real-time video conferencing and video chat, on-demand viewing, storage-based media replay, consumer generated content, and surveillance with fixed or moving cameras [Doc. N16886 *Joint Call for Evidence on video compression with capability beyond HEVC*, Apr. 2017].

3.5.2 Conversion and Coding for HDR/WCG Video

In 2013, MPEG started to explore the area of extended dynamic/color volume. In an initial phase, requirements were gathered, and exploration experiments were initiated [Doc. N14510 *Requirements and explorations for HDR/WCG content distribution and storage*, April 2014]. The exploration experiments covered a wide range of aspects related to the understanding of how HDR/WCG content differs from standard dynamic range (SDR) content and how this affects the attempts to compress it. Several different transfer functions were investigated, different color spaces were explored, and subsampling methods and different bit-depth representations were tested. One of the major challenges was finding reliable and efficient methods for subjective quality assessment of HDR video and collecting HDR video content that could be used as a representative test set for carrying out experiments. It was identified that all the available objective metrics suffered from poor correlation with subjective video quality. The HDR video content used as test material in MPEG was stored either in a display-referred 12-bit SMPTE ST 20841 RGB 4:4:4 format or in a linear light half-float EXR RGB 4:4:4 format. The sequences were all represented with BT.2020 color primaries, which were graded for up to 4000 cd/m^2 using either the P3 color gamut or the BT.709 color gamut [71].

Efficient representation of color pixel in an UltraHD video format reduces the amount of data by optimizing the use of integer code values and by restricting the range of light and color that can be represented for distribution or storage purposes. To compress HDR video content stored in floating-point values, three main approaches have been proposed:

1. **Derivation of metadata** to reconstruct HDR content from LDR or vice versa. Only one stream (original HDR or its tone-mapped LDR version) is then compressed and transmitted with the metadata embedded in the bitstream. In the

case of transmitting HDR content, it is first perceptually encoded and then compressed using a video codec.
2. **Using a perceptual curve**, for example, the PQ/HLG that transforms HDR content (floating-point data) into LDR content of high bit depth. The generated content is directly compressed using a single high-bit-depth codec. At the decoder side, the HDR content is reconstructed using the inverse of the perceptual curve. These techniques need only one codec and require a bit depth of 10 to 14 bits for a high-quality HDR content reconstruction.
3. **Scalable coding techniques** require two codec instances—LDR version and a second one to compress the missing data between the tone-mapped version and the original HDR one.

HEVC specification is already usable for HDR/WCG video distribution applications. HEVC versions comprise a set of coding tools and metadata relevant for increasing the dynamic range and widening the color gamut of UltraHD format: support of higher bit-depth signals (with the definition of a consumer profile, Main 10), BT.2020 color gamut for UltraHD format, PQ (*Perceptual Quantizer*, SMPTE ST 2084) and transfer functions (OETF) that can be used for HLG (*Hybrid Log-Gamma*, ARIB STD-B67) correction with RGB and YCbCr color components, one SEI message providing descriptive information on **color volume** (*the color primaries, white point, luminance range*) of the content display, three SEI messages embedding processing parameters for an efficient adaptation at rendering side (*tone mapping*, *knee function*, *color remapping*). In the context of scalable coding, HEVC contains two tools, the color gamut scalability and bit-depth scalability, allowing the support of HDR and WCG with backward compatibility with standard dynamic range SDR BT.709 HD format [ITU-R BT.709-6 *Parameter values for the HDTV standards for production and international programme exchange*, June 2015].

Fast-Track HEVC HDR10 Extension

MPEG has launched in June 2015 a fast-track standardization process to enhance the performance of the HEVC Main 10 profile for HDR/WCG video that would lead to the **HDR10** extension in October 2016 [Doc. N15084 *Requirements and use cases for HDR and WCG content coding*, Feb. 2015]. To achieve this, immediately following its meeting in February 2015, MPEG issued a CfE (*Call for Evidence*) of new tools that may improve the performance of HEVC when used to encode high dynamic range and wide color gamut video [Doc. N15083 *Call for Evidence for HDR and WCG video coding*, Feb. 2015]. A set of anchors targeting broadcast/OTT bitrates are provided using the HEVC Main 10 codec. After a successful call for evidence for High Dynamic Range (HDR), the technical work starts in the video subgroup with the goal to develop an architecture as well as core experiments (CE). Verification tests conducted in-between the April and May 2016 [Doc. X1018 *Verification test report for HDR/WCG video coding using HEVC Main 10 Profile*, June 2016. Doc. N16506 *Revised verification test report for HDR/WCG video coding using HEVC Main10 Profile*, Oct. 2016].

The proponents of the CfE were working together to create a test model, which is called the **ETM** (*Exploratory Test Model*). The test model was finalized in late 2015 [Doc. W0092 *Description of the exploratory test model for HDR/WCG extension of HEVC*, Feb. 2016]. The ETM process was designed as a normative post-processing step, called the *reshaper*. The reshaper was designed out of loop (i.e., after the HEVCMain 10 decoding step), and the output of the reshaper was viewable HDR video. Optionally, the reshaper could be bypassed to output SDR video for a backward-compatible configuration (*Mode 1*). In the nonbackward-compatible configuration (*Mode 0*), this output would be nonviewable, but would potentially provide improved HDR compression efficiency in return.

The HDR end-to-end system shown in Fig. 3.18 consists of four major stages: *pre*-encoding processes, encoding process, decoding process, and *post*-decoding processes. At the HEVC codec boundary AA', input and output are fixed-point nonlinear signal in 10-bit, narrow range, perceptual quantizer ST2084, subsampling 4:2:0, nonconstant luminance Y'CbCr representation. At the system boundary BB', input and output are floating-point linear RGB signal with BT.2020 color primaries. Differences when coding HDR data are as follows:

- gamma transfer function more nonlinear, several knock-on effects,
- subsampling can give luminance artifacts (avoided by subsampling *Luma Adjustment* procedure),

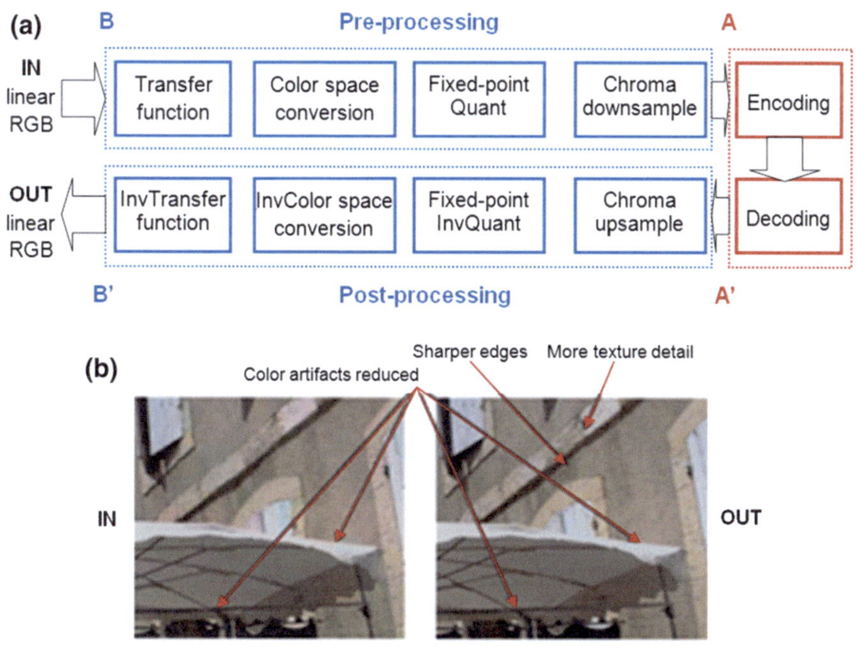

Fig. 3.18 **a** HEVC HDR10 video coding and conversion, **b** examples of typical coding artifacts and image improvements (color, edges, and texture)

- chroma values cluster around 0 value more than for SDR data (counteracted by encoder *Chroma QP offset* optimization), and
- more bits are spent in dark regions of image (counteracted by encoder *Luma DeltaQP* optimization).

Common test conditions (CTC) defines conversion practices and software reference configurations to be used in the context of experiments for HDR/WCG video coding [Doc. W1020 *Common test conditions for HDR/WCG video coding experiments*, Feb. 2016]. CTC are desirable to conduct experiments in a well-defined environment and ease the comparison of the outcome of experiments: test sequences, quantization parameter values, encoder configuration files, and the pre- and post-processing options to be used. The test method adopted for verification test is ITU-T P.910 DCR (*Degradation Category Rating*). The six test sequences are proposed to use. Bitstreams will be generated using HEVC Main 10 Profile, BT.2020 container, ST2084 transfer function, and NCL Y'CbCr color space conversion. All the test sequences have the following characteristics: resolution (1920 × 1080 progressive), original (not coding) color format (RGB 4:4:4), coding format (10 bit 4:2:0), and container (BT.2020 with gamut BT.709/P3D65 depending on the content). The conclusion reached in Feb. 2016 is as follows:

- no new profiles needed for HDR (without backward compatibility considerations),
- guideline development for 4:2:0 10 bit with PQ, and
- further work on backward compatibility.

On June 2015, it was decided that a technical report should be written on the topic of conversion and coding of HDR video using the Main 10 profile for Y'CbCr based on the nonconstant luminance color space and the SMPTE ST2084 transfer function [Doc. N16505 MPEG-H Part 14, Text of ISO/IEC PDTR 23008-14 *Conversion and coding practices for HDR/WCG*, Oct. 2016]. The technical report provides guidance on processing of high-dynamic-range (HDR) video with the purpose to provide a reference for recommended practice operation of HEVC when used for compressing HDR video:

- Processing steps for converting linear light, RGB, 4:4:4 video into ST2084, Y'CbCr, 4:2:0 video before encoding.
- Processing steps for converting ST2084, Y'CbCr, 4:2:0 to linear light, RGB, 4:4:4 after decoding.
- Some high-level recommendations for compression with HEVC are also included in this document. It is recommended to adjust the bit-distribution between chroma and luma, for example, by setting chroma QP offset. It is further recommended to adjust the bit-distribution between dark samples and bright samples, for example, by setting delta QP such that blocks with a high average luma value are assigned lower QP than blocks with a low average luma value. It is also recommended to take into account the activity (var) of a block when setting delta QP for the block.

Although the focus of these guidelines is primarily on 4:2:0 Y'CbCr 10-bit representations, these guidelines may also apply to other representations with higher bit depth or other color formats, such as 4:4:4 Y'CbCr 12-bit video. In addition, this document provides some high-level recommendations for compressing these signals using either the AVC or HEVC video coding standards. A description of post-decoding processing steps for converting these NCL Y'CbCr signals back to a linear light, 4:4:4 RGB representation is also included. The another technical report [Doc. N16508 MPEG-H Part 15, Working Draft 1 of TR: *Signalling, backward compatibility and display adaptation for HDR/WCG video*, Oct. 2016] complements and extends work in the conversion and coding guidelines. Specifically, this report expands on the application of ICtCp, HLG, and SEI messages in the coding of high dynamic range and wide color gamut video.

An evaluation of backward-compatible single layer HDR transmission reports the performance of two possible scenarios for distributing HDR content employing a single layer [Doc. W0106 *Evaluation of backward-compatible HDR transmission pipelines*, Feb. 2016]. One of them compresses HDR10 content, with color conversion and tone mapping performed at the decoding stage to generate SDR, while the other scenario compresses tone-mapped SDR content with inverse tone mapping and color conversion at the decoding side to generate HDR. Additional metadata can be derived to improve the quality of the color and tone mapping. Combining the findings with those from [Doc. M37318 *Evaluation of PQ versus tone mapping for single layer HDR video compression*, 2015] it is concluded that the first scenario (Fig. 3.19a) is preferable over that of the second (Fig. 3.19b).

Long-Track JVET Experiments

The second track standardization process addresses long-term requirements of a future video coding (FVC) standards. As a first step, JVET addressed in January 2017 the *Call for Evidence* (CfE) to interested parties which are in possession of technology providing better compression capability for HDR/WCG video [Doc. N16886 *Joint Call for Evidence on video compression with capability beyond HEVC*, Apr. 2017]. This preliminary CfE will be updated in early April 2017, and responses to the CfE will be evaluated in July 2017. The evaluation of the submissions to the CfE will be done using the *Expert Viewing Protocol* based on recommendation ITU-R BT.2095-0 with JVET participants serving as expert viewers. In addition, proponents are required to submit an input contribution with documentation of weighted PSNR values (at least average of frame wPSNR for each sequence and encoding point, separate for luma and chroma components), tPSNR-Y, deltaE100, and PSNR-L100. Metric definitions and coding conditions are provided in the JVET common test conditions and evaluation procedures for HDR/WCG video. It is requested to also provide the Bjøntegaard Delta-Rate [Doc. VCEG-AI11 *Improvement of BD-PSNR model*, July 2008] for each metric [Doc. N16890 *Common test conditions and evaluation procedures for HDR/WCG video coding*, Apr. 2017].

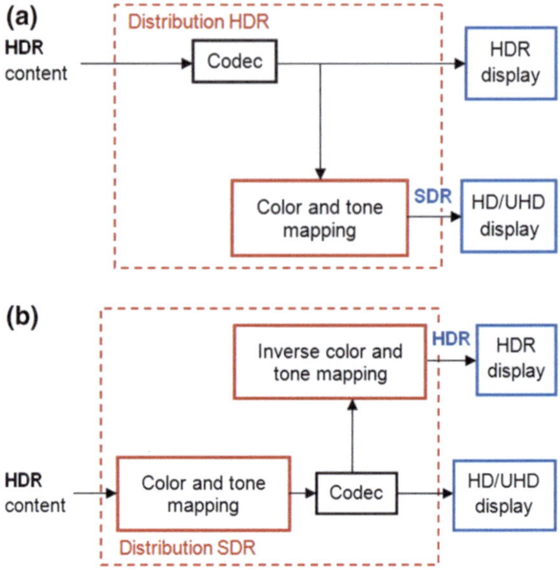

Fig. 3.19 Backward-compatible single layer HDR transmission: **a** distribution of HDR content and derivation of SDR content at the display stage, and **b** distribution of SDR content and derivation of HDR content at the display stage

3.5.3 Projection Conversions and Coding for 360° Video

On June 2016, MPEG established an *ad hoc* group on virtual reality (MPEG-VR/I) which conducted a survey on virtual reality. The understanding of the virtual reality (VR) potential is growing but the market fragmentation due to lack of appropriate standards on storage and delivery format for such content is becoming one of the strong concerns by the industry. The market is rapidly adapting to Virtual Reality (VR) to provide immersive experiences that go beyond what even an UltraHD can offer. Interactivity between the user and the content, high-quality immersive video (HDR, increased spatial resolution), and efficient delivery over existing networks are required. Therefore, MPEG plans to standardize optimized video coding technologies for VR, delivery mechanism, the application format, and other relevant technologies. Based on survey feedback results, MPEG aligned its standardization roadmap with the expected deployment timelines [Doc. N16542 Summary of the *Survey on Virtual Reality*, Oct. 2016].

Since there are no existing standards for end-to-end VR services shown in Fig. 3.20, the lack of interoperability is a significant challenge. An initial specification for 360° video and virtual reality services will be ready by the end of 2017 and is referred to as the MPEG-A OMAF [ISO/IEC 23000 Part 20 *Omnidirectional Media Application Format*, Jan. 2017]. A standard addressing video and audio coding for six degrees-of-freedom (6DOF) where users can freely move around is

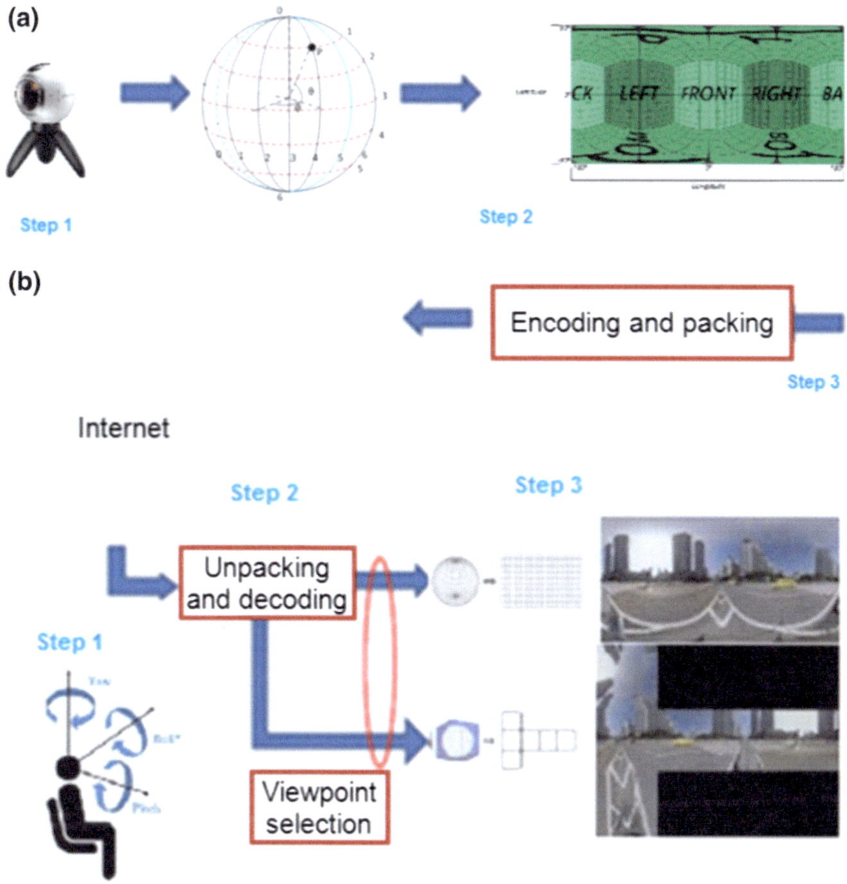

Fig. 3.20 Conversion and coding 360° video in VR end-to-end system: **a** multi-camera array captures video, image stitching to obtain spherical video, pre-processing spherical video to 2D plane projection and 2D video encoding, packing and delivery, and **b** unpacking and decoding, post-processing 2D plane to sphere projection given specific *viewport*, rendering and display

on 5-year roadmap. OMAF currently includes equi-rectangular projection as a projection format but it might include support of additional projection formats with a generalized extension mechanism during its further development. It includes signaling of necessary metadata for interoperable rendering of 360 degree monoscopic and stereoscopic audiovisual data, selection of audiovisual codecs for this application, and the technologies for storage of data in the ISO base media file format (ISOBMFF). The HEVC standard has been chosen as video codec because of its tiling capabilities and MPEG-H 3D audio has been chosen because of its capability of immersive audio representation [72, 74]. The standard will include technologies for the delivery of OMAF content with MPEG-DASH and MMT at a later stage.

As the size of 360° video is becoming a major bottleneck for VR applications and services, projection formats and projection conversions significantly impact on coding efficiency. The scope of future video coding technology consideration performed by the JVET (*Joint Video Exploration Team*) includes camera-view content, screen content, VR/360 video and HDR video. The 360° based end-to-end system consists of four major stages acquisition, *pre*-processing and encoding, decoding and *post*-processing, and rendering with the following characteristics:

- more pre- and post-processing are being applied from camera to display,
- such processing has become more closely related with the codec, and
- in addition to high compression efficiency in directional geometry/coding schemes, the FVC also needs to provide more functionality support (ease of perspective extraction without loss of coding efficiency).

Recently, JVET issued the *Call for Evidence* (CfE) to study the potential need to include 360° video coding technologies in the future video coding standard. [Doc. N16886 *Joint Call for Evidence on video compression with capability beyond HEVC*, Apr. 2017]. This preliminary CfE was issued in early April 2017, and responses to the CfE were evaluated in July 2017. For subjective 360° video evaluation, 2D rectilinear *viewports* will be extracted from the 360° × 180° omnidirectional video, using bilinear interpolation, similar to the default viewport extraction used in the 360 *Lib* software [Doc. N16888 *Algorithm descriptions of projection format conversion and video quality metrics in 360Lib*, Apr. 2017]. The 2D rectilinear viewports will be viewed on ordinary monitors, following the method described for SDR content. Dynamic rectilinear viewports are expected to be used, in which the *yaw* and *pitch* angles may change for each frame in the sequence. The particular dynamic viewports used for evaluation of each sequence will be selected after the submission of YUV files. Proponents are required to submit an input contribution with documentation of multiple objective metrics [Doc. N16891 *Common test conditions and evaluation procedures for* 360 *video*, Apr. 2017].

Evaluation Procedures and Projection Formats

Evaluation procedures for 360° video coding are based on common test conditions and software reference configurations. Common test conditions (CTC) are desirable to conduct experiments in a well-defined environment and ease the comparison of the results. HM HEVC Version 16.15 reference software is agreed to be used for most experiments. For 360-specific coding tools proposal, JEM Version 5.1 software is recommended. The *360Lib* software package is used for packing format manipulation, projection format conversion and video quality metrics computations. High fidelity test materials are provided in YUV4:2:0 format representing 360° video in ERP (*Equi-rectangular projection*). According to the testing procedure specified by Fig. 3.21, prior to the encoding those materials are converted to one of the projection formats (projections) representing 360 video: **ERP** (*equi-rectangular projection*), **ISP** (*icosahedral projection*) [JVET-E0029 Jan. 2017], **CMP** (*Cube Map Projection*), **OHP** (*Octahedron Projection*), **TSP**

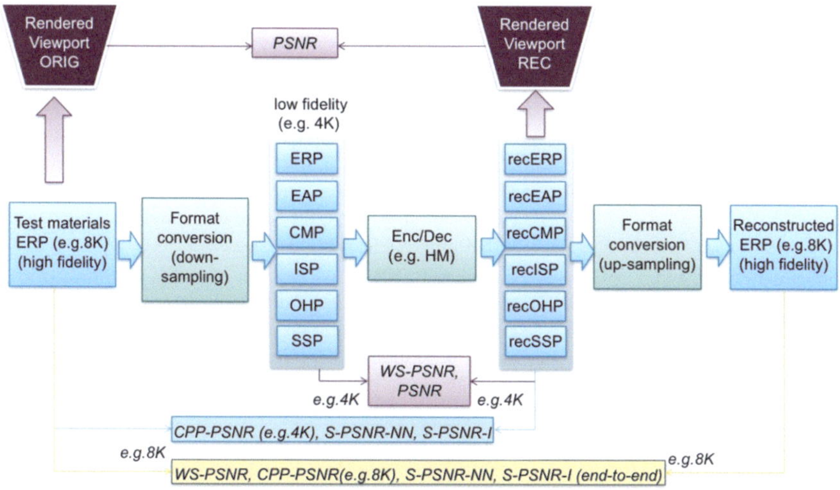

Fig. 3.21 Processing chain for 360° video in CTC testing procedure: end-to-end distortion measurement, cross-format distortion measurement, and coding distortion measurement

(*Truncated Square Pyramid Projection*) [JVET-D0071 Oct. 2016], **SSP** (*Segmented Sphere Projection*) [JVET-E0025 Jan. 2017], and **EAP** (*Equal-Area Projection*) [JVET-D0021 Oct. 2016]. Additionally to PSNR specific quality metric for spherical video listed below will be used: **WS-PSNR** [JVET-D0040 Oct. 2016], **CPP-PSNR** [JVET-D0027 Oct. 2016], **S-PSNR-I**, **S-PSNR-NN** (w/o interpolation) [JVET-D0021 Oct. 2016]. In total 11 different metrics will be reported: both PSNR reported by codec and WS-PSNR between input and output of the codec, CPP-PSNR, S-PSNR-I, and S-PSNR-NN between decoded output and input high fidelity ERP, CPP-PSNR, S-PSNR-I, S-PSNR-NN, and WS-PSNR between original and reconstructed ERP in highest resolution, and PSNR for two view ports.

Referring to Fig. 3.21, for a given input 360° video, projection format conversion is applied first to convert source projection format into coding projection format. In the CTC for 360° video, the coding projection format is in lower resolution than that of the source projection format. For example, 8 K source video is coded in 4 K resolution. After coding, the reconstructed signal in coding projection format is back converted to the source projection format at the source resolution. In the CTC, original video sequences are all provided in the ERP format in either 8 or 4 K.

Spherical Quality Metrics

Four spherical quality metrics are implemented in *360Lib* for 360 video quality evaluation (Table 3.6): weighted to spherically uniform PSNR (WS-PSNR), spherical PSNR without interpolation (S-PSNR-NN), spherical PSNR with interpolation (S-PSNR-I), and PSNR in *Crasters Parabolic Projection* format

Table 3.6 Objective quality metrics

PSNR	Conventional PSNR calculation with equal weight for all samples
Weighted to spherically uniform PSNR (**WS-PSNR**)	The distortion at each sample position is weighted by the area on the sphere covered by the given sample position. All samples on the 2D projection plane are used in WS-PSNR calculation. The two inputs to the metric calculation must have the same resolution and projection format
Spherical PSNR w/o interpolation (**S-PSNR-NN**)	Calculate PSNR based on a set of points uniformly sampled on the sphere. To find the sample value at the corresponding position on the projection plane, nearest neighbor rounding is applied. The two inputs to the metric calculation can have different resolutions and/or projection formats
Spherical PSNR with interpolation (**S-PSNR-I**)	Calculate PSNR based on a set of points uniformly sampled on the sphere. To find the sample value at the corresponding position on the projection plane, bicubic interpolation is applied. The two inputs to the metric calculation can have different resolutions and/or projection formats
CPP-PSNR	Apply another projection format conversion to convert the two inputs into the CPP domain, and calculate PSNR in CPP domain. The two inputs to the metric calculation can have different resolutions and/or projection formats

(CPP-PSNR) [75]. In order to evaluate viewport quality, viewport-based PSNR is also supported in 360 *Lib*.

Calculation of 360-video **objective** quality metrics is performed at different stages in the CTC testing procedure, summarized as the following three categories:

- **End-to-end distortion measurement:** WS-PSNR, CPP-PSNR, S-PSNR-I, and S-PSNR-NN are calculated between the original signal in source projection format and the reconstructed signal in source projection format. The end-to-end distortion considers both projection format conversion errors (including forward and backward projection format conversion) and coding errors.
- **Cross-format distortion measurement:** CPP-PSNR, S-PSNR-I, and S-PSNR-NN are measured between the original signal in source projection format and the reconstructed signal in coding projection format. Partial (only forward) projection format conversion errors and coding errors are measured.
- **Coding distortion measurement:** WS-PSNR and PSNR are measured between the input to the codec and the output of the codec. Only coding errors are measured, and projection format conversion errors are not measured.

Additionally, the CTC includes viewport quality evaluation using viewports generated from the original signal in the source projection format and the reconstructed signal in the coding projection format. Figure 3.22 shows the internal 3D

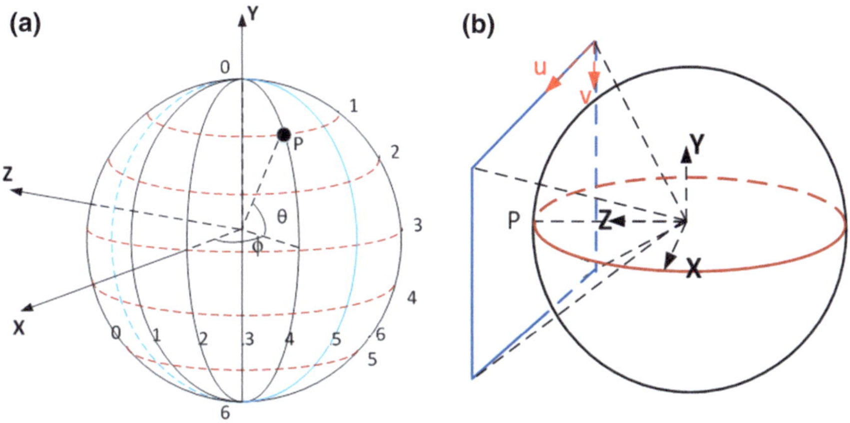

Fig. 3.22 **a** Internal 3D XYZ coordinate definition used to describe the 3D geometry of each projection 360 format, and **b** example of a viewport generation with rectilinear projection

XYZ coordinate definition used to describe the 3D geometry of each projection format representation. Starting from the center of the sphere, X-axis points toward the front of the sphere, Y-axis points toward the top of the sphere, and Z-axis points toward the right of the sphere. The sphere can be sampled with longitude (ϕ) and latitude (θ). The longitude ϕ in the range $[-\pi, \pi]$ is known as *yaw*, and latitude θ in the range $[-\pi/2, \pi/2]$ is known as *pitch*, where π is the ratio of a circle's circumference to its diameter. Rotation of this coordinate system is supported.

Subjective Testing Method

A subjective testing method is included for comparing 360 video projection formats using an HEVC codec [Doc. N16892 *Subjective testing method for comparison of 360 video projection formats using HEVC*, Apr. 2017]. The test sequences are defined in the JVET *Call for Evidence* 360 video section. Five sequences are included: *SkateboardInLot*, *ChairliftRide*, *KiteFlite*, *Harbor*, and *Trolley*. The video evaluation procedure for dynamic view ports is described in CfE [Doc. N16886 *Joint Call for Evidence on video compression with capability beyond HEVC*, Apr. 2017].

The call for evidence requested responses for use cases of video coding technology in three categories: standard dynamic range (SDR), high dynamic range (HDR), and 360° omnidirectional video. The evaluation of the responses at the July 2017 MPEG meeting included subjective testing of the video quality produced by candidate video coding technology. Two responses were received in the SDR category, two responses in the HDR category, and four in the 360° category. The results indicate that for a considerable number of test cases, a significant gain over HEVC had been demonstrated, with comparable subjective quality at 40-50% less bitrate compared to HEVC for the SDR and HDR test cases. In single cases, even higher rate savings could be observed. The substantial benefit was also shown for several 360° video test cases. It has thus been concluded that evidence exists of compression technology that may significantly outperform HEVC that could be used to develop a new standard. As a consequence, MPEG has proceeded toward issuing a formal Call for Proposals (CfP), expected to be issued in October 2017 [73].

Acknowledgements This book chapter was partially supported by COST Action IC1105—3D-ConTourNet.

Sections 3.2 and 3.3 were supported by the Ministry of Education, Youth and Sports (MEYS) of the Czech Republic no. LD15020 (QOCIES) and by the BUT project no. FEKT-S-17-4426. The research described in these sections was financed by Czech Ministry of Education in frame of National Sustainability Program under grant LO1401. For research, the infrastructure of the SIX Center was used. Section 3.4 was supported by National Science Centre, Poland according to the decision DEC-2012/05/B/ST7/01279.

Section 3.5 was supported by the Ministry of Education, Science and Technological Development of the Republic of Serbia under Grant TR-32034, and Secretary of Science of APV under the Grant 142-451-2484/2017-01/01.

References

1. Sin-Yi, J., Chang, N.Y.-C., Chin-Chia, W., Cheng-Hei, W., Kai-Tai, S.: Error analysis and experiments of 3D reconstruction using a RGB-D sensor. In: IEEE International Conference CASE. Taipei (2014)
2. Belhaoua, A., Kohler, S., Hirsh, E.: Estimation of 3D reconstruction errors in a stereovision system. In: Proceedings Modeling Aspects in Optical Metrology. Mnich, Germany (2009)
3. Kyt, M., Nuutinen, M., Oittinen, P.: Method for measuring stereo camera depth accuracy based on stereoscopic vision. In: Proceedings of SPIE 7864. San Francisco, California, USA, (2011)
4. Knight, J., Reid, I.: Active visual alignment of a mobile stereo camera platform. In: IEEE International Conference on Robotics and Automation Proceedings (ICRA). San Francisco, CA (2000)
5. Resko, B., Baranyi, P.: Stereo camera alignment based on disparity selective cells in the visual cortex. In: IEEE 3rd International Conference on Computational Cybernetics (ICCC) (2005)
6. Chang, C., Chatterjee, S.: Quantization error analysis in stereovision. In: 26th Conference on Signals, Systems & Computers. Pacific Grove, CA (1992)
7. Fooladgar, F., Samavi, S., Soroushmehr, S.M.R.: Geometrical analysis of altitude estimation error caused by pixel quantization in stereo vision. In: 20th Iranian Conference on Electrical Engineering (ICEE). Tehran, Iran (2012)
8. Belhaoua, A., Kohler, S., Hirsh, E.: Error evaluation in a stereovision-based 3D reconstruction system. EURASIP J Image Video Process **2010**, 1–12 (2010)
9. Kamencay, P., Breznan, M., Jarina, R., Lukac, P., Zachariasova, M.: Improved depth map estimation from stereo images based on hybrid method. Radioengineering **21**(1), 70–78 (2012)
10. Chang, W., Cho, K., Ryu, W., Lee, S.-Y.: Error cost function for mirror-based three-dimensional reconstruction. Electron. Lett. **50**(16), 1134–1136 (2014)
11. Fooladgar, F., Samavi, S., Soroushmehr, S.M.R., Shirani, S.: Geometrical Analysis of Localization Error in Stereo Vision Systems. IEEE Sens. J. **13**(11), 4236–4246 (2013)
12. Zhao, W., Nandhakumar, N.: Effects of camera alignment errors on stereoscopic depth estimates. Pattern Recogn. **29**(12), 2115–2126 (1996)

13. Ding, X., Xu, L., Wang, H., Wang, X., Lv, G.: Stereo depth estimation under different camera calibration and alignment errors. Appl. Opt. **50**(10), 1289–1301 (2011)
14. Bolecek, L., Ricny, V.: Influence of stereoscopic camera system alignment error on the accuracy of 3D reconstruction. Radioengineering **24**(2), 610–620 (2015)
15. Craig,J.: Introduction to Robotics: Mechanics and Control, 3rd edn. Pearson, Prentice Hall (2004)
16. Wheatstone, C.: Contributions to the physiology of vision I: on some remarkable and hitherto unobserved phenomena of vision. Phil. Trans. R. Soc. (Biol.) **18**(13), 371–375 (1838)
17. Angueira, P., Vega, D.L.L., Morgade, J., Velez, M.M.: Transmission of 3D Video over Broadcasting. In: Zhu, C., Zhao, Y., Yu, L., Tanimoto, M. (eds.) 3D-TV system with depth-image-based rendering, pp. 299–344. Springer, Heidelberg (2012)
18. Lebreton, P., Barkowsky, M., Raake, A., Callet, P.L.: 3D Video In: Möller, S., Ra-Ake, A. (eds.) Quality of Experience Advanced Concepts, Applications and Methods, pp. 299–313. Springer, Heidelberg (2014)
19. Liu, Y., Yang, J., Chu, R.: Objective evaluation criteria for shooting quality of stereo cameras over short distance. Radioengineering **24**(1), 305–313 (2015)
20. Merkle, P., Muller, K., Wiegand, T.: 3D Video: acquisition, coding, and display. IEEE Trans. Consumer Electro. **56**(2), 946–950 (2010)
21. Slanina, M., Kratochvil, T., Ricny, V., Bolecek, L., Kaller, O., Polak, L.: Testing QoE in different 3D HDTV technologies. Radioengineering **21**(1), 445–454 (2012)
22. Polak, L., Kufa, J., Zach, O., Kaller, O., Bolecek, L., Slanina, M., Kratochvil, T.: Study of advanced compression tools for stereoscopic video by objective metrics. In: 26th international conference on Radioelektronika. Kosice, Slovakia (2016)
23. Vetro, A., Tourapis, A.M., Müller, K., Chen, T.: 3D-TV content storage and transmission. In: IEEE Trans Broadcast Spec Issue 3D-TV Horizon: Contents System Visual Percept, **57**(2), 384–394 (2011)
24. (2012) Digital Video Broadcasting (DVB); Frame Compatible Plano-stereoscopic 3DTV. ETSI TS 101 547, v1.1.1
25. (2013) Features of Three-Dimensional Television Video Systems for Broadcasting. ITU-R BT 2160–4, v1.1.1
26. Projector guide, Simple Guide in Process of choosing a Projector. Available online at: http://projector-guide.com/3d-dlp/side-by-side-three-d
27. Sound&Vision.: 3D Broadcast Formats. Available online at:http://www.soundandvision.com/content/3d-broadcast-formats#L5R1U1cc60v48GWT.97
28. Minoli, D.: 3DTV content capture, encoding and transmission building the transport infrastructure for commercial services. T&F Group, Boca Raton (2010)
29. Livolsi, B.: What Does "3D Ready" Mean? Dispelling the Myths about 3D Projection. Available online at: http://www.projectorcentral.com/what_does_3d_ready_mean.htm. (2010)
30. Merkle, P., Smolic, A., Muller, K., Wiegand, T.: Multi-view video plus depth representation and coding. In: 14th International Conference on ICIP. San Antonio, Texas (U.S.A.) (2007)
31. Müller, K., Merkle, P., Wiegand, T.: 3D video representation using depth maps. Proc. IEEE **99**(4), 643–656 (2011)
32. Vetro, A.: Frame compatible formats for 3D video distribution. In: 17th international conference on ICIP. Hong Kong, People's Republic of China (2010)
33. Smolic, A., et al.: Coding algorithms for 3DTV—a survey. IEEE Trans. Circuits Syst. Video Technol. **17**(11), 1606–1621 (2007)
34. Bing, B.: Next Generation Video Coding and Streaming. Wiley, New York (2015)
35. Su, G.-M., Lai, Y.-C., Kwasinski, A., Wang, H.: 3D Visual Communications. Willey, UK (2013)
36. Aflaki, P., Hannuksela, M.M., Hakkinen, J,, Lindroos, P., Gabbouj. M.: Subjective study on compressed asymmetric stereoscopic video. In: 17th International Conference on ICIP. Hong Kong, People's Republic of China (2010)
37. (2014) The International Telecommunication Union (ITU-T); Advanced video coding for generic audiovisual services. ITU-T Rec. H.264

38. VideoLAN Organization.: x264 free software library. Available online at:http://www.videolan.org/developers/x264.html
39. Fraunhofer, H.H.I.: H.264/AVC Software Coordination. Available online at: http://iphome.hhi.de/suehring/tml
40. Vetro, A., Wiegand, T., Sullivan, G.J.: Overview of the stereo and multiview video coding extensions of the H.264/MPEG-4 AVC Standard. In: IEEE Proceedings, **99**(4), 626–642 (2011)
41. VideoHelp – Forum.: Guides, Software. FRIM 3D-MVC Encoder/Decoder 1.26. Available online at: http://www.videohelp.com/software/FRIM
42. Muprhy, C.: Multiview Video Coding: H.264 Annex H (JMVC). Available online at: https://github.com/cmurphy/JMVC
43. Sullivan, G.J., Ohm, J.R., Han, W.J., Wiegand, T.: Overview of the high efficiency video coding (HEVC) standard. IEEE Trans. Circuits Syst. Video Technol. **22**(12), 1649–1668 (2012)
44. Fraunhofer, H.H.I.: High Efficiency Video Coding (HEVC). Available online at: https://hevc.hhi.fraunhofer.de
45. Sullivan, G.J., et al.: Standardized extensions of high efficiency video coding. IEEE J. Sel. Topics Signal Process **7**(6), 1001–1016 (2013)
46. Fraunhofer, H.H.I.: Multiview High Efficiency Video Coding (MV-HEVC). Available online at: https://hevc.hhi.fraunhofer.de/mvhevc
47. FFmpeg.: Cross-Platform Solution to Record, Convert and Stream Audio and Video. Available online:http://ffmpeg.org/download.html
48. Cheng, E., Burton, P., Burton, J., Joseski, A., Burnett, I.: RMIT3DV: Pre-Announcement of a Creative Commons Uncompressed HD 3D Video Database. In: 4th international workshop on QoMEX. Melbourne, Australia (2012)
49. Domanski, M., Grajek, T., Klimaszewski, K., et al.: Poznan Multiview Video Test Sequences and Camera Parameters. ISO/IEC JTC1/SC29/WG11 MPEG 2009/M17050. Xian, China (2009)
50. (2008) Subjective video quality assessment methods for multimedia applications. ITU-T Rec P 910
51. Zach, O., Slanina, M.: A matlab-based tool for video quality evaluation without reference. Radioengineering **23**(1), 405–411 (2014)
52. Wang, Z., Bovik, A.C., Sheikh, H.R., Simoncelli, E.P.: Image quality assessment: from error visibility to structural similarity. IEEE Trans. Image Process **13**(4), 600–612 (2004)
53. Pinson, M.H., Wolf, S.: A new standardized method for objectively measuring video quality. IEEE Trans on Broadcast **50**(3), 312–322 (2004)
54. Yasakethu, S.L.P., et al.: Quality analysis for 3D video using 2D video quality models. IEEE Trans. Consum. Electro **54**(4), 1969–1976 (2008)
55. Saygli, G., Goktug, C., Tekalp, A.M.: Evaluation of asymmetric stereo video coding and rate scaling for adaptive 3D video streaming. IEEE Trans. Broadcast. **57**(2), 593–601 (2011)
56. Sullivan, G.J., Boyce, J.M., Chen, Y., Ohm, J.-R., Segall, C.A., Vetro, A.: Standardized extensions of high efficiency video coding (HEVC). IEEE J. Selec. Topics Signal Process. **7** (6), 1001–1016 (2013)
57. Müller, K., Merkle, P., Wiegand, T.: 3D video representation using depth maps. Proc. IEEE **99**(4), 643–656 (2011)
58. Lu, Yu., Wang, Qing, Ang, Lu, Sun, Yule: Response to call for evidence on free-viewpoint television: Zhejiang University. ISO/IEC JTC1/SC29/WG11, MPEG2016/m37608. San Diego, US (2016)
59. 3D HEVC reference codec available online https://hevc.hhi.fraunhofer.de/svn/svn_3DVCSoftware/tags/HTM-13.0
60. Domański, Marek, Dziembowski, Adrian, Grzelka, Adam, Kowalski, Łukasz, Mieloch, Dawid, Samelak, Jarosław, Stankiewicz, Olgierd, Stankowski, Jakub, Wegner, Krzysztof: [FTV AHG] technical description of Poznan University of technology proposal for call for

evidence on free-viewpoint television. ISO/IEC JTC1/SC29/WG11, MPEG2016/m37893. San Diego, US (2016)
61. Müller, K., Vetro, A.: Common Test Conditions of 3DV Core Experiments Joint Collaborative Team on 3D Video Coding Extension Development of ITU-T SG 16 WP 3 and ISO/IEC JTC 1/SC 29/WG 11 7th Meeting: Doc. JCT3 V-G1100, San José, US, Jan. 2014
62. Domański, M., Grajek, T., Klimaszewski, K., Kurc, M., Stankiewicz, O., Stankowski, J., Wegner, K.: Poznań multiview video test sequences and camera parameters. ISO/IEC JTC1/SC29/WG11 MPEG Doc. M17050, Xian, China, Oct. 2009
63. Rusanovskyy, D., Aflaki, P., Hannuksela, M.M.: "Undo Dancer 3DV sequence for purposes of 3DV standardization. ISO/IEC JTC1/SC29/WG11 MPEG Doc. M20028, Geneva, Switzerland, Mar. 2011
64. Tanimoto, M., Fujii, T., Fukushima, N.: 1D parallel test sequences for MPEG-FTV. ISO/IECJTC1/SC29/WG11 MPEG Doc. M15378, Archamps, France, Apr. 2008
65. Ho, Y.S., Lee, E.K., Lee, C.: Multiview video test sequence and camera parameters. ISO/IECJTC1/SC29/WG11 MPEG Doc. M15419, Archamps, France, Apr. 2008
66. Domański, M., Dziembowski, A. Kuehn, A., Kurc, M., Łuczak, A., Mieloch, D., Siast, J., Stankiewicz, O., Wegner, K.: Poznan Blocks—a multiview video test sequence and cam-era parameters for Free Viewpoint Television. ISO/IEC JTC1/SC29/WG11 Doc. M32243, San Jose, USA, Jan. 2014
67. Big Buck Bunny test sequence available online http://www.bigbuckbunny.org/
68. Zitnick, C.L., Kang, S.B., Uyttendaele, M., Winder, S., Szeliski, R.: High-quality video view interpolation using a layered representation. ACM Trans. Graph. **23**(3), 600–608 (2004)
69. Bjøntegaard, G.: calculation of average psnr differences between RD-curves. ITU-T SG16, Doc. VCEG-M33, Austin, USA, Apr. 2001
70. Alves, G., Pereira, F., daSilva, E.A.B.: Light field imaging coding: Performance assessment methodology and standards benchmarking. In" Proceedings IEEE International Conference on Multimedia & Expo Workshops (ICMEW), 2016
71. Milovanovic, D., Kukolj, D.: Recent advances in UHD video coding technology: High Dynamic Range and Wide Color Gamut. In: Assuncao, P.A., Vanam, R. (eds.) IEEE COMSOC Multimedia Communications Technical Committee, MMTC Communications—Frontiers, Special issue on Ultra-high definition video communications, vol. 11(1), pp. 50–55, Jan. 2016
72. Milovanović, D., Kukolj, D., Bojković, Z.: Recent advances on 3D video coding technology: HEVC standardization framework, Chapter 4 in Connected media in the future Internet era (Kondoz, A., Dagiuklas, T. (eds.)), Springer-Verlag, pp. 77–106 (2016)
73. Ström, J., Samuelsson, J.: Progress report from MPEG. SMPTE Motion Imaging J. **125**(7), 80–84 (2016)
74. Samelak, J., Stankowski, J., Domański, M.: Adaptation of the 3D-HEVC coding tools to arbitrary locations of cameras", International Conference on Signals and Electronic Systems, ICSES 2016, Kraków, Poland, September 5–7 2016, pp. 107–112
75. Zakharchenko, V., Choi, K.P., Park, J.H.: Quality metric for spherical panoramic video. In: Proceedings 9970 Optics and Photonics, SPIE Optical Engineering + Applications, San Diego, 2016. pp. C1–9
76. Grewl, P.K., Viswanath, K.S., Golnaraghi, F.: Minimization of position uncertainty using 3-D stereo imaging technique for the real-time positioning of a handheld breast tissue anomaly detection probe. In: Fourth International Conference on ICCCNT. Tiruchengode, India, (2013)

Chapter 4
Efficient Depth-Based Coding

Carl James Debono, Marek Domański, Sérgio M. M. de Faria, Krzysztof Klimaszewski, Luís F. R. Lucas, Nuno M. M. Rodrigues and Krzysztof Wegner

Abstract This chapter addresses predictive coding methods for depth maps that are required for virtual view synthesis. In multi-view immersive systems, virtual views play a very important role in the overall quality experienced by the users. Despite the fact that depth maps are not viewed by the users, their accuracy has a significant impact on the quality of the corresponding synthesized views. This is mostly due to the geometry information of the scene they represent, which enables reconstruction of any viewpoint lying between two camera views. The indirect impact of depth map quality on synthesized images and video suggests that the optimization of coding algorithms for view synthesis can improve the quality of experience of the user. In this context, the chapter discusses the compression of depth maps using standard coding techniques. A method for compressing the depth map with the use of the standard coding technique of advanced video coding is provided. A depth map quality metric is discussed as this is of paramount importance when allocating

C. J. Debono (✉)
Department of Communications and Computer Engineering,
University of Malta, Msida, Malta
e-mail: c.debono@ieee.org

M. Domański · K. Klimaszewski · K. Wegner
Chair of Multimedia Telecommunications and Microelectronics,
Poznań University of Technology, Poznań, Poland
e-mail: domanski@et.put.poznan.pl

K. Klimaszewski
e-mail: kklima@multimedia.edu.pl

K. Wegner
e-mail: kwegner@multimedia.edu.pl

S. M. M. de Faria · L. F. R. Lucas · N. M. M. Rodrigues
Instituto de Telecomunicações and Politécnico de Leiria, Leiria, Portugal
e-mail: sergio.faria@co.it.pt

L. F. R. Lucas
e-mail: luis.lucas@ipleiria.pt

N. M. M. Rodrigues
e-mail: nuno.rodrigues@co.it.pt

P. A. Assunção and A. Gotchev (eds.), *3D Visual Content Creation, Coding and Delivery*, Signals and Communication Technology,
https://doi.org/10.1007/978-3-319-77842-6_4

bitrates between the texture and the depth videos. Furthermore, experiments regarding compression efficiency and bitrate allocation strategy are described and their results presented.

4.1 Introduction

The multi-view video plus depth (MVD) representation provides a solution to reduce the amount of data that needs to be transmitted in three-dimensional television (3DTV) and free-viewpoint television (FTV) applications. The information in the depth video supplies the geometry necessary to reconstruct any viewpoint that lies between any two camera views. The depth data is composed of large homogeneous areas representing objects that are at the same distance from the camera and sharp edges that indicate changes in the depth.

These characteristics distinguish the depth video from the texture video. The standard codecs are optimized for the texture video and therefore might not be optimal for the depth information. This suggests that better encoding of the depth data can be obtained by exploiting the different characteristics of the data.

This chapter first discusses predictive depth map coding for efficient virtual view synthesis. The depth map is never viewed by the user but it is important for the quality and accuracy of the synthesized views. This suggests that optimizing the coding algorithm for view synthesis can improve the quality of experience of the user. The chapter then discusses the compression of depth maps using standard coding techniques. A method of compressing the depth map with the use of standard coding technique of advanced video coding (AVC) is provided. A depth map quality metric is discussed as this is of paramount importance when allocating bitrates between the texture and the depth videos. Experiments regarding compression efficiency and bitrate allocation strategy are described and their results are presented. Finally, some conclusions are drawn.

4.2 Depth Map Coding for Efficient Virtual View Synthesis

The predictive depth coding (PDC) algorithm [1, 2] has been recently proposed for efficient intra coding of depth maps in the context of the multi-view video plus depth representation format. PDC presents an alternative coding paradigm based on a highly predictive framework combined with a flexible block partitioning scheme. In alternative to transform-based residue coding, a straightforward linear approximation residue coding scheme is used. Reported experiments have shown that PDC is able to achieve a higher rate-distortion performance compared to the current state-of-the-art 3D-HEVC encoder for depth map intra coding. In this section, the PDC algorithm is described and some experimental results comparing the performance of PDC with 3D-HEVC standard are presented.

4.2.1 Algorithm Overview

At its core, the PDC algorithm uses a block-based hybrid coding approach based on intra prediction and residue coding. The depth map is partitioned into non-overlapping blocks of 64 × 64 pixels that can be further partitioned into smaller sub-blocks using a flexible partitioning scheme. Each sub-block can be predicted using directional prediction or, alternatively, a new constrained depth modelling mode (CDMM), as illustrated in Fig. 4.1.

The used directional intra prediction method is based on one of the methods defined in the HEVC standard [3–5]. However, PDC proposes significant improvements including adaptive mode pruning and efficient prediction direction signalling. The block may alternatively be encoded using CDMM, designed for explicitly signalling the edges that are difficult to predict by directional prediction. These edges are typically observed in the bottom-right region of the block, which is harder to predict by directional prediction using left and top neighbouring block

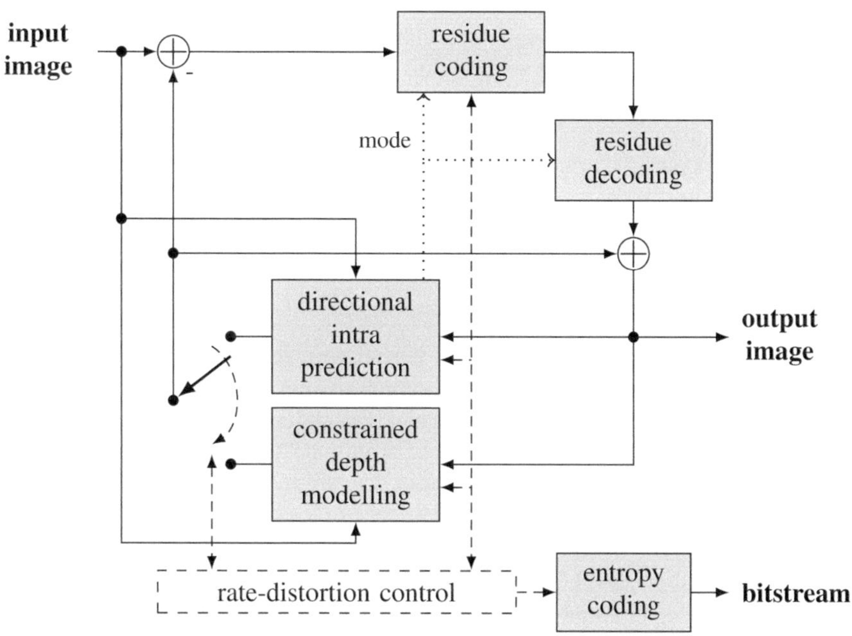

Fig. 4.1 Block diagram of the PDC algorithm for intra depth map coding

samples. CDMM allows to explicitly signal an approximation of the edges in the block as well as surrounding smooth areas. For residue coding, a linear approximation approach that depends on the chosen prediction direction is used. The approximation linear coefficients are transmitted using a depth lookup table (DLT), as proposed in 3D-HEVC. DLT performs a mapping of the valid depth values into index values, being advantageous, for instance, when the range of available values in depth map is pre-quantized. On the encoder side, most of the possible combinations of block partitioning and coding modes are examined and the best one is selected according to a Lagrangian rate-distortion cost. The context adaptive m-ary arithmetic coding (CAAC) [6] is then used for entropy coding, including the symbols that represent the flexible block partition, directional prediction modes, constrained depth modelling mode and residue coding.

4.2.2 Flexible Block Partitioning

In PDC, each block may be partitioned using a flexible scheme based on bintree and quadtree methods. The bintree partitioning [7] recursively divides the input block, either in the vertical or horizontal directions, down to the 1×1 size, as illustrated in Fig. 4.2. Some block sizes with a high ratio between horizontal and vertical dimensions (ratios larger than 4, e.g. 64×1) are not used, lowering overall computational complexity, because of their negligible impact on the coding.

In order to further reduce the encoder's computational complexity, the bintree block partitioning scheme is combined with quadtree partitioning. Three quadtree levels are defined at block sizes 16×16, 32×32 and 64×64. The four partitions, generated by each quadtree partitioning, are processed in a raster scan order. For each available quadtree level, the bintree partitioning is used down to a predefined minimum block size.

4.2.3 Directional Intra Prediction

When combined with the flexible block partitioning scheme, the directional prediction framework provides an efficient representation of depth map edges. The directional intra prediction includes the planar, DC and 33 angular prediction modes [3]. PDC further improves the prediction of depth map signals, by using predefined and adaptive reduction of directional modes, to minimize signalling requirements.

As discussed before, PDC allows a large set of block sizes for prediction. Since some block sizes are very small or narrow, some directional intra prediction modes (e.g. the ones with adjacent directions) may be redundant, giving similar prediction results. Therefore, to save unnecessary calculations and overhead bits, PDC uses a predefined reduced set of available directional modes, mainly for smaller block sizes.

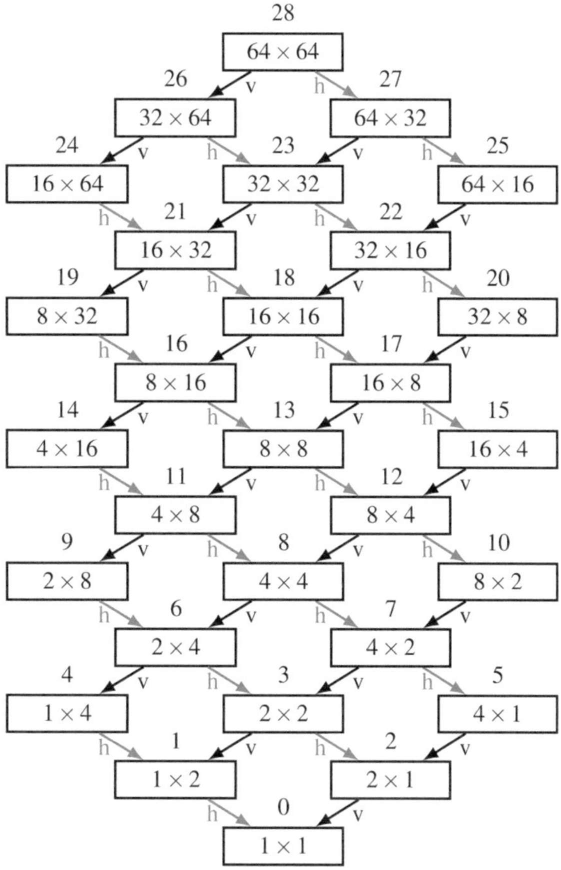

Fig. 4.2 Possible block sizes in PDC and their respective label numbers. Arrows indicate the direction of block splitting

The adaptive reduction of directional modes is another improvement of PDC algorithm, which has similar coding advantages. It exploits the large amount of smooth areas present in depth maps. In these areas, various directional modes may produce the same predicted samples. Adaptive reduction of directional modes is used to avoid this, based on the reference samples in the block neighbourhood. Three groups of directional prediction modes are defined, depending on their reference samples. Group 1 holds all the directions that generate a prediction signal, based exclusively on the top and left block neighbourhood including the top-left pixel, that is angular modes 10 to 26 plus the planar mode. When these reference samples are constant, the modes of group 1 are disabled. As illustrated in Fig. 4.3, modes 10 to 26 tend to produce the same prediction output when the left and top block neighbouring samples are constant. The remaining modes may produce different prediction results if edges are present in down-left or top-right block neighbouring regions (see right side of Fig. 4.3). DC mode can be chosen in place of the disabled modes of group 1 since it produces the same result. When the samples of the neighbour left and down-left regions are constant, the angular modes

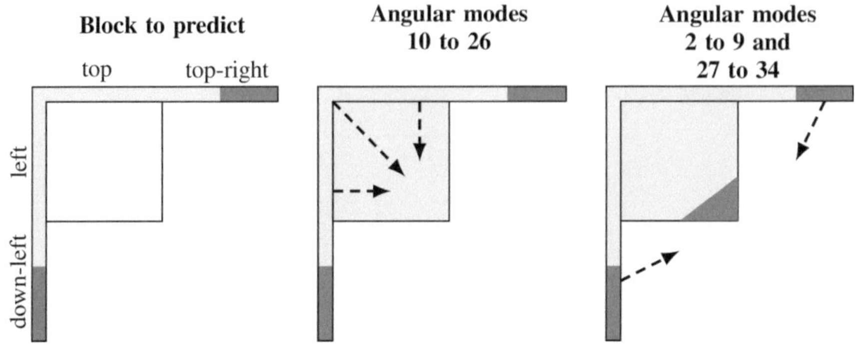

Fig. 4.3 Modes of group 1 (angular 10 to 26) only use left and top neighbouring samples to generate block prediction. Middle figure shows that modes of group 1 produce the same prediction output when the top and left block neighbouring samples are constant

2 to 9 are disabled. In this case, angular 10 mode (horizontal) can substitute these modes. Finally, Group 3 contains angular modes 27 to 34, which depend on the top and top-right neighbour regions, and can be substituted by angular mode 26 (vertical).

4.2.4 Constrained Depth Modelling Mode

CDMM facilitates the intra directional prediction process, by providing an alternative method that explicitly encodes depth map edges that are difficult to predict. Such edges are often present in the bottom-right region of the block. Directional prediction framework reasonably predicts straight edges coming from the left or top block neighbourhood. However, when this is not the case, for example in the right block of Fig. 4.4, it becomes difficult to predict.

CDMM uses a technique similar to the Wedgelet depth modelling mode used in 3D-HEVC [8], but several restrictions were applied to its design, to make it more

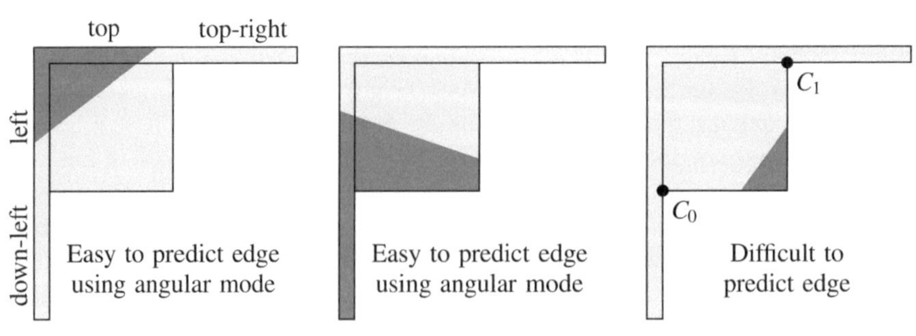

Fig. 4.4 Example of simple edge prediction (left and middle) and difficult edge prediction (right)

efficient in the context of the PDC algorithm. It divides the block into two partitions, which are then approximated by constant values. The block partitioning occurs between two points of the right and bottom margins of the predicting block. In the second restriction, the line drawn between the two selected points should be parallel to the anti-diagonal defined by the down-left and top-right block corners (line defined between points C_0 and C_1 in Fig. 4.4).

The restriction on the block partitioning slope avoids testing many block partitions with different slopes reducing computation time. Furthermore, using a unique partition slope associated with the block size allows for no bitstream overhead in its transmission. The main disadvantage of this restriction is the reduced flexibility to approximate depth map edges. However, the PDC algorithm is able to alleviate this issue by combining CDMM with the flexible block partitioning scheme, which results in anti-diagonal lines with different slopes.

CDMM block partitioning generates two partitions, whose depth values are approximated by using a constant value. For $P1$ partition, an approximate coefficient is derived using the block's neighbourhood. On the other hand, the constant approximation of the $P2$ partition is explicitly transmitted to the decoder.

4.2.5 Residual Signal Coding

The flexible block partitioning scheme combined with the intra prediction framework provides a very efficient prediction of edges, resulting in a smooth residue. Because of this, PDC uses a simple linear modelling method to encode the residue. Figure 4.5 presents the schematic of the PDC residue coding method. Four approximation models are available: constant, horizontal linear and vertical linear, as well as a special case of null residue. Depending on the chosen prediction mode,

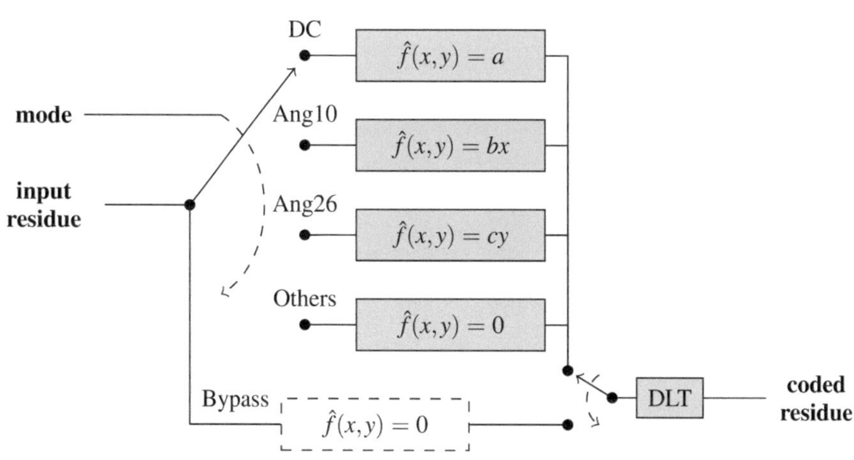

Fig. 4.5 The PDC residue coding method

which is known to both encoder and decoder, one of the available residue approximation models is transmitted. For a more efficient rate-distortion coding, linear non-null residue approximation models can be bypassed by setting a binary flag. In order to better encode pre-quantized depth maps, the DLT algorithm [8] is used to encode linear approximation coefficients.

4.2.6 Rate-Distortion Performance

The intra depth map coding performance of the PDC and 3D-HEVC algorithms is evaluated based on MPEG common test conditions document for 3D video core experiments [9]. 3D-HEVC results were produced using reference software version HTM-13.1, in all-intra configuration. For a fair comparison of depth coding performance, the contour depth modelling mode of 3D-HEVC was disabled, in order to disable inter-component prediction. Performance of depth map coding is evaluated in terms of PSNR quality of resultant virtual views generated based on the decoded depth data and the original texture views, against the reference virtual views, generated based on the original uncompressed depth and original texture views. The average luminance PSNR quality of six intermediate views placed between the positions of the encoded depth maps is used. For the purpose of view synthesis, the state-of-the-art software for linear camera arrangement implemented in HTM software is used [10]. The sum of the bitrate used to encode the three depth map views is considered in the evaluation process.

Figure 4.6 presents the average Bjontegaard Delta Bitrate [11] results of PDC relative to 3D-HEVC, for eight different test sequences, using two configurations with different distortion metrics: the sum of square errors (SSE) and the view synthesis optimization (VSO) [8]. These results show the advantage of the PDC algorithm over the 3D-HEVC approach, when using SSE and VSO distortion metrics. The average bitrate savings of PDC over 3D-HEVC using all-intra

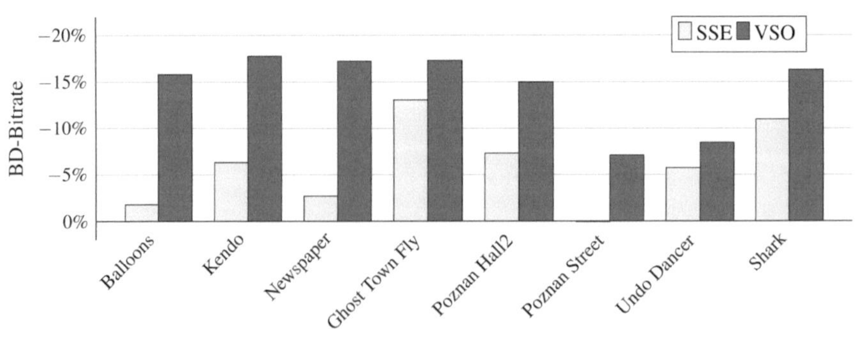

Fig. 4.6 Average BD-BR values of PDC relative to 3D-HEVC using SSE and VSO distortion metrics

configuration and SSE distortion metric are approximately 6%. The average PDC gain is superior when VSO method is used in both algorithms, achieving 14.3% of bitrate reduction for the same quality of virtual views.

4.3 Depth Compression Using Standard Coding Techniques

Sending the depth data along the texture data (normal video frames) for a video sequence is a must in different modern applications, described elsewhere in this book. Such an approach requires that available bandwidth is divided between the texture and depth data. Both kinds of data need to be compressed, be it a joint compression, or completely independent compression for texture and depth.

The most straightforward approach towards the compression of depth maps, represented as a sequence of monochromatic images, is to feed the depth map sequence to a standard video encoder. The same encoder can be, therefore, used for texture and depth compression. This approach has some obvious drawbacks, since standard coding methods, that are used in hybrid coders, are not very well suited for depth map compression. The properties of depth maps make the hybrid coders less suitable for use for their compression [12]. This occurs mostly because of the poor performance of the transform coding for sharp edges, which are found in abundance in depth maps. Moreover, the process of lossy coding performed by hybrid coders introduces offsets in values of depth maps. This can be destructive for depth map quality, while is acceptable for texture.

While all the aforementioned drawbacks have to be kept in mind, still, the encoding of depth maps using a standard video coder is a valid and widespread technique for encoding depth maps.

This section deals with some aspects of such encoding, mostly the bitrate distribution between depth and texture data.

4.3.1 Bitrate Distribution

Encoding textures and depths involves deciding on the way the available bitrate is distributed between texture and depth. It is usually assumed that the depth can be encoded using significantly less bits than allocated to encode the corresponding texture. This comes from the fact that depth is conventionally represented as a monochromatic image (so only a single channel needs to be encoded) with limited textures (usually objects are rather flat therefore their corresponding depth is a simple gradient-filled surface, such areas can be efficiently encoded). On the other hand, the prominent edges in depth maps consume many bits to encode. Another limiting factor is the necessity to avoid significant changes in depth values, which would easily result from high compression ratios. In order to choose the appropriate bitrate ratio between depth and texture, one must decide on the optimization criterion used. For standard video coders, this will usually be the quality of the reconstructed data. This approach, however, is not appropriate for depth, as will be discussed in the next section.

4.3.2 Depth Map Quality

When deciding on the strategy of bitrate distribution, one has to choose the quality metric for depth. Unfortunately, the most widespread objective metric, the PSNR, is unsuitable for depth. The reason for this unsuitability is that depth data are not presented to the viewer, but, instead, are being used in the process of synthetic view generation (so-called depth image based rendering—DIBR). Certain parts of the depth map contribute much more significantly to the resulting virtual view image than others, and, therefore, the quality metric needs to be weighted with the weights appropriate for each pixel separately. In fact, there exist areas in depth maps, where the distortions in depth map will not influence the virtual view synthesis process at all. Unfortunately, PSNR metric does not provide this kind of information.

Thus, the more appropriate metric of the quality of depth seems to be the metric based on the quality of the synthesized view with the use of a given depth map. This turns out to be the most straightforward way of assessing the quality metric, as it inherently takes into account different influence of the values of depth map in different areas. Such a method was adapted by Moving Pictures Experts Group (MPEG) of the International Standardization Organization (ISO/IEC) in the process of evaluating methods of processing and compression of the multi-view video with depth [13]. Adopted method assesses the performance of depth compression method by evaluating the quality of virtual views in terms of PSNR. But even such a simple method has its variants. Because PSNR metric needs a reference image, one has to provide the image to refer to for each frame of virtual view.

The first method is to compare the virtual view generated with the use of compressed depth map to the real view captured by a camera situated in the same position and orientation as were set for a virtual camera, as shown in Fig. 4.7. This approach provides results that are more consistent with the subjective evaluations of the virtual view quality [12], but requires additional video data to be obtained during the capture process, which is not always possible. Another drawback of this method is that the quality evaluation is influenced by the quality of virtual view synthesis algorithm, and thus, comparing results obtained with different view synthesis software is difficult.

Another method is to compare the virtual view generated with the use of compressed depth map to the virtual view generated with the use of uncompressed depth data, as shown in Fig. 4.8. This way the influence of the view synthesis algorithm is diminished, but the results are slightly less consistent with the already mentioned subjective test results. This is the only method that is suitable for

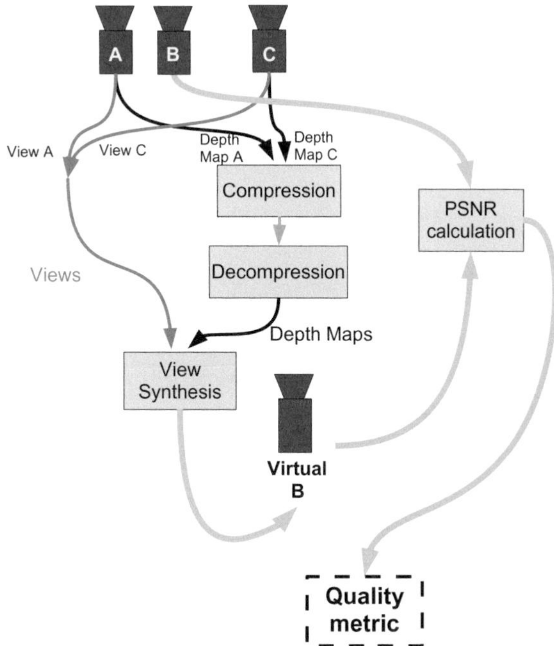

Fig. 4.7 Depth quality assessment using real view as a reference

assessing the quality of dense virtual views, since it is physically impossible to place any real camera for providing the reference view. Also for already recorded sequences, this can be the only feasible metric, since no additional camera data can be provided in such a case.

The method accepted by MPEG relies on this second version of depth quality estimation algorithm. This method is also used for assessing the quality of the overall video plus depth compression. This is therefore the optimization criterion for estimating the optimal or near optimal ratios of bitrate devoted to texture and depth bitstreams in compression.

4.3.3 Bitrate Distribution Between Texture and Depth

Upon agreeing on the way of assessment of the quality of depth, the proper ratio of bitrates devoted to texture and depth can be estimated. Usually, the depth is perceived as the kind of data that can be easily compressed, with high compression ratios. The comparison of bitrates for three exemplary sequences with depth, Book Arrival [14], Newspaper [15] and Pantomime [16] for a given value of quantization parameter index (QP) is given in Fig. 4.9. It can be seen that the bitrate of depth is for all the tested sequences visibly lower than the bitrate for texture. Therefore, usually the bitrate devoted to the transmission of the depth data is limited to about

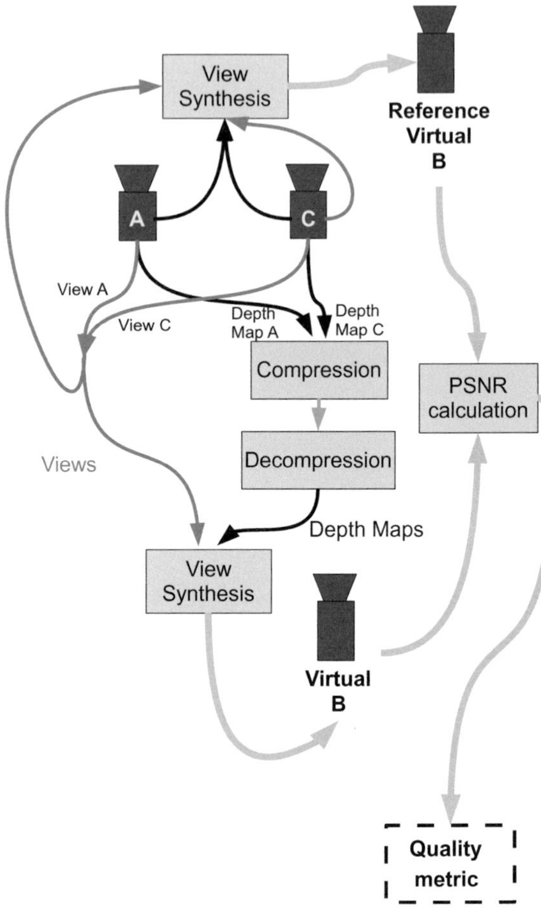

Fig. 4.8 Depth quality assessment using synthetic view as a reference

10–20% of the total bitstream. This rule of thumb may not, however, be the most efficient bitrate distribution. In fact, as it turns out, such a simple approach is not valid for a significant number of cases, as usually depth requires larger bitrate than expected.

The exemplary curves showing the PSNR value of the virtual view for a certain test sequence can be seen in Fig. 4.10. The reference for the PSNR metric was here the real view from a camera. The quantization parameter index for texture is denoted QP, while for depth it is denoted QD. Results for only one representative sequence are given. The dashed horizontal line is the reference quality obtained with uncompressed data. It can be seen that there are different combinations of QP and QD indices that produce the same bitrate, but they can significantly differ in quality of synthesized view. The same holds true for the bitrate ratio—as seen in Fig. 4.11. The quality is highly dependent on the QP—the index of quantization parameter for texture. The dependency on the QD is much less noticeable for the

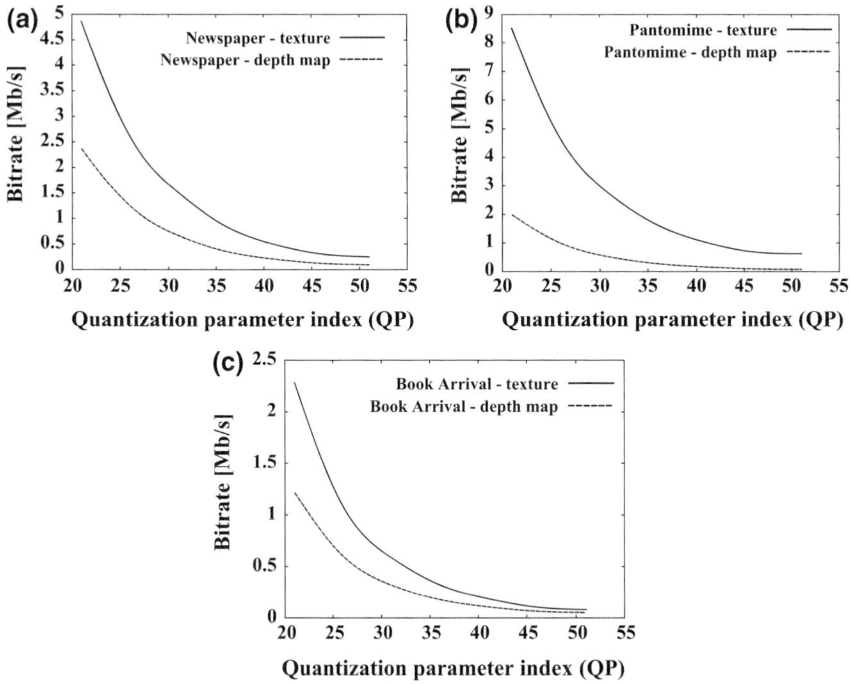

Fig. 4.9 Comparison of bitrates of bitstreams for texture and depth for three different test sequences **a** *Newspaper*, **b** *Pantomine*, and **c** *Book Arrival*

Fig. 4.10 Synthesis quality for different quantization parameter indices for texture (QP) and depth (QD)

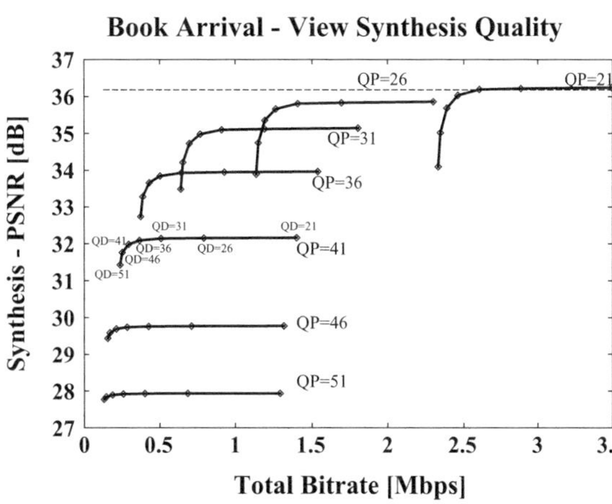

part of the considered data points. However, there is a certain threshold, above which the QD value starts to significantly influence the quality of the virtual view. It can be therefore concluded that in order to achieve the highest coding performance,

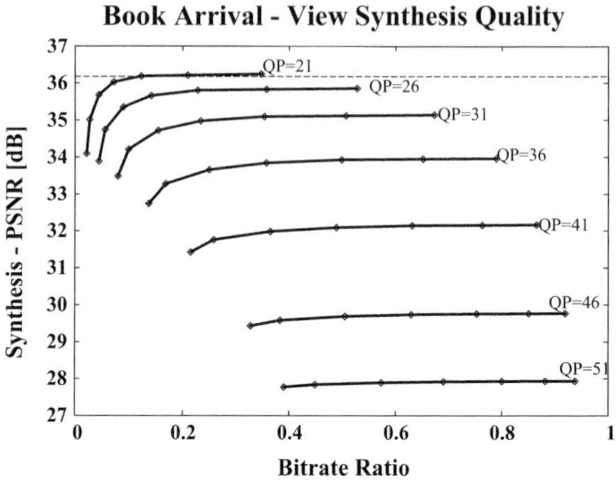

Fig. 4.11 Synthesis quality for different bitrate ratios between texture and depth

the compression should be done so that the bitrate is maintained at the close-to-optimal level.

This optimal bitrate ratio can be estimated using more experimental data. The dense experimental data are shown in Fig. 4.12. One can clearly see the envelope of the family of curves—the bold continuous line above all dashed lines. Dashed lines depict the results for different QP and QD pairs that were tested. The envelope consists of the testpoints with different QP and QD values. It is also interesting to see the bitrate ratios for the points on this optimal curve. The bitrate ratio is shown in Fig. 4.13. Figure 4.13 shows that to get optimal performance, one sometimes needs to assign more than 30% of bitstream to depth coding. Another observation is that the bitrate ratio gets bigger for lower bitrates—this means, that for lower total bitrates one has to assign more bitrate to the depth data, since any further

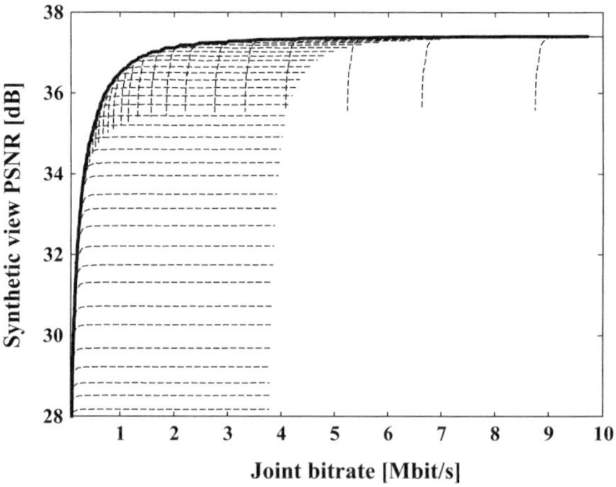

Fig. 4.12 The optimal quality envelope as a function of the total bitrate

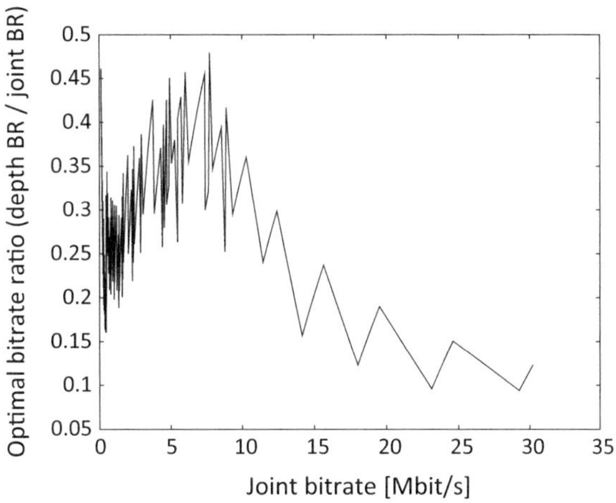

Fig. 4.13 Optimal bitrate ratio for a specific representative test sequence

degradation of depth reconstruction would significantly reduce the overall quality of the virtual views.

In order to operate on the optimal curve (i.e. assign the proper bitrate to depth and texture for a given total bitrate), one needs to adjust the QP and QD indices. Unfortunately, there is no linear dependency between the values of QP and QD laying on the optimal quality curve.

The analysis of experimental data, provided in [17], is concluded in an approximate formula used to estimate the value of QD based on the value of QP index. There are two formulas, and the choice depends on the reference used for PSNR quality estimation. For comparison with a view synthesized using the uncompressed data, the formula is given below.

$$\mathrm{QD} = \begin{cases} \lfloor -0.0216 \times \mathrm{QP}^2 + 2.6872 \times \mathrm{QP} - 29.376 \rfloor, & \mathrm{QP} > 16 \\ 11, & \mathrm{QP} \leq 16, \end{cases}. \tag{4.1}$$

Compression results obtained with the use of Eq. (4.1) for a specific representative test sequence are shown in Fig. 4.14. It can be seen that the performance (solid line) is at close-to-optimal level, compared to the dashed lines for all possible pairs of QP-QD values.

4.3.4 *Coding Depth with Reduced Resolution*

Another observation about coding depth maps with standard coding techniques is that it is usually more beneficial, in terms of compression efficiency versus virtual view quality, to decimate the depth map and obtain the depth map with reduced resolution. This reduced resolution depth map is then compressed, and after decompression, the original size is restored. The experiments have shown, that during the decimation, several aspects need to be observed.

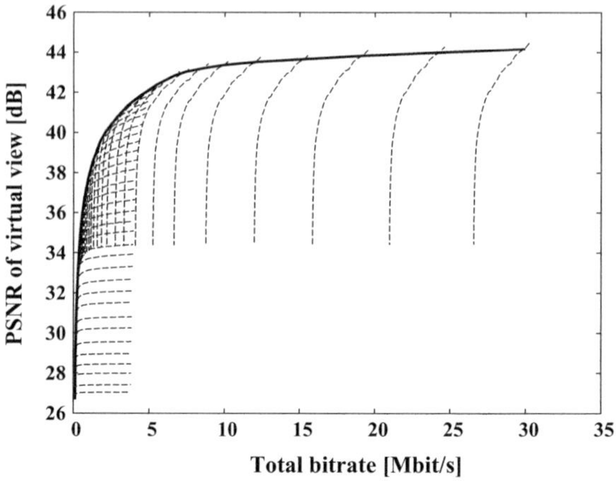

Fig. 4.14 The results obtained with the use of Eq. (4.1)

First—it is not possible to use average value for depth pixels. Doing so would introduce non-existent depth levels to the scene and may cause significantly visible artefacts in virtual views. Therefore, only the values that are already present in the depth map can be used in the decimated version. The most appealing choice is to use the maximum value of the neighbouring pixels during decimation. This can be justified by the observation that this maximum value corresponds to the object closest to the camera and unlikely to be occluded. It will, however, occlude the neighbouring objects in virtual view. This is therefore the object that will, most probably, be seen in the virtual view and it is important to preserve its original depth value.

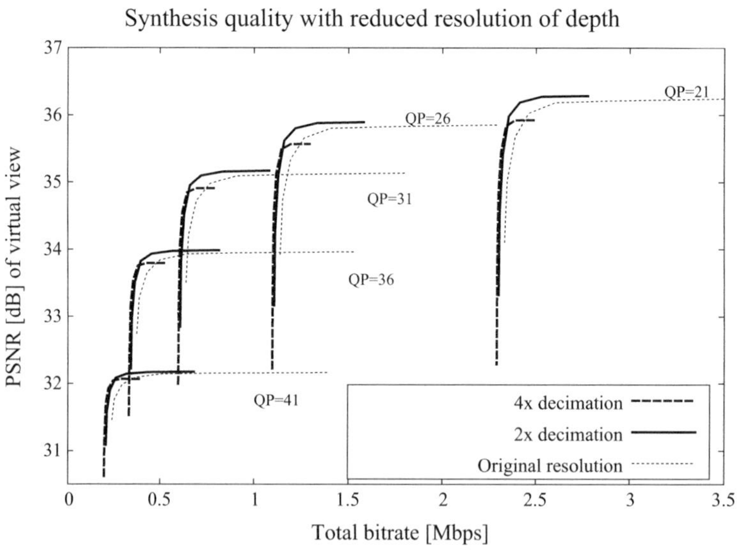

Fig. 4.15 The quality of the virtual view for decimated depth maps

The results of the compression of texture and depth, with the depth being reduced in resolution twofold and fourfold, are shown in Fig. 4.15. It is noticeable that only the twofold reduction improves the compression, while the fourfold reduction decreases the overall quality.

4.4 Conclusion

This chapter has covered predictive depth map coding solutions optimized for view synthesis. Furthermore, it has shown that standard video coding techniques can be used for depth compression, despite their fundamentally different properties than texture, for which the coding methods were developed. The motion compensation and transform coding of a hybrid coder perform sufficiently well also for depth maps. Although better suited methods are being developed, still the most convenient and straightforward method of compressing the depth map is to use standard video coding software.

Acknowledgements Section 4.3 was supported by National Science Centre, Poland according to the decision DEC-2012/05/B/ST7/01279.

References

1. Lucas, L., Wegner, K., Rodrigues, N., Pagliari, C., Silva, E., Faria, S.: Intra predictive depth map coding using flexible block partitioning. IEEE Trans. Image Process. **24**(11), 4055–4068 (2015)
2. Lucas, L., Wegner, K., Rodrigues, N., Pagliari, C., Silva, E., Faria, S.: Intra depth-map coding using flexible segmentation, constrained depth modeling modes and simplified/pruned directional prediction. Joint Collaborative Team on 3D Video Coding Extension Development of ITU-T SG 16 WP 3 and ISO/IEC JTC 1/SC 29/WG11, Jul. 2014
3. ITU-T and ISO/IEC JTC 1/SC 29 (MPEG). High efficiency video coding. Recommendation ITU-T H.265 and ISO/IEC 23008–2, 2013
4. Sullivan, G., Ohm, J., Han, W., Wiegand, T.: Overview of the high efficiency video coding (HEVC) standard. IEEE Trans. Circuits Syst. Video Technol. **22**(12), 1649–1668 (2012)
5. Müller, K., Schwarz, H., Marpe, D., Bartnik, C., Bosse, S., Brust, H., Hinz, T., Lakshman, H., Merkle, P., Rhee, F., Tech, G., Winken, M., Wiegand, T.: 3D high-efficiency video coding for multi-view video and depth data. IEEE Trans. Image Process. **22**(9), 3366–3378 (2013)
6. Witten, I., Neal, R., Cleary, J.: Arithmetic coding for data compression. Commun. ACM **30** (6), 520–540 (1987)
7. Francisco, N., Rodrigues, N., Silva, E., Carvalho, M., Faria, S., Silva, V., Reis, M.: Multiscale recurrent pattern image coding with a flexible partition scheme. In: Proceedings of the 15th IEEE International Conference on Image Processing, October 2008
8. JCT3 V-J1005, "3D-HEVC Test Model 10." Joint Collaborative Team on 3D Video Coding Extension Development of ITU-T SG 16 WP 3 and ISO/IEC JTC 1/SC 29/WG 11 Oct. 2014
9. JCT3 V-G1100.: "Common Test Conditions of 3DV Core Experiments." Joint Collaborative Team on 3D Video Coding Extension Development of ITU-T SG 16 WP 3 and ISO/IEC JTC 1/SC 29/WG 11 Jan. 2014

10. ISO/IEC JTC1/SC29/WG11 MPEG2013/M31520. Wegner, K., Stankiewicz, O., Tanimoto M., Domański, M.: "Enhanced View Synthesis Reference Software (VSRS) for Free-viewpoint Television 2013
11. Bjøntegaard, G.: "Calculation of Average PSNR Differences Between RD-curves." ITU-T SG 16 Q.6 VCEG, Doc. VCEG-M33 Apr. 2001
12. Klimaszewski K., Wegner K., Domański M.: Distortions of synthesized views caused by compression of views and depth maps. In: Proceedings of the 3DTV-Conference 2009—The True Vision Capture, Transmission and Display of 3D Video, Potsdam, Germany, 4–6 May 2009
13. "Call for Proposals on 3D Video Coding Technology". ISO/IEC JTC1/SC29/WG11 MPEG Doc. N12036, Geneva, Switzerland March 2011
14. Feldmann I., Müller M., Zilly F., Tanger R., Müller K., Smolic A., Kauff P., Wiegand T.: "HHI Test Material for 3D Video." ISO/IEC JTC1/SC29/WG11 MPEG Doc. M15413, Archamps, France April 2008
15. Ho Y.-S., Lee E.-K., Lee C.: Multiview video test sequence and camera parameters. ISO/IECJTC1/SC29/WG11 MPEG Doc. M15419, Archamps, France April 2008
16. Tanimoto M., Fujii T., Fukushima N.: 1D parallel test sequences for MPEG-FTV. ISO/IECJTC1/SC29/WG11 MPEG Doc. M15378, Archamps, France April 2008
17. Klimaszewski K., Wegner K., Domański M.: Video and depth bitrate allocation in multiview compression. InL Proceedings of the 21st International Conference on Systems, Signals and Image Processing, IWSSIP 2014, Dubrovnik, Croatia, 12–15 May 2014

Chapter 5
Error Concealment Methods for Multiview Video and Depth

Sérgio M. M. de Faria, Sylvain Marcelino, Carl J. Debono, Salviano Soares and Pedro Amado Assunção

Abstract The different media representation formats and coding techniques currently used to deliver 3D visual information across diverse networks require specific approaches and methods to minimise the perceptual impact of data loss, that may occur along the communications path. This chapter addresses this type of problem by presenting recent advances in error concealment methods, expanding conventional techniques used for 2D video to multiview (MVC) and multiview video-plus-depth (MVD) coded formats. The methods described in the chapter exploit the specific characteristics of multiview formats to achieve highly efficient error concealment performance and, consequently, to improve the perceptual quality delivered to end users, in the presence of transmission losses. In the case of MVC, besides spatial and inter-frame, inter-view correlations are also exploited, while in MVD, the most efficient methods use both the texture (view) and depth information to improve the error concealment performance. The most relevant contributions in this field are described in detail, where the performance of these

S. M. M. de Faria (✉) · S. Marcelino · P. A. Assunção
Instituto de Telecomunicações, Leiria, Portugal
e-mail: sergio.faria@co.it.pt

P. A. Assunção
e-mail: amado@co.it.pt

S. M. M. de Faria · P. A. Assunção
Politécnico de Leiria, Leiria, Portugal

C. J. Debono
Department of Communications and Computer Engineering, University of Malta, Msida, Malta
e-mail: c.debono@ieee.org

S. Marcelino · S. Soares
Douro/ECT Engineering Department, Universidade de Trás-os-Montes e Alto Douro, Vila Real, Portugal
e-mail: salblues@utad.pt

S. Soares
IEETA, University of Aveiro Campus, Aveiro, Portugal

P. A. Assunção and A. Gotchev (eds.), *3D Visual Content Creation, Coding and Delivery*, Signals and Communication Technology,
https://doi.org/10.1007/978-3-319-77842-6_5

advanced solutions is discussed along with comparisons between different methods and benchmarking.

5.1 Introduction

The existence of different representation formats and coding techniques for 3D visual information requires specific approaches and methods to minimise the perceptual impact of data loss, that may occur in delivery services through error-prone networks, such as those generally used in multimedia communications. Video compression of multiview video exploits spatial, temporal and inter-view redundancies to reduce the huge amount of captured data. When such highly compressed streams are delivered over practical networks, transmission errors can cause loss of some of the data packets with significant impact on the quality of experience (QoE) provided to end users. The missing information does not simply affect an image area in the current frame, but due to the various coding dependencies exploited during the compression process, any spatial, temporal or inter-view content that predicts the values from this area will be also erroneous. Therefore, the errors will propagate in space, time and in-between views until the dependencies are interrupted.

In order to provide good quality of experience (QoE) in 3D multiview services over networks, the lost information needs to be reconstructed using methods capable of minimising the impact of the resulting impairments. To overcome the errors over communications channels, error control strategies, such as forward error correcting (FEC) codes, can be applied. However, when the FEC codes fail other algorithms are needed at the decoder side to limit the error effects and estimate the missing video content. The former techniques are known as error resilience coding, which restricts the propagation of the errors to a certain extent, while the latter are classified as error concealment (EC), which estimate the missing information from the received data, including stream syntax elements. In this case, depending on the type of coded data affected by losses, spatial and temporal information may be used to recover missing image regions to reduce the impact of lost data. Conventional EC techniques, normally used in 2D video, may also be implemented in robust decoders of multiview (MVC) and multiview video plus depth (MVD).

In this chapter, more efficient error concealment methods are described. These exploit the specific characteristics of multiview formats, which allow the use of new types of information in order to improve the error concealment performance and, consequently, the perceptual quality delivered to end users. As described in the following sections, in the case of MVC, besides spatial and inter-frame, inter-view correlations may also be efficiently exploited, due to the high similarity between frames captured from different viewpoints at the same time instant. In the case of MVD, besides the different viewpoints, for each texture (view) frame there is also a corresponding depth map, thus texture and depth information can be jointly used in order to improve error concealment performance. Based on the previous underlying principles, the chapter describes the most recent advances in error concealment methods for MVC and MVD, performance evaluation, discussion of results and benchmarking using known methods as reference. The efficiency of these advanced solutions is discussed based on the quality of virtual views as this is the ultimate performance indicator for the reconstruction efficiency of error concealment methods in MVC and MVD robust decoders.

5.2 Error Concealment for Multiview Video

In general, error concealment is a post-processing operation applied after the actual decoding process to fill in any missing content, by exploiting the error-free information that reaches the receiving side. Due to the nature of the coding algorithms, any missing blocks in images affected by errors always have some level of correlation with other blocks located in their spatial, temporal and inter-view neighbourhood. This has been the underlying principle behind single-view concealment methods, which have been inherited for multiview solutions along with the addition of inter-view neighbourhood exploitation. This section describes the most important methods that have been used in the past and also recent research contributions to improve the quality of multiview video delivery over error-prone networks, by using efficient error concealment methods.

5.2.1 Basic Methods Using Neighbouring Regions

Spatial correlation methods are based on the similarity between pixels located in neighbouring areas of the same image. A possible approach to implement a spatial EC method is to store the location of all erroneous and dropped slices in an error map during the decoding process [1]. To recover such missing content, the EC algorithm starts from the top and bottom edges in the vertical plane and from the left and right edges in the horizontal plane of the frame and moves to the centre of the lost slice. Spatial EC methods are particularly useful when transmission errors corrupt pixels within an intra-coded frame, because no temporal or inter-view neighbours that can be used to recover the missing regions. In these cases, error concealment relies exclusively on the information available in the spatial neighbourhood. A weighted average of the closest available boundary pixels values is performed to fill in the missing pixels. The values of the weights are found depending on the inverse distance between the pixel to conceal and the reference, as shown in Fig. 5.1. Missing pixels closer to boundary generally correlate better with the pixels on the boundary and therefore result in better quality than others, which are further away. The missing pixel value is interpolated using:

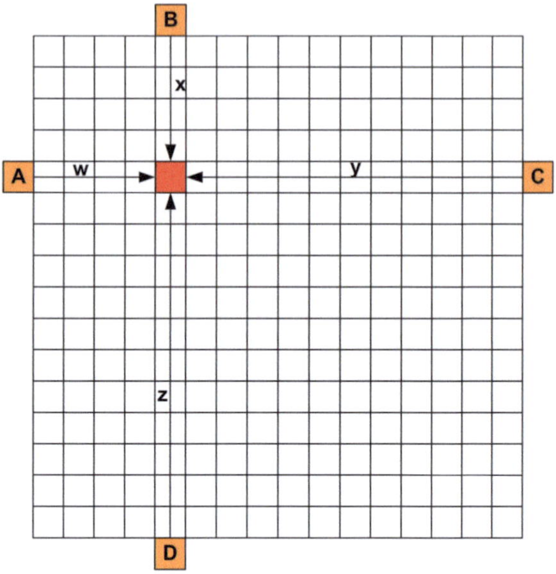

Fig. 5.1 Spatial concealment

$$\frac{(wA + xB + yC + zD)}{(w + x + y + z)} \tag{5.1}$$

where A, B, C and D represent the reference pixel values and w, x, y, and z represent the pixel distances, as shown in Fig. 5.1.

In the case of inter-predicted frames, there are more candidates to be used by the error concealment process. In addition to the spatial candidates discussed above, the missing pixels can be estimated from temporal and inter-view neighbours. Temporal concealment methods estimate the motion vector of the missing data computed from the motion vectors of the spatial neighbourhood. These motion vectors provide pixel candidates in the previous frame. If the average of these motion vectors represents a very slow moving or stationary content, the lost data is filled using the collocated pixels in the reference frame. If on the other hand there is high movement in the scene, a motion vector needs to be selected from the neighbouring data and assigned to the lost blocks. The result represents a good estimate given that motion of nearby content is statistically highly correlated. Figure 5.2 represents the concept behind this technique where mv_l, mv_t, mv_r and mv_b represent the left, top, right and bottom motion vector, respectively. The neighbouring blocks can be divided into smaller partitions, allowing to increase the number of candidate vectors.

This method provides a list of motion vector candidates that can be used to recover the lost blocks. To select the best match from the list, the error at the boundary for each candidate replacement is computed, and the one presenting the

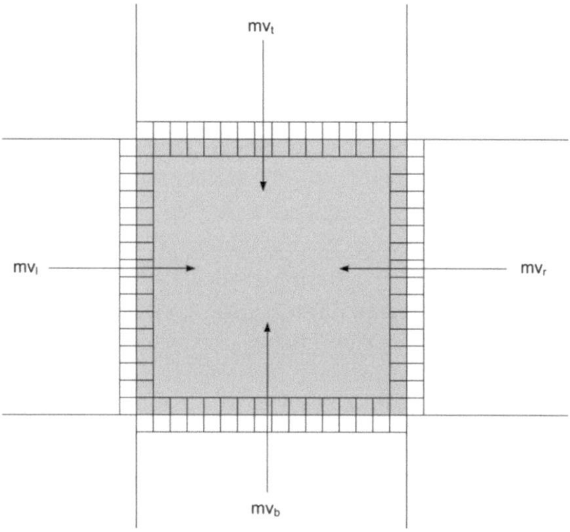

Fig. 5.2 Motion vector candidates for temporal concealment

smallest error is selected. The calculation is done using the boundary matching algorithm (BMA) [2]:

$$d_{sm} = \underset{\mathrm{dir} \in \{\mathrm{left,top,right,bottom}\}}{\mathrm{minarg}} \left\langle \left(\sum_{j=1}^{16} \left| Y_{\mathrm{ref}}(mv_{\mathrm{dir}})_j - Y_{\mathrm{rec}j} \right| \right) / N \right\rangle, \tag{5.2}$$

where d_{sm} represents the Luma (Y) error, mv_{dir} represents the motion vector being considered, Y_{rec} is the reconstructed Luma value at the boundary of the frame, Y_{ref} is the reference Luma value on the other side of the boundary and N is the average number of pixel elements. The summation to 16 is for a 16 × 16 block. This error represents a measure of smoothness between the correctly decoded blocks and the ones replacing the lost ones. The zero-motion vector also forms part of the list of candidates to cater for the possibility that the corrupted content was a SKIP. If the current frame is bi-predicted, then the vectors to be used will be selected from a list of candidate vectors. When only one of motion vectors is available, then the selection is trivial. When both are available, the forward prediction motion vector is used.

The multiview standard requires that one of the views is compatible with single-view coding and therefore presents no inter-view dependencies. Hence, frames in this view simply use the same single-view techniques, which rely on the spatial and the temporal concealment methods previously discussed. The other views can have inter-view dependencies, which can be either dependent on another view or on two other views. This situation is similar to the temporal dependencies, but instead of motion vectors they now use disparity vectors. Therefore, the candidate disparity vectors from the neighbourhood of the missing content can be used to find replacements in the other views [3]. Note that, in order to allow random access, the anchor frames do not exploit temporal redundancy, but still exploit inter-view redundancy. Hence, only the disparity vectors are candidates in such cases. Other frames can use both temporal and inter-view concealment. A decision algorithm should be used to determine the best reference frame for concealment, depending on the error generated when comparing the boundaries.

These techniques form the basis of all error concealment methods found in the literature. Improvements are based on using different combinations of replacement candidates, different error measurement solutions to determine the smoothness, and different matching algorithms, such as the outer boundary matching algorithm [4]. The results obtained from error concealment strongly depend on the type of sequence and, typically, favour slow moving content with few errors. Further work is still needed to improve the results of concealment and understand its impact in high-definition 3D viewing.

5.2.2 *Recent Advances in EC for Multiview Video*

Based upon the basic techniques described above, there are different approaches to obtain improved results and better visual quality when multiview video suffers from network losses. The following methods are relevant examples of recent advances in this field.

A spatial EC method for MVD, which uses the depth information to restore the corresponding texture image based on thresholding of the depth map, is described in [5] by A. Ali et al. This approach shows high potential, since the main edges of the depth map define the limit between the background and foreground, which can also be used in the reconstruction process of the corrupted texture. However, the validation of the method is not very strong because a simple error pattern was used with small error rates, which makes it difficult to evaluate the accuracy and to validate the performance as a general result. Moreover, since the MVD format is used and the quality of the images synthesised from the recovered texture images was not evaluated, the actual impact of the recovered texture images on the overall quality is not fully known.

A full-frame EC method for MVC was proposed by Liu et al. [6] using an approach that takes into consideration the underlying principles based on motion similarities between frames in the temporal domain. Nevertheless, in this case the redundancy between adjacent views is used. As the EC method described in [7], a motion field is computed based on the previously decoded frames. Liu et al. proposed a similar approach, but taking advantage of the MVC video characteristics, where the motion information of the adjacent views is used to recover the corrupted frames. The motion field of the adjacent views is based on a simplified global disparity model, where the motion vectors (MVs) from adjacent views are considered similar, simply taking into account the camera displacement. When a frame from a non-base view is lost, each lost MB is recovered by copying all the block partitions and respective MVs from the adjacent view. When motion information is not available in the adjacent view, as for intra-coded MBs, spatial error concealment is performed by setting the lost MB being concealed as a skip in *P*-coded pictures or direct mode, in the case of *B*-coded pictures [8]. However, the authors only compared the proposed method with temporal replacement method (TR), where gains up to 2.6 dB and 0.97 dB, on average, were achieved for the tested sequences. Despite the fairly good results, this method still has room for improvement, mainly due to simplistic EC approach that uses the same global disparity for all recovered MBs. This might be an advantage in terms of computational complexity, but may reduce the EC accuracy when compared to the case where individual disparity values would be used for each recovered MB.

Using a different approach, Chen et al. in [9] also proposed a full-frame error concealment method for stereoscopic video using a frame difference projection based on the disparity (DFDP) of a stereo pair. Figure 5.3 shows an example of how the temporal similarity between frames (inter-frame) and also between views (inter-view) is exploited. The method assumes that the left view is the base view, allowing this stream to be independently decoded without the need of other views. In the case of the right view, the correlation between views was taken into account, meaning that the adjacent views are needed in the decoding process. Chen et al. proposed a method that is targeted for non-base encoded views, where inter-view information is also exploited.

The left view is defined by l, the right view by r and t defines the time instant. A pixel from $f_{(r,t)}$ of a corrupted frame can be obtained by motion and disparity vectors, one due to motion activity ($\overrightarrow{m_l}$ and $\vec{m}_r$) and the other due to disparity between frames ($\overrightarrow{d_t}$ and $\vec{d}_{(t-1)}$).

Assuming that objects in the scene do not change significantly along time, nor between frames from adjacent views, it is reasonable to consider that the MVs $\overrightarrow{m_l} \approx \vec{m}_r$ and $\overrightarrow{d_t} \approx \vec{d}_{t-1}$, indicating that both inter-view and inter-frame correlations

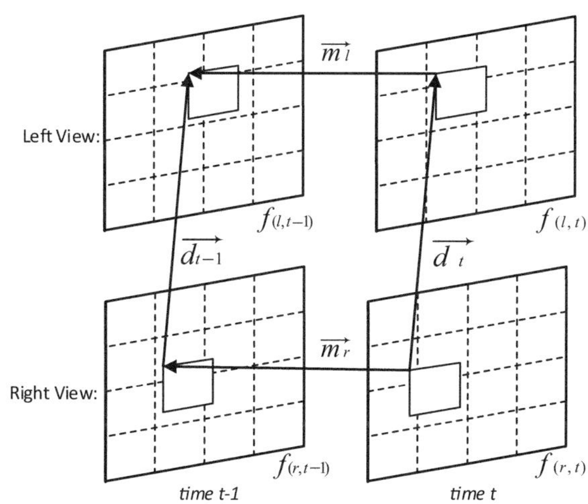

Fig. 5.3 Inter-view and inter-frame correlations

are high. Chen's DFDP method is based on this assumption and comprises three main functions: (1) *Change Detection*, (2) *Disparity estimation* and (3) *Frame difference projection*.

In the first function, *Change Detection*, a temporal change detection is performed by computing the absolute frame differences for all pixels between a certain temporal instant t and $t-1$ in the left view l. The resulting matrix Δf, defined by Eq. (5.3), represents the corresponding frame difference, which is then filtered with a mean filter, followed by thresholding [10]. In order to detect the moving objects, pixels belonging to the foreground and background are separately identified. This filtered matrix is represented by $M_{(l,t-1\to t)}(x,y)$ and defined by Eq. (5.4), where x and y are the corresponding pixel coordinates. The threshold T is computed by an iterative algorithm, as described in [9].

$$\Delta f_{(l,t-1\to t)}(x,y) = \left| f_{(l,t)}(x,y) - f_{(l,t-1\to t)}(x,y) \right| \tag{5.3}$$

$$M_{(l,t-1\to t)}(x,y) = \begin{cases} 1, & \Delta f_{(l,t-1\to t)} \geq T \\ 0, & \text{otherwise} \end{cases} \tag{5.4}$$

In the second function, *Disparity estimation,* the horizontal disparity estimation is computed between the stereo pair l and r (a parallel camera arrangement is used). It is assumed that a disparity vector can be decomposed into two components, the global and the local disparity. The global disparity $d_{\text{global}}^{(t-1)}$ is computed for the regions where temporal changes occur, as defined by Eqs. (5.5) and (5.6). The objective is to compute the disparity d, where the absolute differences between frames belonging to left (l) and right (r) views are smaller.

$$E_{\text{global}}^{t-1} = \sum_{M_{(r,t-1\to t)}(x,y)} \left| f_{(l,t-1\to t)}(x,y) - f_{(r,t-1\to t)}(x-d,y) \right| \tag{5.5}$$

$$d_{\text{global}}^{(t-1)} = \arg\min_{d} E_{\text{global}}^{t-1} \tag{5.6}$$

The local disparity $d_{\text{local}}^{(t-1)}$ is computed using an $m \times m$ window, as defined by Eqs. (5.7) and (5.8). To compute the local disparity, an 8×8 window was used in a search range defined by $d \in [-20, 20]$.

$$d_{\text{local}}^{(t-1)} = \arg\min_{d} E_{\text{local}}^{t-1}(x,y,d) \tag{5.7}$$

$$\boldsymbol{E}_{\text{local}}^{t-1}(\boldsymbol{x},\boldsymbol{y},\boldsymbol{d}) = \sum_{\gamma=-\frac{m}{2}}^{\frac{m}{2}} \sum_{\varepsilon=-\frac{m}{2}}^{\frac{n}{2}} \left| \boldsymbol{f}_{(l,t-1\to t)}(\boldsymbol{x}+\gamma,\boldsymbol{y}+\varepsilon) - \boldsymbol{f}_{(r,t-1\to t)}\left(\boldsymbol{x}+\gamma-\boldsymbol{d}-\boldsymbol{d}_{\text{global}}^{(t-1)},\boldsymbol{y}+\varepsilon\right) \right| \tag{5.8}$$

After obtaining the global and local disparity components, the final disparity values $d^{(t-1)}(x,y)$ are expressed as:

$$d^{(t-1)}(x,y) = d_{\text{global}}^{(t-1)} + d_{\text{local}}^{(t-1)} \tag{5.9}$$

In the third and final function of this EC method, frame difference projection is performed, based on the inter-frame and inter-view correlation, knowing that MVs $\overrightarrow{m_l} \approx \vec{m}_r$ and $\overrightarrow{d_t} \approx \vec{d}_{t-1}$. The change detection map $M_{(r,t-1\to t)}(x,y)$ and the temporal frame difference $\Delta f_{(r,t-1\to t)}(x,y)$ of the right view can be computed based on the left view $M_{(l,t-1\to t)}(x,y)$ and $\Delta f_{(l,t-1\to t)}(x,y)$. It is considered that the lost frame is the right view from temporal instant $t(f_{(r,t)})$, which is recovered using pixels from $f_{(r,t-1)}$ together with the temporal distances $\Delta f_{(r,t-1\to t)}(x,y)$, as defined by

$$\boldsymbol{\Delta f}_{(r,t)}(\boldsymbol{x},\boldsymbol{y}) = \boldsymbol{f}_{(r,t-1\to t)}(\boldsymbol{x},\boldsymbol{y}) + \boldsymbol{\Delta f}_{(r,t-1\to t)}(\boldsymbol{x},\boldsymbol{y}) \tag{5.10}$$

To validate the method, Chen et al. compared with two other techniques [11, 12], which also exploit the correlation between views to recover the lost regions. The proposed method is able to achieve good results and to surpass the best reference methods by an average luminance peak signal-to-noise ratio (PSNR) of 1.42 dB. But, it would be also interesting to compare the proposed method with other popular EC methods tailored for 2D video, such as temporal replacement (TR) or motion vector extrapolation (MVE), in order to conclude more clearly the accuracy of the proposed method over other techniques.

Chung et al. proposed an EC in [13], which is similar to the previous one [9]. This approach also exploits the correlation between different views, in order to extract MVs from uncorrupted views to the one being concealed. The novel idea of this work is to consider the occlusions between views [9]. Figure 5.4 shows how the occlusions between the views are detected. For example, a scene composed by points *a, b, c, d, e, f* and *g*, which are represented in two distinct frames at the same temporal instant from two views, e.g. a stereo pair. Considering an object that is located at a certain distance from the background represented by pixels *a* and *b*,

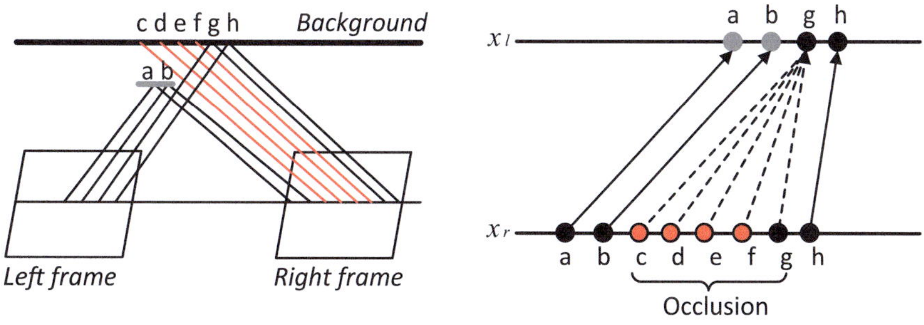

Fig. 5.4 Detection of occluded region between views

some pixels of the background are not visible by none of the views. These pixels belong to the occluded region, represented by *c, d, e* and *f*. Since the EC cannot be accurate for these pixels, as presented in [9], a further method to fill this region must be used.

After recovering the lost region corresponding to the non-occluded areas, the hole filling is performed not only on the occluded areas, but also in other regions where temporal EC was not successfully performed. In some cases, some MVs can point to the same pixels, resulting in empty regions that were not concealed. The non-concealed regions are filled, by checking the motion activity of neighbours previously recovered. Using a 5×5 window around the empty pixel, and if the motion intensity from such window is below a predefined threshold, temporal replacement (TR) is used. When the motion activity is above the threshold, the holes are filled with the spatially nearest available pixels. Comparing to the implementation in [9], on which the method of Chung is based, the authors reported an average PSNR gain of 0.4 dB. In the case of individual frames, the PSNR gain over the method in [9] can be over 1.7 dB, which is a quite significant improvement. Since the average PSNR advantage over the best reference methods is not very high (0.4 dB), the proposed method would be better validated if subjective tests were made, in order to verify if such a small PSNR difference has some effect on the perceived quality by the viewers.

An EC algorithm for MVC, based on the FMO H.264/AVC resilience tool, was proposed by Micallef et al. in [3] and [14]. Since the adopted EC techniques cannot deal with full-frame loss, the method relies on the assumption that when errors occur, only portions of the images are lost (e.g. slices). This method also exploits both inter-view and inter-frame correlations, but in this case, the similarities are not exploited by computing a disparity for the corrupted pixels. A more specific approach is used, less computationally expensive because the disparity vectors (DVs) are extracted from the MVC bitstream. Based on these DVs, EC techniques that were primarily developed for 2D video can also be adopted, such as the ones presented in [15], by adding the DVs to the set of candidates. The MV or DV producing the smaller distortion at the boundaries of the lost region is chosen to recover the corrupted blocks. Depending on the MVC coding structure, DVs might not be available for error concealment. In case of the first view (view 0), which is typically the base view, DVs are not available and EC is performed in the same manner as 2D video. In the other views, anchor frames only have available DVs and not MVs, because in such frames only inter-view prediction is exploited in the coding process. In some view frames, both MVs and DVs are available for EC. As mentioned before, Micallef et al. tested these EC methods for MVC using FMO and also another coding scheme that uses a fixed slice size of 150 bytes. Using FMO or the fixed slice size, the reconstruction accuracy was clearly higher than the accuracy in the scheme with fixed slice size. Comparing a reference method (i.e. FC) with the proposed method, PSNR gains over 2 dB were achieved, proving its effectiveness, though the diversity of EC reference methods used for benchmarking is limited to consolidate the gains obtained from the use of disparity information (i.e. DVs).

In the work described in [16], Stankiewicz et al. proposed an EC algorithm for multiview video, using the MVD format. The major novelty of this method is the use of depth-image-based rendering (DIBR) in EC, where the synthesis of a virtual view is used to recover the lost areas. In addition to DIBR, also intra-based and temporal techniques are used. In the DIBR process, it is assumed that the depth maps are available at the decoder by either being transmitted through different channels or generated on-site. First, the lost regions are recovered using a combination of the inter-view (DIBR) and temporal techniques, but since these might not be able to recover all pixels of the lost regions, the remaining areas are filled using an intra-technique. Regarding, the inter-view and temporal techniques, these two are used to recover all the missing regions. After performing this task, only one of the methods is chosen based the estimated accuracy of the EC technique. Although using demanding simulation conditions with high packet loss rates (PLR), up to 50%, only a very specific loss scenario was considered, where the corrupted frames are not used as reference, thus the error propagation does not exist. Since the test scenario is not very realistic, the ability of this EC method to mitigate the negative effects of error propagation is not known. The quality of the proposed EC method was assessed by comparing it with two other reference methods, temporal replacement (TR) and another temporal EC method, similar to the one described in [17]. The image quality was measured through a set of subjective tests and it was found that the proposed EC method achieves better results than the reference methods for almost all sequences. Only for one test sequence, the method did not achieve the best result, which was justified by the inconsistency in the lightning environment between cameras, which severely affects the accuracy of inter-view concealment.

Another EC algorithm, proposed by Xiang et al. [18] for stereoscopic video, is based on an autoregressive model (AR) [19]. This method starts by acquiring the motion information (MVs) and also the DVs of the corrupted regions. Then, the AR coefficients are computed based on the spatial correlations using both the previously acquired MVs and DVs. The final step is to apply the AR model on all pixels of the lost regions, using weighted interpolation of the selected prediction directions. Note that each of the MVs and DVs is refined using BMA, then the best MV and DV is chosen for the recovery process. Each pixel of the lost region is computed by using a weighted interpolation of the pixels that belong to a window with size $(2 \times R)$. R is the radius of this window, centred on the pixel located at the point given by MV or DV from the reference frame (temporal-correlated or inter-view frame).

Figure 5.5 shows an example of pixels selection used to compute the weighted interpolation. A cross $\times$ in the lost MB of the current frame defines the lost pixel, while a circle ◯ defines the surrounding pixels in the reference frames that are contained inside the region defined by $2R$. Pixels defined by ◯ are used in the weighted interpolation, with an associated weight α that is computed by the proposed AR model. The authors reported PSNR gains up to 1.28 dB in comparison to the error concealment method implemented in JM H.264/AVC reference decoder, for the base view (without inter-view redundancy to exploit). For the second view,

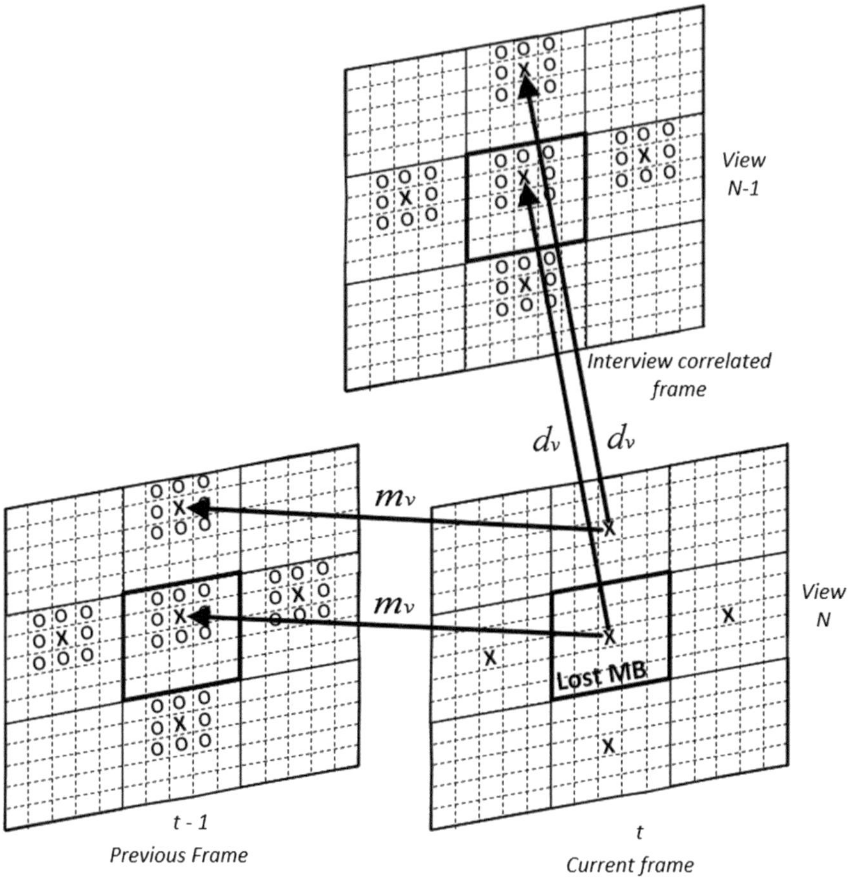

Fig. 5.5 Temporal and inter-view EC model

where inter-view redundancy exists, the PSNR gain is higher, up to 3.32 dB, revealing the importance of using DVs to achieve an accurate EC.

In the work described in [20], Yan et al. proposed an EC technique for MVD which has many similarities with his previous work [21], where the main focus is to recover corrupted texture by taking advantage of additional information given by the depth maps. Hybrid motion vector extrapolation (HMVE) and pixel-based motion vector extrapolation (PMVE) EC techniques are implemented with the addition of depth support. First, a set of extrapolated MVs for each pixel of the lost region is computed using HMVE and PMVE. Then, the depth value of each MVs point is checked based on the assumption that objects present in the scene with similar motion have also similar depth values. When depth maps are lost, no specific EC is performed for depth maps, therefore the conventional PMVE or HMVE is used. Since some MVs might point to the same pixels, some other lost pixels might not be recovered. In such cases, a simple weighted interpolation of the nearest available pixels is performed, using pixels that were already recovered using the previous techniques. The use of depth map information to recover corrupted texture in both PMVE and HMVE EC algorithms is reported as being a significant advantage. Without the use of depth, the HMVE achieves better quality than PMVE. With the addition of depth maps, the advantage of the HMVE method is even more significant. Like other EC techniques described in this chapter, the quality of the synthesised images is not evaluated. Therefore, it is not possible to assess the effect the recovered depth and texture on the synthesis of virtual views.

An error concealment method, based on view synthesis, was proposed by Doan et al. in [22] to recover corrupted colour images, while for corrupted depth maps, the error concealment method is described in [23]. For the sake of simplicity, the authors considered an application where two views with the corresponding texture and depth maps are used. The left view is considered to be error free. This approach is also based on the assumption that slices containing eighty macroblocks are randomly lost. Regarding the EC of the texture, which is the main focus of this method, when a missing region of the right view is detected, the corresponding pixels of such regions are synthesised through DIBR, in order to find the matching pixels of the left view. In the synthesis process, some pixels may not be synthesised if the matching pixels belong to occluded regions. If the number of unsuccessful synthesised pixels is higher than a predefined threshold, the corrupted region is recovered using conventional methods [23]. Otherwise, the following steps are performed in order to choose a temporal or synthesis based error concealment method: (i) first, inter-view MV prediction is performed by computing the DVs, using the available depth of the corresponding corrupted texture image. Since such DVs are block-based, they may be somehow inaccurate at the pixel level. To overcome this problem, a block partitioning scheme is adopted, using the disparity motion field computed from the depth maps. It allows MB partitions from 16×16 pixels down to 8×8 pixels, and the partition size is chosen according to the one that produces less deviation from the disparity MV field; (ii) second, view synthesis based selection is performed by using not only the typical temporal MVs (zero-MV, neighbour-MVs) but also the DVs computed in the previous step and the co-located depth MVs. The best MV/DV is selected by computing the distortion between the predicted block and the corresponding synthesised texture. Besides considering the distortion between predicted block and synthesised texture, the distortion at the boundary of the recovered region is also taken into account in order to achieve a more accurate spatial smoothness; (iii) finally, a selection is performed between a temporal error concealment (TEC), using the MVs/DVs of the previous steps, or simply by synthesising the missing regions. This is performed by selecting the technique that produces less distortion at the boundaries of the missing region, computed using BMA. V. Doan et al. compared the proposed method with BMA, DMVE and also the technique described in [23]. Compared to the best reference method [23], the results show that the proposed method is able to achieve improvements up to 2.19 dB, for a 20% PLR. To validate these results, and to allow comparison to other methods, the quality of the synthesised views using the recovered texture and depth maps should have been evaluated.

The methods previously described follow different approaches, but the goal is essentially the same, i.e. to recover missing image regions in multiview video prone to network losses during transmission. In the case of MVD, depth maps are also transmitted along with the corresponding texture views. Since the depth data is substantially different from texture views, errors or data loss in depth has a completely different impact on the 3D video quality. Therefore, specific error concealment methods must be used for depth data, as described in the next section.

5.3 Methods for Error Concealment of Depth Maps

Most of the work developed so far in the field of depth map error concealment for MVD is based on the use of temporal information extracted from video-plus-depth decoded streams. Due to the low bit rates of coded depth data most common data loss scenarios involve the loss of full depth frames, but depending on the coding/packetisation method, also single blocks or groups of blocks may be lost (bursts). As mentioned before, despite the fact that depth maps are not directly displayed, they play a very important role in the overall quality of the synthesised views, significantly affecting the synthesis quality. Therefore, in error-prone environments, it is crucial to use EC methods that can mitigate the effects of errors in depth maps that would lead to inaccurate synthesis. In this section, a review of the most important EC methods for depth maps is presented, having in mind that some of these publications do not refer exclusively to depth map EC, but also to recover corrupted texture. Depth map EC algorithms are strongly influenced by the existing techniques for 2D and 3D video that were previously described in this chapter. As in 2D and 3D, some techniques rely only on the available spatial information to recover the lost regions, while other techniques rely on temporal and inter-view information. Finally, other techniques rely on the combination of all approaches.

The research work described in [24] exploits the redundancy in motion information between texture images and their corresponding depth maps to recover lost regions, in both texture images and depth maps. However, despite the good objective results, the depth contours are not preserved and some blocking artefacts are noticeable in the depth map, mainly at the edges between the foreground and the background. As pointed out before, this leads to poor quality in the synthesised views. In [23], it is assumed that in presence of depth map errors, if the corresponding texture image region is free of errors, then, it can be used to recover the missing depth. Although good objective results are reported, the impact on the quality of the synthesised images of other views was not evaluated. In another recent work [25], the authors propose an EC method for both texture and depth, but additionally using temporal information. In many of the referenced publications, the impact of the error concealment accuracy of texture and depth maps in the synthesis process is not evaluated, which is rather important since one of the main advantages of using MVD is the capability of synthesising other virtual view points.

An EC method for 3D video transmission that is used on both texture images and depth maps was proposed by Yan et al. in [26]. A BMA technique is used together with the available motion information to recover both corrupted depth maps and texture images. As in the other EC techniques with a similar approach, the BMA technique is used to select the best MV candidate, but in this case, the MV candidates are obtained not only from the uncorrupted spatially and temporal neighbouring regions but also from the corresponding depth map. However, the use of motion information from depth maps to texture and vice versa can be questionable. To exploit redundant motion information between texture and depth, the MV sharing process should not be done indiscriminately, because not all MVs of the texture are suitable to be used for the depth, and vice versa. This is much more evident when using depth MVs in texture, because these MVs are much more likely to be inaccurate, while the texture MVs are much more correlated with depth maps. This is mainly due to the fact that many objects in the scene might have motion, but at the same depth level and in this type of region the depth does not change significantly. Typically, the depth MVs that are more similar to those of the texture are those located at the objects' edges. At the edges of an object, sharp changes in depth maps values occur, and then MVs at such locations are more likely to have high motion correlation with texture motion. As mentioned before, Yan et al. used depth MVs as additional candidates in the BMA technique to recover texture errors. It would be clearer if it is known how frequently the depth MVs are used, when compared to texture MVs of neighbouring regions. Despite the good results of the EC accuracy in the texture video, the authors did not evaluate the quality of the reconstructed depth maps. Furthermore, in these experiments, only one video sequence was tested. Since the proposed method is intended to be used with MVD, the quality of the synthesised views using the recovered texture and depth should have been also evaluated. Therefore, it is very difficult to get a precise idea of its accuracy for synthesised images. The method developed by Liu et al. in [23] also exploits the temporal correlation between texture and depth. This EC method for depth maps is based on a similar approach as described in [24], where the MVs from the texture are directly used to recover the lost regions in the depth maps. It is assumed that depth does not change dramatically over time, and then using MVs from texture is an effective approach. However, such assumption is not totally assertive for very fast-moving areas and also in motion over the z-axis, i.e. perpendicular to the camera plane.

A full-frame EC method is described in [24, 27] by Hewage et al. to be applied for depth maps in MVD based on the SVC (Scalable Video Coding) coding architecture, where the depth maps are encoded as an enhancement layer. In this method, to recover lost texture or depth frames, a similar approach as previously described by Yan et al. in [26] is used. The MVs from the corresponding depth maps or texture are used to conceal the error effects in the respective corrupted depth map or texture. Since this is intended to be used for entire depth frame loss, BMA is not used to refine the MVs, as in [26]. In this case, depth or texture MVs are used directly in the error concealment process. To improve the EC performance of the depth maps, MVs from texture images are also used in the enhancement layer and then to limit the bit rate increase, an upper limit of 25% of the corresponding texture bit rate is imposed. In the performance evaluation, the quality of synthesised image was not measured, only the PSNR of depth maps, which is not the best metric because the quality of synthesised images does not follow the PSNR of the corresponding depth maps. That is why it is important to evaluate error concealment performance of depth maps by assessing the quality of the corresponding synthesised views. Nevertheless, the available results show that sharing motion information between texture and depth maps can be quite effective. Hewage et al. presented an example in [24] where the proposed method is able to achieve good results when compared with TR methods. However, since the depth maps are not used directly for display, subjective tests and/or assessment of the synthesised image are needed to consolidate such results.

Another method for MVD using SVC encoding is presented in [25], where the texture images were encoded as the base layer and the depth maps were encoded in the enhancement layer. The authors rely on the assumption that due to the SVC coding scheme, when a texture region is lost, the co-located region of the depth map has high probability of being also corrupted, due to the interlayer correlations. The proposed technique starts by recovering any possible errors in the texture images and then, by exploiting the correlations between texture and depth, the depth map is recovered in a second step. Besides using temporal information from the temporally adjacent regions, the spatial neighbouring regions of the lost areas can also be used by a block matching technique, in order to find the best blocks for reconstruction. The algorithm classifies the lost areas into two separate groups. The first group includes the lost blocks that have at least one surrounding block in the spatially neighbouring regions. The second group includes either blocks with corrupted neighbours or blocks that have already been concealed. In the case of missing regions, classified as belonging to the second group, EC is performed by using temporal replacement TR from the last decoded frame. In the case of the lost regions belonging to the first group, Fig. 5.6 shows the data involved in the reconstruction. By using the available blocks on top, bottom, left and right of the lost block, a search procedure is carried out to find four MVs that correspond to these four neighbour blocks. The search is done on the previously decoded frame, which is used as the reference frame with a 32×32 pixel search window. The best MV is chosen by computing a border continuity metric (BCM), as described in [28]. After choosing the best MV, the lost block is recovered by motion compensation.

Regarding the EC of the depth map, a lost block is also classified into two groups, by determining whether the lost region is co-located (first group) or not (second group) with the lost region in the texture image. If the lost depth region belongs to the first group, EC is performed by using the best MV found in the texture video. The method relies on the correlation that exists between texture and the corresponding depth maps. In the case where the lost depth regions belong to the second group, TR is also used for EC. This EC method has many aspects in common with other solutions presented in this chapter, such as performing block matching search in the neighbourhood of the lost regions using the correlations

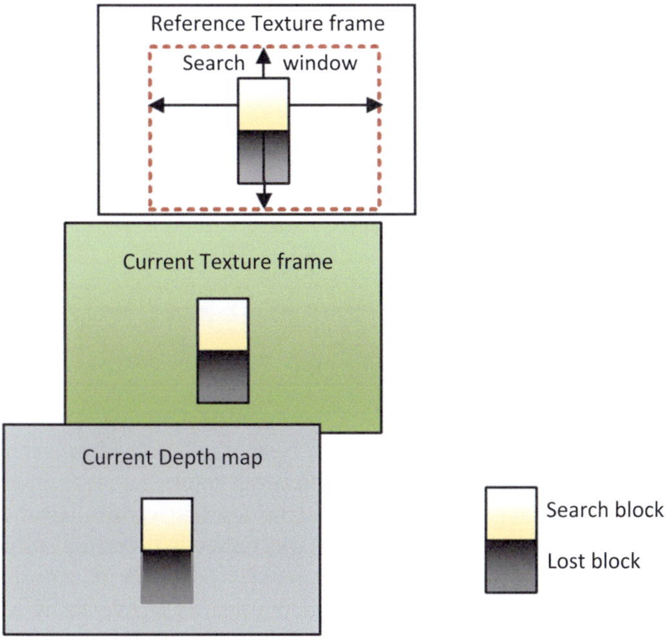

Fig. 5.6 Hewage et al. texture and depth error concealment

between texture and depth. In this respect, the relative performance was not evaluated in comparison to more sophisticated temporal methods, such as MC, MVE or PMVE. Regarding the SSIM, the results show that the highest difference between the proposed method and the TR is less than 0.4%, which might reveal that SSIM is not a significant metric to evaluate the recovered texture and depth in this scenario. Performance evaluation of depth error concealment through image and video synthesis is not reported.

For stereoscopic video, chung et al. [29] proposed a temporal EC method to deal with data losses in both texture images and depth maps of MVD. Missing regions of depth maps are recovered by exploiting the correlations between texture and depth of the same view and from the adjacent view, depending whether the lost frame belongs to the base view or to the second view. The prediction structure shown in Fig. 5.6 is used, assuming that the base view is the left (C_L) one and the non-base view is the right one (C_R). The group of pictures (GOP) length is eight pictures ($T_i, i = 0, 1, \ldots, 7$). D_L and D_R define the left and right depth maps, respectively. In the encoding process, besides temporal correlation (motion compensation prediction, MCP) also disparity compensation prediction (DCP) is used. In the base view, only MCP is used and in the non-base view, MCP and DCP are used. Since texture and depth maps are separately encoded, MCP and DCP are only used within the same type of data, as shown in Fig. 5.7. The actual error concealment approach

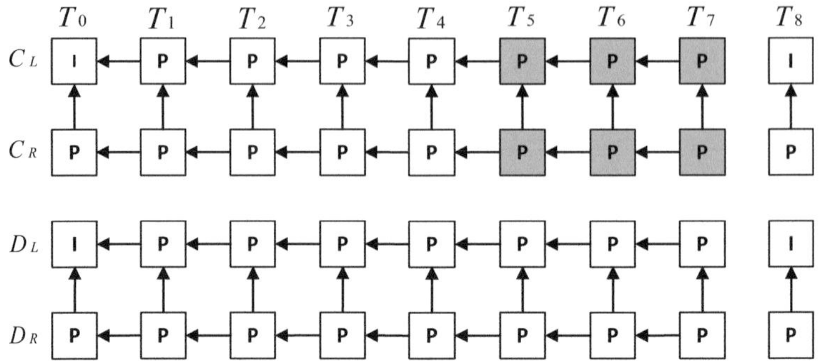

Fig. 5.7 chung et al. [29] stereoscopic coding scheme

depends whether a lost frame belongs to the base or non-base view and also on the availability of the corresponding texture and depth frame.

When a left depth map is lost, the EC algorithm first verifies whether the corresponding texture image of the same view was correctly decoded. In this case, its MVs are directly copied for the depth map, in order to recover the lost depth frame. If the texture image is also lost, then the depth map is recovered by using Pixel based motion vector extrapolation (PMVE) [30]. In the case where the right view is lost (non-base view), EC is performed by first checking the availability of the corresponding depth map of the same view. If the depth map is available, pixels of the lost right texture frame are synthesised using the depth map associated to the left texture image. If the corresponding depth map is not available, the occluded pixels are detected by using both texture and depth from the right view. Then, the non-occluded ones are first recovered by using the inter-view similarities [13]. The remaining occluded pixels are further recovered using PMVE. Finally, the right depth frames errors are recovered directly using MVs from the corresponding right texture frame. For pixels that have been encoded with DCP, the temporal MV is not available and its recovery is done as described in [13]. To evaluate the proposed EC method, Chung et al. compared it with four other reference methods described in [9, 27, 26, 31]. Random full-frame loss was simulated for both texture and depth maps in both views using an error percentage of 10%. Since this EC method is intended to be used for both texture and depth, the authors evaluated the PSNR of texture video sequences and also the intermediate synthesised view using the recovered texture and depth maps. When compared to the reference methods, the proposed method is able to obtain consistent PSNR improvements over all test sequences, on both texture images and also over the corresponding synthesised views. Besides the good error concealment results, it would be enlightening to know what are the negative effects on the synthesis when these errors occur only in the depth maps. Liu et al. in [32] followed a similar approach as previously described. At the encoder side, the occlusions are first detected in the texture regions and the texture MVs of such regions are used in the encoding process of the depth. When a frame is lost, at the decoder side, EC is performed in two main steps: (i) first the non-occluded regions of the lost frame are recovered by synthesising those pixels; (ii) in the second step, the MVs that were embedded in the coding process are used to reconstruct the lost regions that correspond to the occluded areas. The approach of forcing texture MVs of the occluded regions in the depth maps is an innovative idea, but despite texture MVs being highly correlated with depth maps MVs, as mentioned before, this redundancy tends to decrease in the presence of fast-moving regions. Since the texture MVs are used for depth maps coding, the cost is translated into an increased residue due to less accurate motion information. Regarding the texture images, the author reported significant PSNR gains over the reference method used for comparison of results. In the case of the depth maps, also good PSNR results were reported. However, as mentioned before, the depth maps are not used for display and the quality of the synthesised images was not evaluated.

An EC method for MVD to recover entire frame loss for both texture and depth maps was proposed by Lie et al. in [33]. Regarding recovery of depth maps, the proposed algorithm is based on MV sharing [31] and BMA. Typically, BMA is not used in full-frame loss but rather in error concealment of lost regions that have at least some neighbour regions decoded without errors. This method starts by using MV sharing on the co-located regions of the texture image, where MVs exists. In intra-coded texture MBs, the MB is divided into 4×4 pixel sub-blocks and for each of them a MV is computed. These co-located MVs, with the intra-coded texture blocks, are computed using the texture MVs of the neighbouring region of the intra-coded MB. All the MVs recovered in this step are designated as DEC_BF. The subsequent task is to refine DEC_BF with BMA using the information from depth map together with the corresponding texture information. The accuracy of the method was evaluated by separately computing the PSNR of both the recovered texture and depth maps. As mentioned before, using PSNR to evaluate the quality of depth maps might not be the best choice. This problem is more evident when the PSNR differences are smaller. In one of the five sequences, the depth PSNR advantage over the reference methods is up to 2.42 dB for a 15% PLR, but for all the other sequences, the depth PSNR gain is lower than 0.5 dB, which may not be a valid performance indicator.

In [34], Zhang et al. proposed an EC technique for depth maps, which is based on the selection of three other techniques, previously described. The first technique is the weighted spatial interpolation of the four uncorrupted neighbours. The second technique is an inter-frame EC method using TR. Finally, the third technique is MV sharing, as described in [31]. The method proposed by Zhang et al. starts to recover each missing block using these three methods. Then the winning EC method is selected by verifying the similarity of the recovered depth block with the corresponding region of the depth maps in the adjacent view of the same temporal instant.

The similarity measure is based on the computation of the disparity MVs obtained from the depth maps. Figure 5.8 shows an example of how this task is performed, where I_L corresponds to the left view I_R to the right view, D_L and D_R correspond to the left and right depth maps, $k = (x, y)$ defines the coordinates of the texture co-located depth block B in the left view, while $k' = (x', y')$ defines the

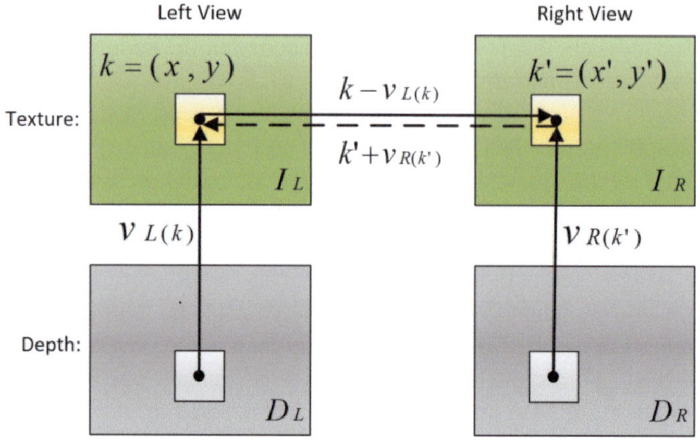

Fig. 5.8 Zhang et al. disparity MVs computation based on depth maps

position of the associated block B in the right view. In the non-occluded regions, the original depth maps, the DVs (disparity vectors)$v_L(k)$ and $v_R(k')$ should be very similar. These DVs are computed based on the corresponding depth maps, as defined by Eq. (10). In this equation, only the horizontal component is shown (vx_l), since it is considered that only horizontal disparity exists due to the 1D arrangement of the camera array.

$$\boldsymbol{vx_l} = \boldsymbol{f}.\boldsymbol{l_t}\left[\frac{\boldsymbol{D_L(k)}}{255}\left(\frac{1}{\boldsymbol{Z_{\min}}} - \frac{1}{\boldsymbol{Z_{\max}}}\right) + \frac{1}{\boldsymbol{Z_{\max}}}\right] \tag{10}$$

In Eq. 10 f defines the focal length, l_t defines the baseline, and 255 is the maximum value for a depth map with 8-bit resolution. $Z_{\min}$ and $Z_{\max}$ are, respectively, the minimum and maximum depth value of the depth map sequence. The winning concealment technique is the one that minimises the distortion between the blocks in texture images, which are computed based on the disparity vectors $v_L(k)$ and $v_R(k')$.

The authors evaluate the accuracy of the proposed method by computing the PSNR of the synthesised views. This is performed by comparing the recovered depth maps of three reference EC methods with the one obtained from the proposed method. The novel aspect of the method is the computation of the disparity vectors using the recovered depth, since this type of approach is not common in the literature. The performance was measured by using HEVC to encode depth data and then loss patterns using a uniform distribution of lost blocks. Additionally, it is used a packetisation scheme capable of ensuring error-free decoded regions surrounding lost blocks, which may not be very realistic. A loss scenario with small blocks is considered (e.g. 8 × 8 pixels), which eases block reconstruction through EC. Thus, the quality gains of the method are modest in comparison to other methods solely based on spatial interpolation techniques.

The various error concealment methods described above mostly rely on three main groups of techniques that can be classified as: spatial domain, inter-view domain and temporal domain. The next subsections describe recent advances in each of these domains.

Spatial domain EC techniques for depth

In the spatial domain, error concealment methods were developed relying on the contours of the corrupted depth map itself, followed by a weighted spatial interpolation [35, 36]. This concept is the core of advanced methods, where two approaches emerge to recover lost contours: the first is based on geometric fitting using *Bézier* curves, while the second exploits the similarities between the depth maps and the associated texture frame contours. In these methods, depth map contours representing sharp transitions between different depth levels are reconstructed using curve fitting techniques based on *Bézier* curves or texture frame contours. First, all contours representing sharp transitions in depth values are extracted from the received depth map. Second, depth lost blocks are classified into two categories: non-edge lost blocks and edge lost blocks. For the non-edge lost blocks, weighted sample interpolation is used to compute values for the missing depth samples, while for edge lost blocks, the missing edge/contour is first reconstructed by using *Bézier* curves [35] or texture contours [36]. Based on the recovered depth contours, depth values inside such lost blocks are also computed using weighted interpolation. In this case, contours are used as boundaries to separate regions with different levels of depth. The proposed algorithm comprises four main steps: Initially, all contours are extracted from the depth map. Then, the contour around each lost area is analysed to find matching endpoints that should be connected together. Finally, based on the matching endpoint pairs obtained in the previous step, an additional pair of control points is computed to reconstruct the contour using a *Bézier* [35] curve or reconstructed by using texture contours. Finally, all lost blocks are reconstructed using weighted sample interpolation. When comparing the performance of these two methods, the average PSNR of the synthesised images is quite similar in the tested sequences, where their variation is not larger than 0.1 dB, but surpassing the reference methods up to 1.91 dB. A variant method was devised [37], which recovers missing contour segments from both texture and *Bézier* interpolation. The results demonstrated that the combination of both methods leads to improved quality of the synthesised views up to 1.01 dB.

Inter-view domain EC techniques for depth

Inter-view domain concealment methods are based on the correlation between depth maps and texture images from different views. Such inherent characteristic of these representation formats allows to exploit similarities between depth and texture views to reconstruct the corrupted depth maps. To this end, block matching using warping functions with geometric transforms proved to achieve quite good performance [38]. In this work, it is assumed that only the depth map of one view is affected by errors resulting in missing blocks/regions. This method is structured in three stages:

1. Weighted interpolation of the lost values using the non-corrupted neighbour values only.
2. Error concealment using warping vectors obtained from block matching using geometric transformations (BMGT) on the texture images. Figure 5.8 shows an example of how a lost depth region is recovered using BMGT. The warping vectors are found by searching in regions of both texture images, co-located with the lost region of the corrupted depth map. The warped quadrilateral mapping is then used to interpolate the lost region, by using values of the non-corrupted depth map.
3. Weighted interpolation of the lost values using the non-corrupted neighbour values and the depth contours, which are reconstructed based on the edge information of the texture image. Details of this method can be found in [36].

In a typical *BMA*, motion is represented by a simple translation of a rectangular area. In the *BMGT* algorithm, an image block is warped to match the best possible representation of a complex motion. The block deformation is performed through image warping, by means of a geometric mapping. When a lost block belongs to a non-homogeneous region, the recovery process is performed by using *BMGT*. This processing sequence is described by the following three steps:

- **Step 1**: The texture image associated with the corrupted depth map (*View 2*) is used as a reference to perform a *BMGT* search, in order to find the matching quadrilateral in the texture image of *View 0* (see Fig. 5.9). The BMGT search is performed using the fast search technique, as described in [38].
- **Step 2**: In this step, taking the best quadrilateral match for the texture image *View 0* in the previous step, the depth map values of *View 0* are used to recover the lost area of *View 2* (inverse mapping). The matching refinement is performed by warping the quadrilateral area until the best possible match is found for the lost region. The mapped depth values are candidates to fill the missing ones in the lost region. The match is verified by evaluating the distortion between depth values of the mapped candidate block (in *View* 0) and the non-corrupted depth values in the neighbourhood of the lost area (in *View* 2). Three rows (top and bottom neighbour values) and three columns (left and right neighbour values) of depth values are used to measure the distortion between the candidate block and the surrounding area of the region to be recovered. In case of error bursts, only top and bottom neighbour values are used, as the left and right neighbours are unavailable. *SAD* is computed to measure the distortion and a predefined threshold *th*, defined as $th = 100 \times N_{\mathrm{pel}}$, is used to decide whether the mapped values from *View* 0 are suitable to recover the lost region of *View* 2. N_{pel} is the number of depth neighbour values used to compute *SAD* and the constant 100 was empirically obtained from the experimental results. If the computed distortion (i.e. *SAD*) corresponding to the best quadrilateral used to recover the lost

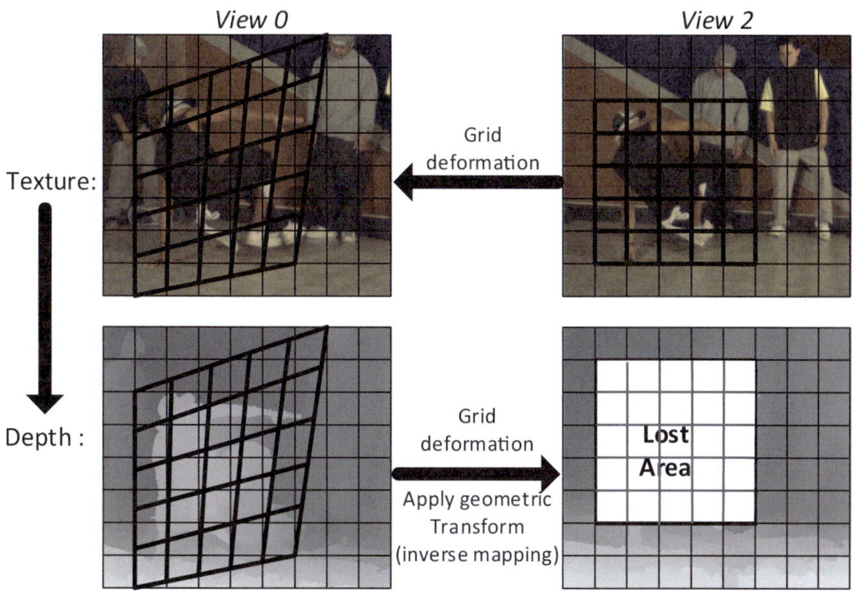

Fig. 5.9 Depth error concealment with BMGT

depth region is larger than *th*, this block is discarded and *BMGT* is not used to recover the lost region.

- **Step 3**: Since each view is obtained from a different camera position, the intensity of the depth map values used to recover the lost region from one view to another may be different from the actual ones. Therefore, after recovering the lost depth values in *Step* 2, an intensity compensation is performed by taking into account the edge information of the recovered block extracted by the *Canny* algorithm.

The results achieved by this method outperform the reference ones used for comparison, especially in sequences where the depth maps contain many depth levels, corresponding to several objects in the scene at different depths. The method shows good efficiency to recover the lost regions, by preserving the objects geometry and depth edges which are of major importance to achieve high quality synthesised images. The results also shown that this method achieves not only higher subjective quality in synthesised images than the reference one but also better PSNR results, that can be up to 5.61 dB for the tested sequences [38].

Temporal domain EC techniques for depth

Regarding temporal domain techniques, geometric similarity between texture and depth maps is exploited by using the motion information extracted from the corrupted depth map itself and also from the texture images using a similar technique, as described in [38]. These temporal techniques are combined with the previous spatial domain techniques, resulting in improved accuracy and better error concealment performance. By exploiting the geometric nature of depth information, a *BMGT* approach jointly used with depth contour reconstruction is devised to achieve accurate interpolation of arbitrary shapes within lost regions of depth maps. The experimental results show that, for different packet loss rates, up to 20%, the depth maps recovered by the proposed method produce virtual views with higher quality than the existing methods based on motion information and spatial interpolation. An average PSNR gain of 1.48 dB is obtained in virtual views synthesised from depth maps using this method [39].

These methods were further investigated in order to efficiently recover lost descriptions in multiple description coded (MDC) depth maps [40, 41]. As the coarse depth version, decoded from a single description, significantly affects the quality of the resulting synthesised images, the research problem tackled in this topic was focused on efficient concealment of missing descriptions when a single one is lost. The method proposed to recover corrupted depth maps is based on a coarser decoded version, which is recovered by applying the spatial/temporal error concealment techniques to the received description, which significantly reduces the negative effects of losing a depth map description. This error concealment method for MDC depth maps can be used when any description is lost. The experimental results show that the proposed method is very efficient, not only when applied to small depth error regions but also in large error areas, even when an entire depth frame is lost. Results obtained with the proposed method show that the use of motion information from the texture clearly improves the reconstruction performance. Furthermore, the combination of temporal and spatial techniques results in an accurate MDC depth map values, significantly improving the quality of the synthesised views, e.g. a PSNR improvement up to 2.29 dB for loss rates of 10%.

5.4 Conclusions

Multiview video delivery using the current coding standards, such as the H.265/HEVC and its multiview extensions, is prone to transmission errors and data loss with strong impact on the quality provided to end users. Joint encoding of texture images (views) and depth maps allow sharing some common information from each side, when parts of the corresponding streams are lost in the networks. This chapter presented several error concealment methods for multiview video and depth maps, that are capable of improving the quality of the visual information delivered to users when transmission errors occur. Besides methods inherited from conventional 2D video, several extensions to stereo and multiview were described, including specific methods for depth maps, which are extremely important for the quality of synthesised views. Overall, the chapter presented the most important recent research in this field.

References

1. Wang, Y.-K., Hannuksela, M.M., Varsa, V., Hourunranta, A., Gabbouj, M.: The error concealment feature in the H.26L test model. In: Proceedings of the International Conference on Image Processing, pp. 729–732, September (2002)
2. Lam, W.M., Reibman, A.R., Liu, B.: Recovery of lost or erroneously received motion vectors. Proc. Int. Conf. Acoustics Speech Signal Process. **5**, 417–420 (1993)
3. Micallef, B.W., Debono, C.J.: Error concealment techniques for multi-view video. In: Proceedings of the IFIP 2010 wireless days, October 2010
4. Thaipanich, T., Wu, P.H., Kuo, C.C.J.: Low-complexity video error concealment for mobile applications using OBMA. IEEE Trans. Consum. Electron. **54**(2), 753–761 (2008)
5. Ali, A., Karim, H.A., Arif, N.A.M., Sali, A.: Depth image-based spatial error concealment for 3-D video transmission. *Res. Dev. (SCOReD), 2010 IEEE Student Conf* 421–425 (2010)
6. Liu, S., Chen, Y., Wang, Y.-K., Gabbouj, M., Hannuksela, M.M., Li, H.: Frame loss error concealment for multiview video coding. Proc. IEEE Int. Symp. Circuits Syst. ISCAS **2008**, 3470–3473 (2008)
7. Belfiore, S., Grangetto, M., Magli, E., Olmo, G.: An error concealment algorithm for streaming video. In: Proceedings of the International Conference on Image Processing - ICIP 2003, vol. 3, pp. 649–52 (2003)
8. Vetro, A., Wiegand, T., Sullivan, G.J.: Overview of the stereo and multiview video coding extensions of the H.264/MPEG-4 AVC Standard. In: Proceedings of the IEEE, vol. 99, no. 4, pp. 626–642 (2011)
9. Chen, Y., Cai, C., Ma, K.-K.: Stereoscopic video error concealment for missing frame recovery using disparity-based frame difference projection. In: Proceedings of the 16th IEEE International Conference on Image Processing—ICIP 2009, pp. 4289–4292 (2009)
10. Gonzalez, R.C.: Digital image processing. Pearson Education India (2007)
11. Bilen, C., Aksay, A., Akar, G.B.: Motion and disparity aided stereoscopic full frame loss concealment method. In: Proceedings of the IEEE 15th Signal Processing and Communications Applications—SIU 2007, pp. 1–4 (2007)
12. Pang L., Yu, M., Yi, W., Jiang, G., Liu,W., Jiang, Z.: Relativity analysis-based error concealment algorithm for entire frame loss of stereo video. In: Proceedings of the 15th IEEE 8th International Conference on Signal Processing, vol. 2 (2006)
13. Chung,T., Sull, S., Kim, C.: Frame loss concealment for stereoscopic video based on inter-view similarity of motion and intensity difference. In: Proceedings of the 17th IEEE International Conference on Image Process—ICIP 2010, pp. 441–444 (2010)
14. Micallef, B.W, Debono, C.J., Farrugia, R.A.: Performance of enhanced error concealment techniques in multi-view video coding systems. In: Proceedings of the 17th IEEE International Conference on Systems Signals Image Process—IWSSIP—2011, pp. 1–4 (2011)
15. Lam, W.-M., Reibman, A. R., Liu, B.: Recovery of lost or erroneously received motion vectors. In: Proceedings of the IEEE International Conference on Acoustics, Speech, and Signal Processing—ICASSP 199, vol. 5, pp. 417–420 (1993)
16. Stankiewicz, O., Wegner, K., Doman M., ski: Error concealment for MVC and 3D video coding. In: Proceedings of the Picture Coding Symposium—PCS 2010, pp. 498–501 (2010)
17. Liu, H., Wang, D., Li, W., Issa, O.: New method for concealing entirely lost frames in H.264 video transmission over wireless networks. In: Proceedings of the IEEE 15th International Symposium on Consumer Electronics—ISCE 2011, pp. 112–116 (2011)
18. Xiang, X., Zhao, D., Ma, S., Gao, W.: Auto-regressive model based error concealment scheme for stereoscopic video coding. Proc. IEEE Int. Conf. Acoustics Speech Signal Processing ICASSP 2011, pp. 849–852 (2011)
19. Wu, X., Barthel, K.U., Zhang, W.: Piecewise 2D autoregression for predictive image coding. In: Proceedings of the International Conference on Image Processing—ICIP 1998., vol. 3, pp. 901–904 (1998)

20. Yan B., Zhou J.: Efficient Frame Concealment for Depth Image Based 3D Video Transmission. *IEEE Trans. Multimedia* **99**(1) (2012)
21. Yan, B., Gharavi, H.: A Hybrid Frame Concealment Algorithm for H.264/AVC. IEEE Trans. Image Process. **19**(1), 98–107 (2010)
22. Doan, V.-H., Nguyen, V.-A., Do, M.N.: Efficient view synthesis based error concealment method for multiview video plus depth. Proc. IEEE Int. Symp. Circuits Syst. ISCAS **2013**, 2900–2903 (2013)
23. Liu, Y., Wang, J., Zhang, H.: Depth image-based temporal error concealment for 3D video transmission. IEEE Trans. Circuits Syst. Video Technol. **20**(4), 600–604 (2010)
24. Hewage, C.T.E.R., Worrall, S.T., Dogan, S., Kondoz, A.M.: A Novel Frame Concealment Method for Depth Maps Using Corresponding Colour Motion Vectors. In: Proceedings of the 3DTV Conference on True Visual—Capture, Transmission and Display 3D Video, pp. 149–152 (2008)
25. Hewage, C.T.E.R., Martini, M.G.: Joint Error Concealment Method for Backward Compatible 3D Video Transmission. In: Proceedings of the IEEE 73rd Vehicular Technology Conference—VTC Spring 2011, pp. 1–5 (2011)
26. Yan, B.: A Novel H.264 Based Motion Vector Recovery Method for 3D Video Transmission. IEEE Trans. Consum. Electron. **53**(4), 1546–1552 (2007)
27. Hewage, C.T.E.R., Worrall, S., Dogan, S., Kondoz, A.M.: Frame concealment algorithm for stereoscopic video using motion vector sharing. In: Proceedings of the IEEE International Conference on Multimedia Expo 2008, pp. 485–488 (2008)
28. Wang, Y., Ostermann, J., Zhang, Y.-Q.: Video processing and communications. vol. 5. Prentice Hall Upper Saddle River (2002)
29. Chung, T., Sull, S., Kim, C.: Frame loss concealment for stereoscopic video plus depth sequences. IEEE Trans. Consum. Electron. **57**(3), 1336–1344 (2011)
30. Chen, Y, Yu, K., Li, J., Li, S.: An error concealment algorithm for entire frame loss in video transmission. In: Proceedings of the IEEE International Conference on Picture Coding Symposium, pp. 15–17 (2004)
31. Bilen, C, Aksay, A., Akar, G.B.: Two novel methods for full frame loss concealment in stereo video. In: Proceedings of the Picture Coding Symposium (2007)
32. Liu, X., Peng, Q., Fan, X., Frame loss concealment for multi-view video plus depth. In: Proceedings of the IEEE 15th International Symposium Consumer Electronics—ISCE 2011, pp. 208–211 (2011)
33. Lie W.-N., Lin, G.-H.: Error concealment for 3D video transmission. In: Proceedings of the IEEE International Symposium on Circuits and Systems - ISCAS - 2013, pp. 2559–2856 (2013)
34. Zhang, X., Zhao, Y., Lin, C., Bai, H., Yao, C., Wang, A.: Warping-driven mode selection for depth error concealment. In: Proceedings of the IEEE Global Conference on Signal and Information Processing—GlobalSIP 2014, pp. 302–306 (2014)
35. Marcelino, S., Assuncao, P., de Faria, S.M.M., Soares, S.: Error recovery of image-based depth maps using Bézier curve fitting. In: Proceedings of the International Conference on Image Processing—ICIP 2011, pp. 2293–2296 (2011)
36. Marcelino, S., Assuncao, P., de Faria, S.M.M., Soares, S.: Lost block reconstruction in depth maps using color image contours. In: Proceedings of the Picture Coding Symposium - PCS 2012, pp. 253–256 (2012)
37. Marcelino, S., Assuncao, P., de Faria, S.M.M., Soares, S.: Efficient depth error concealment for 3D video over error-prone channels. In: Proceedings of the IEEE International Symposium on Broadband Multimedia Systems and Broadcasting—BMSB 2013, pp. 1–5 (2013)
38. Marcelino, S., Assuncao, P., Faria, S.M.M., Soares, S.: Depth map concealment using interview warping vectors from geometric transforms. In: Proceedings of the 20th IEEE International Conference on Image Processing—ICIP 2013, pp. 1821–1825 (2013)
39. Marcelino, S.: Adaptação e Optimização de Qualidade em Serviços Futuros de Vídeo 3D—Depth error concealment for 3D video over error-prone networks. Universidade de Trás-os-Montes e Alto Douro, Vila Real, Portugal (2016)

40. Correia,P., Marcelino, S., Assuncao, P., de Faria, S.M.M., Soares, S, Pagliari, C., da Silva, E.: Enhancement method for multiple description decoding of depth maps subject to random loss. 3DTV-Conference: The True Vision-Capture, Transmission and Display of 3D Video (3DTV-CON), pp. 1–4 (2014)
41. Marcelino, S., Assuncao, P., de Faria, S.M.M., Soares, S.: Robust decoding of MDC Depth Maps for enhanced 3D video over Hybrid Broadcasting Networks. IEEE International Symposium on Broadband Multimedia Systems and Broadcasting, pp. 1–5, Nara—Japan (2016)

Chapter 6
Light Field Image Compression

Caroline Conti, Luís Ducla Soares, Paulo Nunes, Cristian Perra, Pedro Amado Assunção, Mårten Sjöström, Yun Li, Roger Olsson and Ulf Jennehag

Abstract Light field imaging based on a single-tier camera equipped with a micro-lens array has currently risen up as a practical and prospective approach for future visual applications and services. However, successfully deploying actual light field imaging applications and services will require identifying adequate coding solutions to efficiently handle the massive amount of data involved in these systems. In this context, this chapter presents some of the most recent light field image coding solutions that have been investigated. After a brief review of the current state of the art in image coding formats for light field photography, an experimental study of the rate-distortion performance for different coding formats

C. Conti (✉) · L. Ducla Soares · P. Nunes
Instituto Universitário de Lisboa (ISCTE-IUL), and Instituto de Telecomunicações, Lisbon, Portugal
e-mail: caroline.conti@lx.it.pt

L. Ducla Soares
e-mail: lds@lx.it.pt

P. Nunes
e-mail: paulo.nunes@lx.it.pt

C. Perra
Department of Electrical and Electronic Engineering, University of Cagliari, Cagliari, Italy
e-mail: cperra@ieee.org

P. A. Assunção
Instituto de Telecomunicações and Politécnico de Leiria, Leiria, Portugal
e-mail: amado@co.it.pt

M. Sjöström · Y. Li · R. Olsson · U. Jennehag
Mid Sweden University, Sundsvall, Sweden
e-mail: marten.sjostrom@miun.se

Y. Li
e-mail: yun.li@miun.se

R. Olsson
e-mail: roger.olsson@miun.se

U. Jennehag
e-mail: ulf.jennehag@miun.se

P. A. Assunção and A. Gotchev (eds.), *3D Visual Content Creation, Coding and Delivery*, Signals and Communication Technology,
https://doi.org/10.1007/978-3-319-77842-6_6

and architectures is presented. Then, aiming at enabling faster deployment of light field applications and services in the consumer market, a scalable light field coding solution that provides backward compatibility with legacy display devices (e.g., 2D, 3D stereo, and 3D multiview) is also presented. Furthermore, a light field coding scheme based on a sparse set of microimages and the associated blockwise disparity is also presented. This coding scheme is scalable with three layers such that the rendering can be performed with the sparse micro-image set, the reconstructed light field image, and the decoded light field image.

6.1 Introduction

Light field imaging based on a single-tier camera equipped with a micro-lens array (MLA)—simply referred to as light field (LF) in this chapter—has currently risen up as a practical and prospective approach for future visual applications and services. However, successfully deploying actual LF imaging applications and services will require identifying adequate coding solutions to efficiently handle the massive amount of data involved in these systems.

In this context, this chapter overviews some relevant LF image coding solutions that have been recently proposed in the literature. For this, the chapter starts reviewing the state of the art in image coding formats for LF photography in Sect. 6.2. Moreover, since the choice of the used data format strongly influences the LF coding performance, a comprehensive analysis of the rate-distortion performance for different coding formats and different coding architectures applied to LF image coding is presented in Sect. 6.3. In addition to this, aiming at allowing faster deployment of LF applications and services in the consumer market, a scalable LF coding solution that provides backward compatibility with legacy display devices (e.g., 2D, 3D stereo, and 3D multiview) is presented in Sect. 6.4. This display scalable solution makes use of an efficient Inter-layer prediction scheme that when combined with a spatial displacement compensated prediction is able to achieve, in most of the cases, better rate-distortion performance than the non-scalable HEVC solution.

Furthermore, an LF coding scheme based on a sparse set of microimages (MIs) and the associated blockwise disparity is presented in Sect. 6.5. This coding scheme is scalable with three layers such that the rendering can be performed with the sparse MI set, the reconstructed LF image, and the decoded LF image. Moreover, it is shown that this coding scheme improves considerably the coding efficiency with respect to HEVC Intra and is slightly better than the spatial displacement compensated prediction with multiple hypotheses.

6.2 Light Field Image Representation

Since the first approach proposed by Lippman [1] to capture light rays, continuous research and technological developments led to production of LF cameras (a.k.a. plenoptic cameras) that are now available in the consumer market and also for research and scientific applications. Such cameras are mainly characterized by their ability to record not only the light intensity but also the directionality of light rays that reach the camera. This is equivalent to sample the continuous plenoptic function in (6.1), which describes the intensity of light rays passing through any point at a 3D spatial location (x, y, z), i.e., the camera center, from any possible direction (θ, ϕ) with wavelength λ at any instant t.

$$P = P(x, y, z, \theta, \phi, \lambda, t) \tag{6.1}$$

For practical acquisition and representation of light fields, the high dimensionality of the plenoptic function is reduced by assuming that the optical spectrum is monochromatic and the light intensity does not change over the discrete acquisition time of each sample (i.e., a single shot that captures one image at each instant t). Moreover, the LF is not captured for all possible 3D positions in the scene space. Instead, only the light projected onto the 2D camera plane is recorded. This simplification turns the 7D plenoptic function into a 4D representation of light fields, which is commonly used by defining two parallel planes, the camera plane (u, v) and the image plane (s, t). In such 4D model, the light field $L(s, t, u, v)$ defines the intensity of a light ray intersecting both planes [2]. This representation allows visualization of a light field as a (u, v) array of (s, t) images (i.e., different views or perspectives) or as a (s, t) array of (u, v) images (i.e., sub-aperture images of the whole captured scene [3]. Since in currently available LF cameras, 4D light fields are captured as a two-dimensional matrix of tiled 2D MIs, the latter is also the most common representation format used in many application areas and computational algorithms based on the information conveyed by light directionality.

However, such tiled representation may not enable simple and fast access to other types of implicit information embedded in a 4D light field, such as surface reflection and the 3D structure of objects in the scene, i.e., depth. For extracting and processing such type of information the epipolar plane image (EPI) representation is in general more appropriate. An EPI representation of a 4D light field can be thought as a large set of views, where the viewpoints all lie in the common focal plane and the views are projected onto the same image plane I. If P is parameterized with coordinates (s, t) and I with coordinates (x, y), then by fixing a camera coordinate t and image plane coordinate y, the resulting cut in the (x, s) plane is the EPI image. The EPI structure captures 2D views from different viewpoints and encodes the depth information as the slope of line structures in the 2D (x, s) planes.

6.3 Light Field Image Coding Formats

Regardless of the representation format, LF images require a huge amount of data to capture and store the light intensity along with directional information. The number of samples that is necessary to capture the intensity and direction of light rays is much higher than the spatial resolution of conventional 2D images usually rendered in end-user devices.

Due to the inherent redundancy of LF representation, this type of visual data can be easily compressed using conventional image/video coding methods. However, such redundancy is dependent on the data structure that is used in conjunction with each specific coding scheme. For instance, when using standard image/video encoders, which are not specifically tailored for images comprising a lot of MIs with sharp boundaries and highly redundant content, there is a mismatch between such input data structure and the coding units used in most standard compression algorithms. Standard image and video encoders have been used for this purpose, but optimal exploitation of the intrinsic redundancy of LF data requires specific preprocessing. For instance, the correlation between neighboring MIs was exploited in [4, 5] and the correlation in sub-aperture images [6], using three-dimensional transforms was exploited in [7, 8]. Another method based on preprocessing the raw LF in two steps was prosed in [9]. The first step consists in partitioning the raw LF in tiles of equal size and then, in the second step, these tiles are ordered as a pseudo-temporal sequence in order to adapt the data to subsequent HEVC temporal predictive coding. The results show that by exploiting redundancies in the spatial and view angle domain, the HEVC encoding tools are more efficient than JPEG exploiting only spatial redundancies in the whole LF image. Another result of interest is the significant difference between *RD* performance of the tile-based scheme and that of JPEG, which is quite large for high compression ratios (e.g., bpp = 0.1), but much lower for small compression ratios (e.g., bpp = 1). Such results indicate that the benefits of exploiting both the spatial and view angle correlations decrease as the compression ratio also decreases. Thus, for lower compression ratios, exploiting the data redundancy in the two dimensions may result in similar coding efficiency as exploiting redundancy only in the spatial dimension.

For other applications, where the full accuracy of the originally captured LF needs to be preserved, lossless encoding must be used for the entire representation data. This is required for applications with stringent accuracy requirements, such as medical imaging, computer vision for industry, microscopy, etc. For such purpose, LF lossless coding methods have been reported in the literature. For instance, in [10], Perra encodes the non-rectified lenslet image by exploiting the correlation between microimages, like Henlin et al. [11], where the proposed method encodes the sub-aperture images extracted from the rectified lenslet data, exploiting Inter-image correlations by applying different predictors to regions of the same depth. An experimental study on lossless light field coding using standard codecs is presented in [12], using preprocessing techniques to convert the LF data to a format that enables higher lossless compression performance of current standard encoders.

The study analyzes the use of two types of preprocessing techniques that increase the compression efficiency of standard lossless encoders, namely lenslet data rearrangement and color transformation.

6.3.1 Light Field Image Coding Using HEVC

This section presents a performance evaluation study of the coding efficiency attained by the standard high-efficiency video coding (HEVC) using different LF representation formats [13]. To this aim, a data set comprising 12 LF images was captured with the Lytro-Illum camera, which stores the data on LPR files ($\approx$55 MBytes each). This is basically a container format comprising several types of data (the RAW image as captured by the sensor, a thumbnail in PNG format and system settings, among others). The RAW image itself is a 10 bits pack, in GRBG format, with a total resolution of 7728 $\times$ 5368. The RAW files were processed using the "Light Field Toolbox for Matlab", which allows to decode and rectify the captured information using the camera's specific calibration data, comprising a set of white images [14]. The main output of this process is a reconstructed LF corresponding to a 625 $\times$ 434 matrix of MIs, each one capturing the light coming from 15 $\times$ 15 different directions. The data set used in this study is characterized in Table 6.1.

Five data formats were defined to evaluate the standard HEVC coding efficiency, corresponding to different data structures of the same YUV LF. Three of these are organized as still images and encoded using the HEVC still image profile. For the remaining two, LF images are decomposed into sequences of different views in a pseudo-video arrangement encoded using the "Low-Delay *B*", "Low-Delay *P*" and "Random Access" video coding configurations. The following formats were used for the HEVC Still Image Profile.

Table 6.1 Data set used to evaluate the HEVC light field coding performance

	Light field image	Visual content
1	Euro	A 2e coin
2	Bottles	Bottles (1.5 L) on a table
3	Bottle caps	Plastic bottle caps on a grey table
4	Corridor	Corridor in backlight
5	WhiteFlowers	Large green leaves and few pink flowers
6	RedFlowers	Small red and white flowers
7	Park	A few cars at a park exit
8	Garden	Part of a garden with medium-sized trees
9	TrashCans	Large ($\approx$1.80 m) recycling containers
10	Twobottles	Two plastic bottles (1.5 L)
11	People	Four adults at building entrance
12	SkinSpots	Dark spots ($\approx$2 mm) on white skin

Fig. 6.1 Data set used to evaluate the HEVC light field coding performance

- **Light field (Lenslet)**—This is the LF comprising a matrix of MIs obtained with Light Field Toolbox for Matlab, as described in Sect. 9.2. An example can be seen in Fig. 6.1.
- **All-views**—The LF data is rearranged by first extracting the different angular views which are then placed side by side, as seen in Fig. 6.2.
- **Light field filled**—This is similar to Lenslet but the black corner pixels of each MI are filled by extending the left-neighbor pixels (Fig. 6.3).

The pseudo-video formats were obtained by using two different approaches to arrange sequences of views. Both of them result in a pseudo-video sequence such that adjacent views correspond to "temporally adjacent" frames in order to obtain high Inter-frame correlation. In general, this is observed when views have small view-angles, i.e., where disparity is smaller. The two pseudo-video formats used in this study are the following.

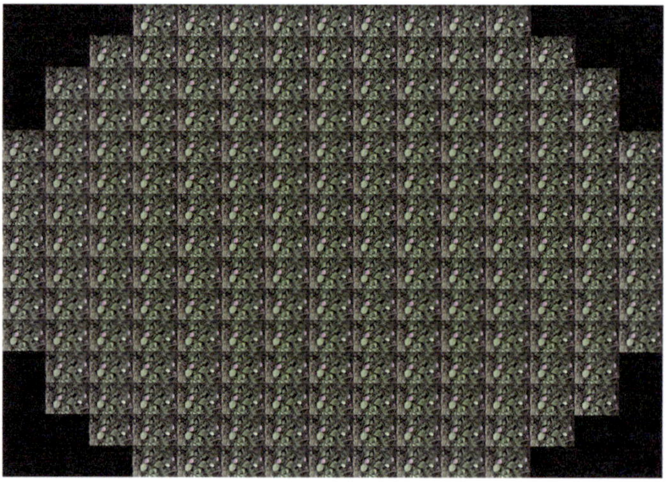

Fig. 6.2 Light field—all views

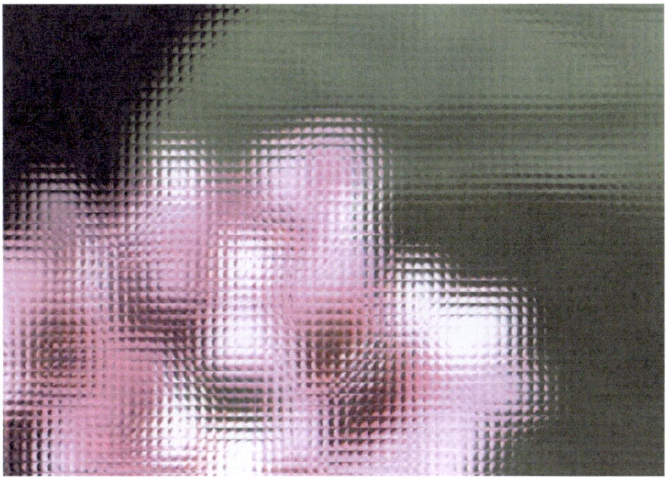

Fig. 6.3 Light field filled

- **Raster**—The pseudo-video sequence is obtained by gathering the views from left to right and top-down, following the scan path shown in Fig. 6.4 (left).
- **Spiral**—The pseudo-video sequence is obtained starting from the central view outwards, following the spiral scan path shown in Fig. 6.4 (right).

For both Raster and Spiral pseudo-video formats, the HEVC configurations used for encoding the light field were the following: All-Intra, Low-Delay *B*, Low-Delay *P*, and Random Access. In the next section, the performance of these coding configurations is evaluated, under test conditions adapted from [15].

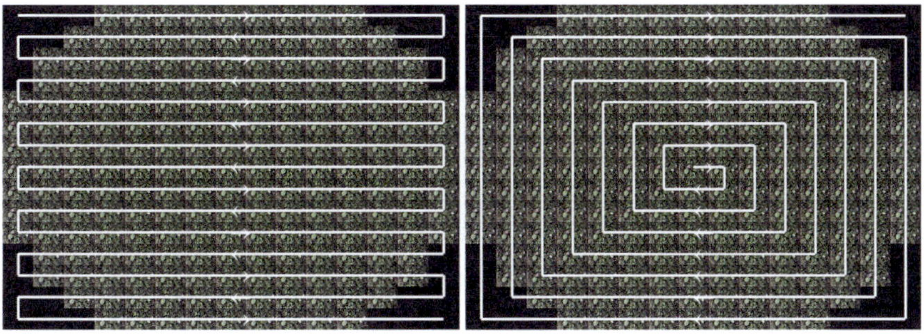

Fig. 6.4 Scan patterns to generate pseudo-video from *all views*: raster (left), spiral (right) (© 2017 IEEE. Reprinted, with permission, from [13].)

6.3.1.1 Coding Efficiency

The coding efficiency obtained from *Bottles*, *People,* and *RedFlowers* is shown in Figs. 6.5, 6.6 and 6.7 for the different configurations referred to above. From these figures, it is quite obvious that different data formats have huge impact on the HEVC coding efficiency. For different arrangements of the LF data and HEVC coding configurations, the PSNR exhibits significant variations, which can be greater than 10 dB at the same rate (bpp). It is worthwhile to note in these figures that the pair (*data format, coding configuration*) does not correspond to a consistent relative coding performance for different visual content. Given the particular structure of the LF data, comprised of a matrix of tiny microimages with dark corners, a consistent worst performance would be expected from the *light field* format. However, while this is true for *People* and *Bottles*, in the case of *RedFlowers,* the *Spiral All-Intra* format is the one achieving the poorest performance. This is most likely due to the fact that the visual content of *RedFlowers* inside the MIs also contain further high-frequency components corresponding to many small leaves of the flowers. Therefore, besides the high-frequency nature of the LF format itself, the coding efficiency is also greatly influenced by the characteristics of the visual content in each MI. On average, the *Bottles* and *People* LF images should not have so much high-frequency content in each MI, which justifies the results shown in the Figures. Further research is necessary to find a valid

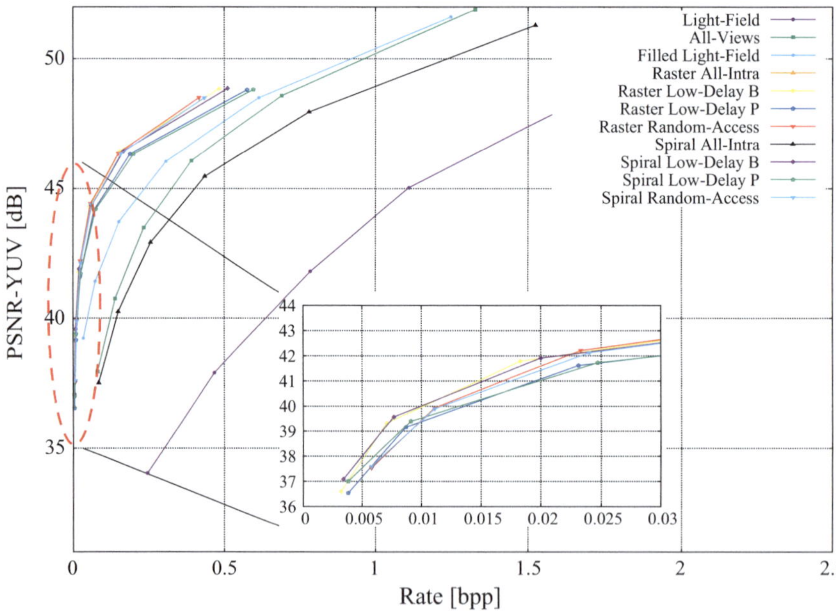

Fig. 6.5 HEVC efficiency for LF image *Bottles* (© 2017 IEEE. Reprinted, with permission, from [13].)

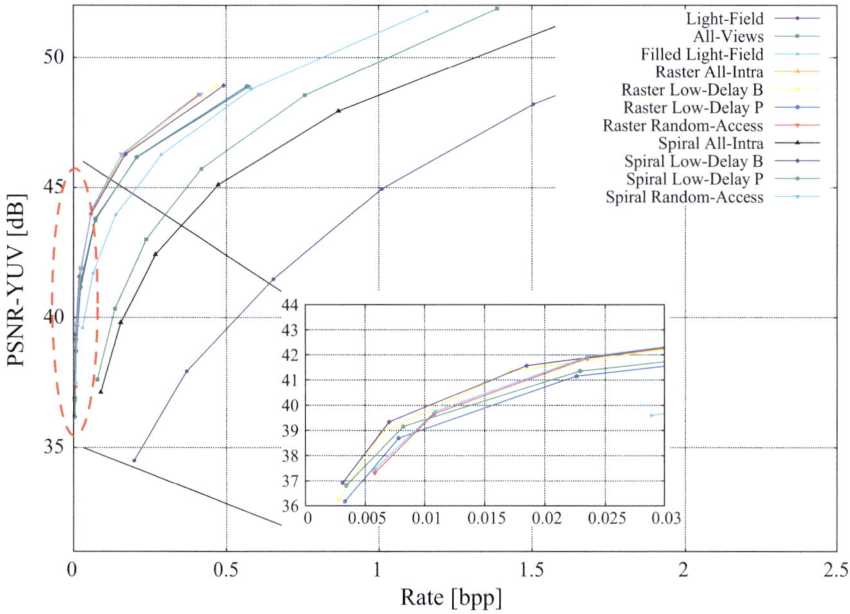

Fig. 6.6 HEVC efficiency for LF image *People* (© 2017 IEEE. Reprinted, with permission, from [13].)

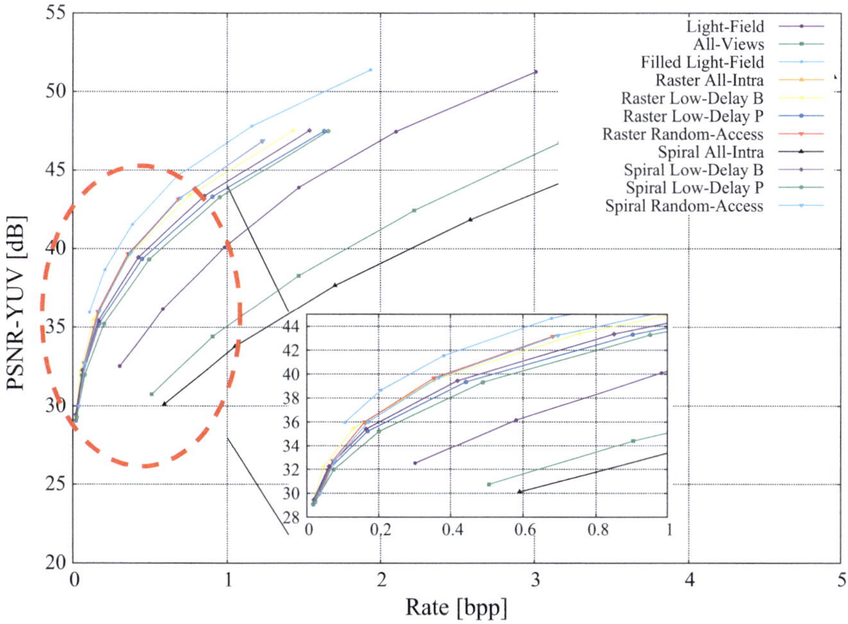

Fig. 6.7 HEVC efficiency for LF image *RedFlowers* (© 2017 IEEE. Reprinted, with permission, from [13].)

threshold for the high-frequency content of MIs, below which coding lenslet light field images might be better than Intra coding of all views (i.e., All-Intra).

Figures 6.5, 6.6, and 6.7 also show a detailed zoom of the lowest rates between 0 and 0.03 bpp. For the pseudo-video formats, one can observe that very low rates are obtained for acceptable levels of PSNR. In this operational region, pseudo-video coding produces very similar results for all data formats, due to the use of high quantization parameters, which contribute to vanish the small differences between adjacent views.

In Figs. 6.5, 6.6, and 6.7, it is clear that organizing the LF data as pseudo-video sequences provides much better performance than still images, as expected. This can also be seen in Table 6.2, where the coding efficiency is shown for the whole set of LF images. Table 6.2 reports the results obtained with quantization parameter $QP = 12$ and $QP = 37$. As expected from the results above, the pseudo-video formats (Raster and Spiral) achieve lower bitrates in comparison with the other formats using the still image profile. The Raster and Spiral scan patterns produce very similar results, which suggests that either one can be used without significant differences in performance.

The results of this simulation study lead to the conclusion that high-efficiency coding of LFs is not only dependent of the encoder configuration but also requires appropriate data rearrangement in order to obtain the best performance. The same coding configuration produces quite different results when using the same input data arranged in a different format. There are also intrinsic signal characteristics of each micro-lens, such as the amount of high-frequency content, that influence the relative coding performance of the various methods. Further research is necessary to find the best LF preprocessing algorithms that are capable of guaranteeing a consistent relative performance across all coding configurations, for any type of content.

6.4 Scalable Light Field Coding for Backward Display Compatibility

In addition to the challenge of proposing efficient coding solutions for handling the huge amount of data involved in LF application systems, another important issue when trying to deliver LF content to end-users is to provide backward compatibility with existing legacy receivers (either 2D, or current stereo or multiview). Dealing with this specific concern is an essential requirement for enabling faster deployment of new LF imaging application services in the consumer market. For enabling this, an efficient scalable LF coding approach is then desirable whereby decoding only the adequate subsets of the scalable stream, 2D or 3D compatible video decoders can present an appropriate version of the LF content. Regarding the scalable coding solution, although simulcast is a possible approach, the bandwidth consumption may not be acceptable, thus demanding a more efficient scalable coding solution.

Table 6.2 RD coding performance comparison for the set of LF images in Table 6.1

	Light field (Lenslet)		All-Views		LF filled		Raster video (Low-Delay *B*)		Spiral video (Low-Delay *B*)	
Sequences	bpp	PSNR	bpp	PSNR	bpp	PSNR	bpp	PSNR	bpp	PSNR
QP = 12										
BottleCaps	3.003	51.03	1.556	51.53	1.732	50.97	0.787	48.00	0.814	48.03
Bottles	2.406	51.58	1.326	51.89	1.246	51.61	0.484	48.85	0.511	48.85
Corridor	2.524	51.62	1.699	51.74	1.506	51.62	0.637	48.57	0.670	48.59
Euro	2.868	51.37	2.254	51.43	1.888	51.41	0.843	48.00	0.906	48.02
Garden	2.688	51.65	2.372	51.99	1.523	51.61	0.902	48.32	0.970	48.33
Park	2.967	51.31	2.155	51.60	1.826	51.35	0.818	48.03	0.891	48.02
People	2.263	51.69	1.388	51.88	1.160	51.77	0.471	48.92	0.492	48.92
RedFlowers	3.009	51.25	4.286	51.43	1.935	51.37	1.431	47.54	1.538	47.53
SkinSpots	3.315	50.94	2.225	51.34	2.074	51.04	0.895	47.60	0.945	47.63
TrashCans	3.394	51.08	1.518	51.60	2.477	50.99	0.750	48.21	0.750	48.22
TwoBottles	2.811	51.24	1.382	51.74	1.391	51.33	0.584	48.42	0.610	48.45
WhiteFlowers	3.250	51.06	3.248	51.25	2.129	51.13	1.167	47.60	1.252	47.62
QP = 37										
BottleCaps	0.309	32.99	0.020	40.43	0.022	39.30	0.002	39.65	0.002	39.80
Bottles	0.244	34.03	0.079	37.95	0.033	39.23	0.003	36.59	0.003	37.09
Corridor	0.239	33.92	0.094	36.20	0.045	37.79	0.004	34.82	0.005	35.43
Euro	0.197	33.25	0.075	34.77	0.054	36.74	0.005	33.57	0.006	33.80
Garden	0.249	33.86	0.141	34.65	0.048	37.89	0.004	33.40	0.004	33.78
Park	0.352	33.16	0.117	35.69	0.039	36.95	0.004	34.27	0.005	34.83
People	0.196	34.49	0.077	37.60	0.029	39.60	0.003	36.25	0.003	36.91

(continued)

Table 6.2 (continued)

	Light field (Lenslet)		All-Views		LF filled		Raster video (Low-Delay *B*)		Spiral video (Low-Delay *B*)	
Sequences	bpp	PSNR	bpp	PSNR	bpp	PSNR	bpp	PSNR	bpp	PSNR
RedFlowers	0.302	32.52	0.506	30.74	0.106	35.95	0.014	29.14	0.019	29.45
SkinSpots	0.402	32.27	0.025	36.39	0.044	36.68	0.002	35.59	0.002	35.77
TrashCans	0.341	32.18	0.064	38.53	0.106	35.02	0.005	37.03	0.006	37.48
TwoBottles	0.382	34.13	0.043	38.96	0.018	40.31	0.002	37.72	0.002	38.18
WhiteFlowers	0.343	32.68	0.243	32.89	0.076	36.23	0.006	31.46	0.007	32.05

In this context, a display scalable architecture for LF coding is presented in Sect. 6.4.1 (as first proposed in [16]) using a three hierarchical layer approach so as to accommodate from the end-user who wants to have a simple 2D version of the LF content to be visualized in a conventional 2D display; to the end-user who wants have a more immersive and interactive visualization by using a more advanced LF display technology, such as an integral imaging display [17–20] or a head-mounted display for augmented and virtual reality [21, 22]. As discussed in Sect. 6.4.2, a preprocessing is necessary for generating the content for each hierarchical layer before coding. Based on this hierarchical coding architecture, Sect. 6.4.3 presents an light field (LF) enhancement codec to efficiently encode the LF content in the highest layer [23]. Finally, Sect. 6.4.4 performs the evaluation of the presented display scalable codec.

6.4.1 Display Scalable Coding Architecture

A display scalable architecture for light field coding (DS-LFC) with a three-layer approach is used here as illustrated in Fig. 6.8. As can be seen, each layer of this scalable coding architecture represents a different level of display scalability:

- **Base Layer (2D Layer)**—The base layer represents a single 2D view, which can be used to deliver a 2D version of the LF content to 2D displays devices. This 2D view is then coded with conventional HEVC [24] Intra coder to provide backward compatibility with a state-of-the-art coding solution. Then, the

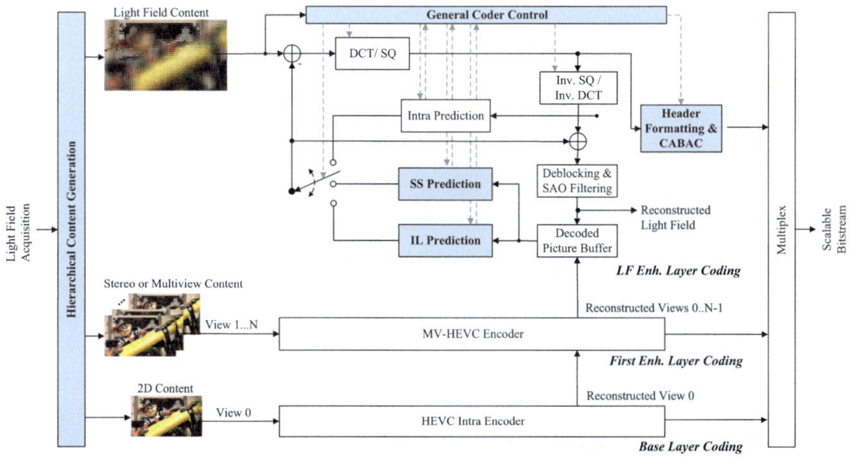

Fig. 6.8 Scalable light field coding architecture using three hierarchical layers for backward display compatibility. The novel and modified blocks are highlighted in blue shaded blocks

reconstructed 2D view is used for coding the higher layers, as illustrated in Fig. 6.8.

- **First Enhancement Layer (Stereo or Multiview Layer)**—This layer represents the necessary information to obtain an additional view (representing a stereo pair) or various additional views (representing multiview content). This is to allow stereo and autostereoscopic devices to play versions of the same LF content. The content in this layer can be then encoded by using a standard stereo or multiview coding solution [25–28], and the reconstructed 2D views are then made available to be used for coding of the LF enhancement layer (Fig. 6.8). In this work, the multiview extension of HEVC, MV-HEVC [28], is adopted. With these solutions [25–28], Inter-view prediction can be used to improve the coding efficiency between the base layer and the first enhancement layer, as well as within the views in the first enhancement layer. However, it should be noticed that efficient prediction mechanisms between the base layer and the first enhancement layer and within the first enhancement layer are not addressed in this chapter since these cases have been extensively studied in the context of MVC [25], and in the 3D video coding extensions of the HEVC [28]. For a good review of these 3D video coding solutions, the reader can refer to [25–28] as well as Chaps. 3 and 4.
- **Second Enhancement Layer (LF Enhancement Layer)**—This layer represents the additional data needed to support full LF display. The content in the LF enhancement layer is then encoded by using the LF enhancement coding solution presented in Sect. 6.4.3, as depicted in Fig. 6.8.

High compression efficiency is still an important requirement for the scalable coding architecture presented in this section. In this context, the scalable coding solution should be able to improve the rate-distortion (RD) coding performance compared to independent compression of the three different layers (the simulcast case).

6.4.2 Hierarchical Content Generation

Generating 2D and 3D multiview content from LF content basically means producing various 2D views with different viewing angles. For this, a particular rendering algorithm needs to be chosen and some information about the acquisition process—such as the MI resolution and MLA structure (i.e., the array packing scheme and the micro-lens shape)—needs to be known at both encoder and decoder sides.

In the work presented in this section, the rendering algorithm proposed in [29] and referred to as basic rendering is adopted for this hierarchical content generation. The idea behind these algorithms is to combine suitable patches from each MI to properly compose a 2D view image. Then, as explained in [29], the process of generating a 2D view image can be controlled by the following two main parameters:

- **Patch Size**—It is possible to control the plane of focus in the generated 2D view image (i.e., which objects will appear in sharp focus) by choosing a suitable patch size to be extracted from each MI. Therefore, during a creative postproduction process, a proper patch size will be selected for generating the content for the first two hierarchical layers. It is worth noting that this decision is limited to the available depth range in the captured LF image.
- **Patch Position**—By varying the relative position of the patch in the MI, it is possible to generate multiple 2D views with different horizontal and vertical viewing angles (i.e., different scene perspectives). It is also worthwhile to note that this choice is also made in a creative manner, and the number of views and their corresponding positions may be based on a target type of display device that will be used for visualization.

In other words, there is a large degree of freedom when defining how to generate the content for the base and first enhancement layers. Therefore, the performance of the scalable coding solution shall be analyzed while taking into account the parameters that control this process.

6.4.3 Efficient LF Enhancement Layer Coding Solution

Since the lower layers of the proposed DS-LFC codec presented in Sect. 6.4.1 are based on the HEVC [24] standard (or on its extension for multiview coding MV-HEVC), the LF enhancement encoder proposed in this section is also based on the hybrid coding techniques of HEVC, as illustrated in Fig. 6.8, so as to modify as few aspects of the underlying architecture as possible. Notice that, although the LF enhancement layer encoder presented in Fig. 6.8 targets LF still image coding, it can be also extended for scalable LF video coding by including also the HEVC Inter-frame coding.

The main blocks of the proposed HEVC-based LF enhancement encoder (highlighted in Fig. 6.8) are explained in the following.

6.4.3.1 Self-similarity (SS) Prediction

The SS prediction [30–32] (Fig. 6.8) is used to exploit the redundancy within the highest enhancement layer and to improve coding efficiency. As can be seen in Fig. 6.9a, a significant cross-correlation exists between neighbor MIs in the LF image captured with an LF camera.

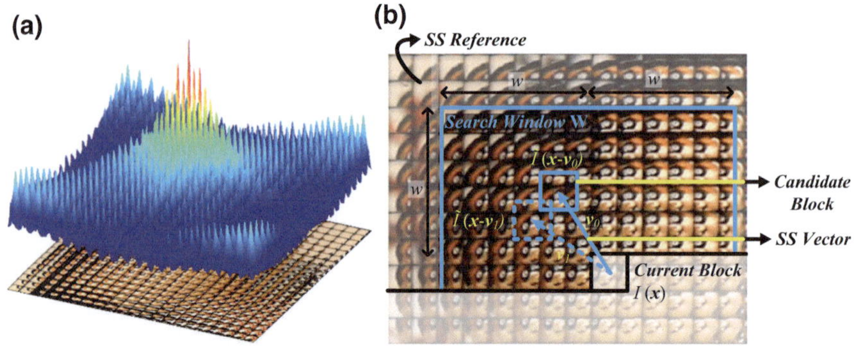

Fig. 6.9 SS prediction: **a** inherent MI cross-correlation in a light field image neighborhood; and **b** SS estimation process (example of a second candidate block and SS vector for bi-prediction is shown in dashed blue line). Reprinted from [33]. Copyright (2017), with permission from Elsevier

Hence, the SS prediction is a spatial displacement compensated prediction [33] which makes use of a block-based matching algorithm to estimate the prediction block with the highest similarity (according to appropriate criteria) to the current block in the previously coded and reconstructed area of the current picture itself (the SS reference, as seen in Fig. 6.9b). This predictor block can be generated from a single candidate block [30, 31] or from a combination of two different candidate blocks [32, 33] (Fig. 6.9b). Hence, the relative position between the current and the "best" candidate block(s) is signaled by one of two SS vector(s), $\boldsymbol{v}_i$, (Fig. 6.9b).

As a result of the SS prediction, the residual information and the SS vector(s) are coded and sent to the decoder.

6.4.3.2 Inter-layer (IL) Prediction

An IL prediction mode can also be used to further improve the LF enhancement coding efficiency by removing redundancy between the LF content and its stereo or multiview version from the enhancement layer underneath.

For this, an Inter-layer reference (ILR) is constructed by using information from the lower layers. This ILR picture can be then used as new a reference frame for employing an IL compensated prediction (see Fig. 6.8) when encoding the LF image. To build an ILR picture, the following information is needed:

- **Set of 2D Views**—The set of reconstructed 2D views obtained by decoding the bitstream in the lower layers is available in the decoded picture buffer, as depicted in Fig. 6.8;
- **Acquisition Parameters**—These parameters comprise information from the LF capturing process (such as the MI resolution and the MLA structure) and also information from the 2D view generation process (i.e., size and position of the

patches). As explained in Sect. 6.4.3.4, this information has to be conveyed along with the bitstream to be available at the decoding side.

Therefore, two steps are distinguished when generating an ILR picture, which are explained in the following.

Patch Remapping

Although most of the LF information is discarded when rendering each view in the hierarchical layer generation block in Fig. 6.8 (see Sect. 6.4.2), it is still possible to reorganize the reconstructed view texture information into its original positions in the LF image. This is the purpose of the patch remapping step. The input for this step is the coded and reconstructed views from the two lower layers, as well as the acquisition parameters used for acquiring these views at the encoder side.

The patch remapping simply corresponds to an inverse process of the rendering algorithm used Sect. 6.4.2. More specifically, it corresponds to an inverse mapping (referred to here as remapping) of the patches from all rendered and reconstructed views to their original positions in the LF image, as illustrated in Fig. 6.10a. A template for the LF image assembles all patches, and the output is referred to as the sparse ILR picture, as seen in Fig. 6.11a.

MI Refilling

This step aims at emulating the significant cross-correlation existing between neighboring MIs so as to fill the holes in the sparse ILR picture (built in the previous step) as much as possible.

Since there is no information about the disparity/depth between objects in neighboring MIs, the disparity is defined in a patch-based manner, by using the patch size parameter that was used in the hierarchical layer generation block (see Sect. 6.4.2). An illustrative example of this process is shown in Fig. 6.10a for only

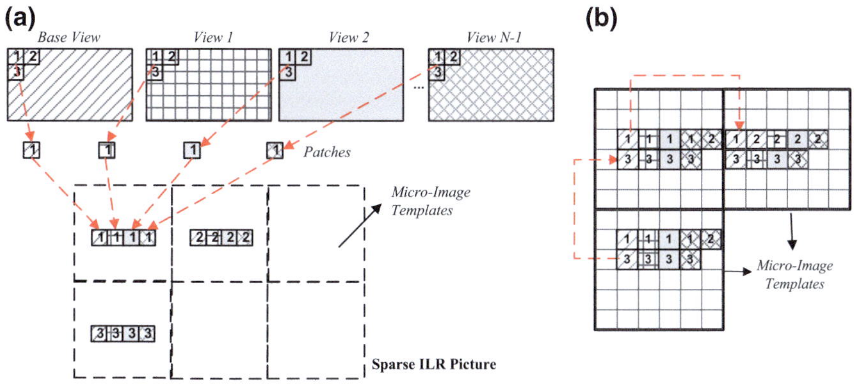

Fig. 6.10 The process to generate an ILR picture to be used in the proposed IL prediction: **a** patch remapping step; and **b** MI refilling step

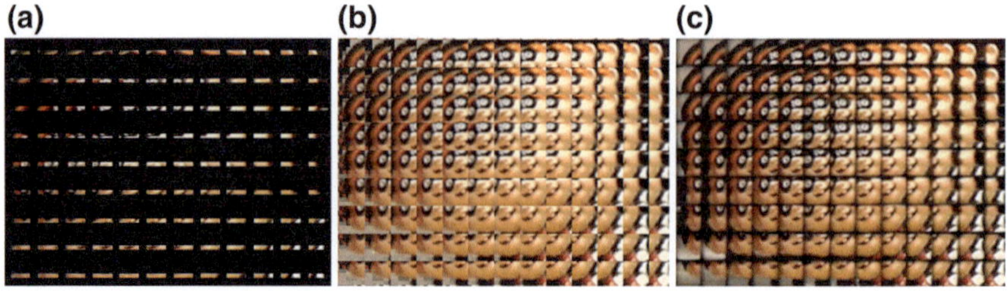

Fig. 6.11 Illustrative example of a portion of an ILR built for the LF image plane and toy (frame 123): **a** the sparse ILR picture; **b** the corresponding complete ILR constructed using the MI refilling algorithm; and **c** the corresponding portion of the original LF image (which is coded in the LF enhancement layer)

three neighboring MIs in the sparse ILR picture. As can be seen, for each MI in the sparse ILR picture, an available set of pixels (see Fig. 6.10a) is copied to a suitable position in a neighboring MI that is shifted by the patch size. Additionally, the number of neighboring MIs where the patch may be copied to depends on the size of the MI and the patch size. Finally, the output of the process is the ILR picture (see Fig. 6.11b).

It is worthwhile to notice that there are still opportunities to enhance the proposed IL prediction (notably, the MI refilling step) and to enlarge the applicability of the proposed DS-LFC solution. A possibility is to incorporate supplementary data (such as depth, ray-space, and 3D model data) into the scalable bitstream. This solution will be further studied in future work.

6.4.3.3 Intra Prediction

HEVC Intra prediction is available as an alternative prediction when selecting the most efficient mode for encoding a CB in the LF enhancement layer (Fig. 6.8). The decision between the different available prediction modes is made in a rate-distortion optimization (RDO) manner [34] as in conventional HEVC [24].

6.4.3.4 Header Formatting and CABAC

Additional high-level syntax elements are carried through the scalable bitstream to support this new type of scalability. These are basically: (i) acquisition parameters that are used to generate the content for the lower layers and are also necessary to build the ILR picture (i.e., MI resolution, MLA structure, size and position of the patches); and (ii) dependency information for signaling the use of the novel reference pictures (SS reference and ILR). Finally, residual and prediction mode signaling data are entropy coded using CABAC.

6.4.4 Performance Assessment

To evaluate the performance of the proposed DS-LFC codec, the following test conditions were considered:

- **Light Field Test Images**—Six LF images with different spatial and MI resolutions are considered to achieve representative RD results. These are (see Fig. 6.12): *Fredo*, *Seagull*, *Laura*, *Demichelis Spark* (first frame of a sequence with identical name), *Robot 3D*, and *Plane and Toy* (frame number 123 from a sequence with identical name). The first three images are available in [35] and the remaining images in [36]. The original tested images were rectified to have all MIs with integer number of pixels, and they were then converted to the Y'CbCr 4:2:0 color format.
- **Hierarchical Content Generation**—To generate the content for the 2D, stereo or multiview layers, the six LF test images were processed using the algorithm Basic Rendering [29] (Sect. 6.4.2). In this process, a set of 9 × 1 regularly spaced 2D views was generated—one for the base layer and the remainder for the first enhancement layer. Additionally, the patch size was chosen to represent the case where the main object of the scene is in focus. Based on the above decisions, the chosen patch sizes and positions for each LF test image are summarized in Table 6.3.

(a) *Fredo* 7104 × 5328 $MI_{resol} = 74 \times 74$

(b) *Seagull* 7104 × 5328 $MI_{resol} = 74 \times 74$

(c) *Laura* 7104 × 5328 $MI_{resol} = 74 \times 74$

(d) *Demichelis Spark* 2850 × 1558 $MI_{resol} = 38 \times 38$

(e) *Robot 3D* 1904 × 1064 $MI_{resol} = 28 \times 28$

(f) *Plane and Toy* 1904 × 1064 $MI_{resol} = 28 \times 28$

Fig. 6.12 Example of a central view rendered from each light field test image (with the corresponding characteristics below each image)

Table 6.3 Test conditions—patch sizes and positions (in pixels) for generating content for the lower hierarchical layers using the DS-LFC solution (for each light field test image in Fig. 6.12)

Test image	Patch size (focus plane)	Patch positions (view's perspectives)
(a)	10	{(−24,0), (−18,0), (−12,0), (−6,0), (0,0), (6,0), (12,0), (18,0), (24,0)}
(b)	9	{(−24,0), (−18,0), (−12,0), (−6,0), (0,0), (6,0), (12,0), (18,0), (24,0)}
(c)	10	{(−24,0), (−18,0), (−12,0), (−6,0), (0,0), (6,0), (12,0), (18,0), (24,0)}
(d)	12	{(−8,0), (−6,0), (−4,0), (−2,0), (0,0), (2,0), (4,0), (6,0), (8,0)}
(e)	4	{(−8,0), (−6,0), (−4,0), (−2,0), (0,0), (2,0), (4,0), (6,0), (8,0)}
(f)	4	{(−8,0), (−6,0), (−4,0), (−2,0), (0,0), (2,0), (4,0), (6,0), (8,0)}

- **Codec Software Implementation**—For these tests, the reference software for the MV-HEVC extension version 12.0 [37] is used as the base software for implementing the proposed DS-LFC codec.
- **Coding Configuration**—The results are presented for four QP-values (22, 27, 32, and 37). The same QP-value was used for coding all hierarchical layers. In the proposed DS-LFC codec, all the views in the lower layers are independently encoded as Intra-frames. Notice that, other configurations for encoding the content in the first layer are still possible, notably, by enabling Inter-view prediction (coding as P or B frames). However, due to the large number of possible test condition combinations, the following sections will focus on analyzing the influence of varying the parameters for generating the content for the lower layers in the performance of the proposed IL prediction. Following this, the LF enhancement layer is encoded as an Inter-B frame.
- **Search Strategy**—Considering both IL and SS prediction, a search range value of 128 is adopted for all tested LF images. The full search algorithm with the HEVC quarter-pixel accuracy is also used.
- **RD Evaluation**—For evaluating the RD performance of the proposed LF enhancement layer encoder, the distortion, in terms of PSNR, of the reconstructed LF image in the LF enhancement layer is considered. The rate is presented in bits per pixel (bpp), which is calculated as the total number of bits needed for encoding all scalable layers, divided by the number of pixels in the LF raw image. Therefore, the BD [38] results are presented in terms of the luma PSNR of the reconstructed LF image in the LF enhancement layer and the corresponding rate in terms of bpp values.
- **Additional Objective Quality Metrics**—Additionally, to analyze the performance in terms of the quality for views synthesized from the reconstructed content in the LF enhancement layer, the distortion is also measured in terms of average PSNR and SSIM values calculated for a set of 3×3 views rendered

from viewpoint positions equally distributed in horizontal and vertical directions. This metric is referred to here as $PSNR_{3\times3Views}$ and $SSIM_{3\times3Views}$. These views are different than the views rendered for the lower layers (except for the central view). The standard deviation for each of these metrics is also used as a dispersion evaluation of the presented average values. For rendering the views, the same algorithm used for generating content for each hierarchical layer is used (i.e., basic rendering or weighted blending [29]).

The next subsections present and analyze the performance of the proposed DS-LFC solution and compare it to the following solutions:

- **DS-LFC (Simulcast)**—This scalable codec corresponds to the benchmark for the simulcast case, where the content from each hierarchical layer is coded independently with the MV-HEVC standard using "All-Intra, Main" configuration [39].
- **DS-LFC (SS Simulcast)**—In this case, the content from the LF enhancement layer was coded with the DS-LFC codec but only enabling the SS prediction and conventional HEVC Intra prediction (without IL prediction). Hence, not only local spatial prediction is exploited (with Intra prediction) but also the nonlocal spatial correlation between neighbor MIs (with SS prediction). Since when using the SS prediction each scalable layer is still coded independently (from each other), the proposed DS-LFC (SS) can be seen as an alternative simulcast coding solution.
- **HEVC (Single Layer)**—In this case, the entire LF image is encoded into a single layer with HEVC using the main still picture profile [24]. Since the proposed DS-LFC codec provides an HEVC-compliant base layer, this solution is used as the benchmark for non-scalable LF coding, and the resulting bit savings are compared to the proposed scalable LF coding solution so as to analyze the cost (in terms of RD performance) of supporting display scalability in the bitstream.

6.4.4.1 Overall DS-LFC RD Performance

To assess the performance of the proposed DS-LFC codec, Table 6.4 presents the RD performance in terms of the Bjøntegaard Delta in PSNR (BD-PSNR) and rate (BD-BR) [38] regarding the benchmarks solutions for all test images in Fig. 6.12.

From these results, the following conclusions can be derived:

- **Comparison with simulcast cases**—The RD performance of the proposed DS-LFC is significantly better than the DS-LFC (Simulcast) for all tested images, with average BD gains of 2.05 dB or 33.71% of bit savings (see Table 6.4). The gains are much more expressive for test images with higher MI resolution, where the BD gain goes up to 3.00 dB with 44.56% of bit savings (for *Seagull*). These gains show the efficiency of the predictive coding tools used in the LF enhancement encoder. Moreover, comparing the DS-LFC (Proposed) solution with the DS-LFC (SS Simulcast), improved RD performance can be attained by taking advantage of the redundancy in all domains (local and nonlocal spatial domain, and Inter-layer domain), leading to average BD gains of 0.35 dB or −6.56%.

- **Comparison with HEVC (Single Layer)**—As shown in Table 6.4, the proposed DS-LFC solution presents better RD performance, in terms of average BD gains (0.90 dB and 13.49%), than the non-scalable HEVC (Single Layer), showing that it is possible to support a display scalable bitstream with no additional bit rate cost. Moreover, for LF images with larger resolution and MI sizes, it is even possible to achieve significant better RD performance with the proposed DS-LFC (with BD gains of up to 2.40 dB and 37.90% of bit savings). On the other hand, for some LF images with smaller resolutions and MI sizes, the scalability is allowed at a cost of some compression efficiency penalty (up to −0.56 dB and 7.54% of penalty). However, it is important to notice that the worse RD performance of the proposed DS-LFC solution is, in this case, also due to the set of 9 × 1 views that are independently encoded as Intra-frames in the lower layers, instead of enabling the Inter-view prediction to improve the RD performance.

6.4.4.2 Quality of Rendered Views

To assess the performance of the proposed scalable coding architecture regarding the quality of rendered views, the RD performance of the DS-LFC (Proposed) is here presented in terms of the $PSNR_{3\times3\text{Views}}$ and $SSIM_{3\times3\text{Views}}$ metrics and compared to the DS-LFC (Simulcast) and HEVC (Single Layer) solutions. The results are illustrated in Figs. 6.13 and 6.14, respectively, for the worst (i.e., for test image Robot 3D) and best case (i.e., for test image Seagull) in terms of DS-LFC (Proposed) RD coding gains.

Table 6.4 BD-PSNR and BD-BR performance of the proposed DS-LFC codec against the benchmarks (for each test image)

Test image	DS-LFC (simulcast)		DS-LFC (SS simulcast)		HEVC (single layer)	
	PSNR (Db)	BR (%)	PSNR (dB)	BR (%)	PSNR (dB)	BR (%)
(a)	2.85	−41.32	0.44	−8.52	2.08	−32.27
(b)	3.00	−44.56	0.43	−9.08	2.40	−37.90
(c)	2.59	−33.05	0.35	−5.86	1.32	−19.99
(d)	1.14	−29.04	0.26	−7.56	-0.19	6.80
(e)	1.18	−13.02	0.26	−3.12	-0.56	7.54
(f)	1.53	−20.58	0.34	−5.22	0.32	−5.13
Average	2.05	−33.71	0.35	−6.56	0.90	−13.49

It was observed that there is a consistent relative RD performance gain using the three different quality metrics. In all cases, the proposed DS-LFC outperforms the simulcast cases with significant gains, showing the advantage of using the proposed IL prediction for improving the RD performance. In terms of the $\mathrm{PSNR}_{3\times3\mathrm{Views}}$ metric (Fig. 6.14a), the RD gains (using the BD metric [38]) of the DS-LFC (Proposed) solution go up to 2.79 dB or 14.82% compared to DS-LFC (Simulcast) and 2.48 dB or 38.62% with respect to HEVC (Single Layer). In the worst case (Fig. 6.14a), supporting a display scalable bitstream using the DS-LFC (Proposed) solution results in a RD performance penalty (using the BD metric [38]) of 0.37 dB or 4.89% of bit saving loss.

Regarding the standard deviation values presented in Figs. 6.13 and 6.14, a more careful analysis of the $\mathrm{PSNR}_{3\times3\mathrm{Views}}/\mathrm{SSIM}_{3\times3\mathrm{Views}}$ results for each rendered views showed that views rendered from viewpoint positions near to the border of the MIs presented larger variation in PSNR/SSIM values. These variations are more significant in the case of *Demichelis Spark*, *Robot 3D* (see Fig. 6.13) and *Plane and Toy* mainly due to the increased vignetting that appears in these images, at the border of each MI.

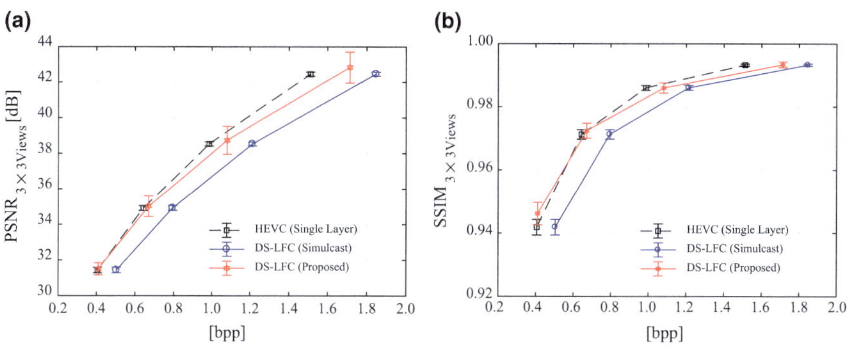

Fig. 6.13 RD performance for a set of rendered views from image *Robot 3D* (Fig. 6.12e) in terms of: **a** $\mathrm{PSNR}_{3\times3\mathrm{Views}}$ versus bpp; and **b** $\mathrm{SSIM}_{3\times3\mathrm{Views}}$ versus bpp

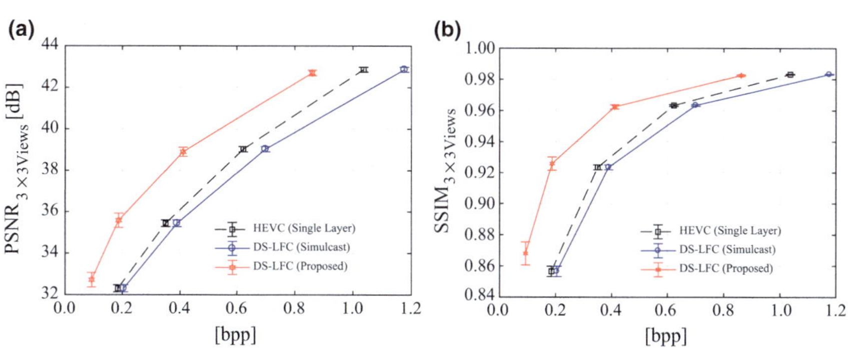

Fig. 6.14 RD performance for a set of rendered views from image *Seagull* (Fig. 6.12b) in terms of: **a** $\mathrm{PSNR}_{3\times3\mathrm{Views}}$ versus bpp; and **b** $\mathrm{SSIM}_{3\times3\mathrm{Views}}$ versus bpp

6.5 Sparse Set of Micro-lens Images and Disparities for an Efficient Scalable Coding of Light Field Images

The information in light field images has a high degree of correlation, as its elements are projections captured from a single scene out of different angles for many positions. In the previous sections of this chapter, this correlation has been modeled in different ways to enable efficient compression. In the present section, the correlation is described by introducing disparity maps in a similar way as depth maps. It is based on a number of articles that describe and evaluate compression of LF that use disparity maps [40, 41] and multi-hypothesis Intra prediction [33, 42] for compression of LF images, both from focused as well as conventional LF cameras [43].

The use of disparity maps to describe the correlation between MIs is particularly suitable when the full LF image is produced by focused LF cameras. The reason is that each MI constitutes a small perspective view of the observed scene with a fair amount of information overlap. The disparities so constitute a shift of pixels between adjacent MIs. This pixel shift is also rather small for data coming from LF cameras, as the distances between the MIs are small and objects are located at intermediate and long distances from the camera.

The principal reason for using the disparity between MIs can also be found in other compression schemes. For example, in an earlier work [44], the scheme arranges light field images into a grid, where images within the grid are recursively predicted from a few Intra coded images. It was later improved by using homography transformations to describe the disparities between views in the light field [45].

The compression scheme using a sparse set of micro-lens images and disparities that is presented in this section consists of three parts:

- **Sparse set of MIs**—The MIs of the original LF image are decimated by selecting every s MI and so constitute a new LF image of subsampled MIs.
- **Disparity maps**—The disparity between adjacent MIs is described by a best value of pixel shift. These disparities so constitute two maps, one describing the horizontal, one the vertical pixel shifts.
- **Refinement by Inter and Intra prediction**—The two previous parts, sparse set of MIs and the disparity maps, enable the prediction of a LF image of full resolution. The third part contains a refinement to obtain a high-quality LF image by predicting from this first LF image of full resolution.

6.5.1 Scalability

The three parts of the compression scheme constitutes a successive refinement of the final LF image reconstruction, and therefore is the basis for a scalability built into the compression scheme. The *first layer* includes a decimated image with a lower angular and spatial resolution. Image reconstruction from this layer can be useful for thumbnails or presentations on smaller displays and devices with lower computational power. The first layer so forms a scalability with respect to resolution. The *second layer* includes additional information so that a LF image with full spatial and angular resolution can be reconstructed, although with a reduced image quality than the original. The second layer so forms a scalability with respect to quality. The *third layer* adds further information that enable a full resolution LF image with the highest possible images quality for the selected compression ratio.

6.5.2 Displacement Intra and Inter Prediction Scheme

Intra prediction schemes are efficient compression methods for images that contain correlated information. This is demonstrated in Sect. 6.4, in which self-similarity prediction for LF images is used. Likewise, the block-copying mode (BC) [46] was introduced into the HEVC codec in order to compress screen contents that contain plenty of repeating patterns in text and graphics. Both self-similarity and HEVC-BC are single-hypothesis Intra predictions, i.e., they find a best block match in the already decoded part of the image and keeps the disparity vector to describe the prediction candidate block, from which the current block is predicted.

The *displacement Intra prediction scheme* was proposed and investigated in [33, 42]. It employs a multi-hypothesis Intra prediction by subdividing the already decoded part of the image into two areas, from each of which a best candidate for block prediction is searched. These two blocks are candidates to be used as a reference for the Intra prediction. There is also a third candidate, which is the mean of two blocks found by searching near the first two blocks. The best match of the three candidates is used for the Intra prediction. The scheme works well both for LF images from focused as well as conventional LF cameras [43].

In the case of light field video coding, i.e., a sequence of LF images, the Intra prediction scheme can also incorporate predictions from previous or future frames to search for good candidates. Thereby, the previous or future frames are loaded into the reference picture list, and the rate-distortion optimization of the codec selects the best prediction mode among the Inter prediction, the displacement Intra and the original HEVC Intra. The combination of these modes provides more possible prediction candidates for an efficient compression. See Fig. 6.15.

6.5.3 Encoding

A schematic description of the encoding of LF images using the sparse set of micro-lens images and disparities is given in Fig. 6.16. The scheme is subdivided into three layers that constitute the basis of the scalability.

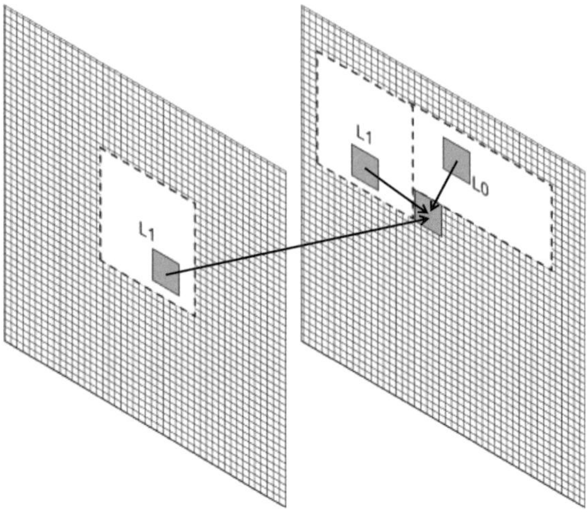

Fig. 6.15 Multi-hypothesis prediction in displacement Intra and Inter prediction scheme. Each block may be predicted from previously decoded areas in the same frame (L0 and L1), and from previous frames (L1). L0 and L1 are reference list 0 and 1, respectively

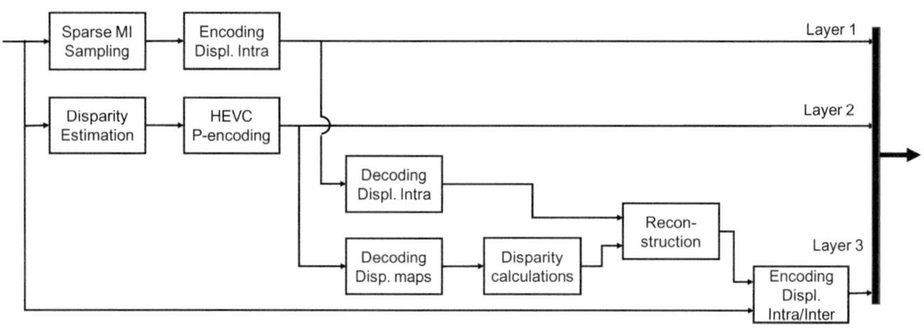

Fig. 6.16 Schematic overview of decoding for sparse set of MIs and disparities. The LF image is encoded in three scalability layers. Layer 1 decimates the LF image to a fewer number of MIs, and is encoded by the displacement Intra prediction scheme. Layer 2 estimates horizontal and vertical disparities between all adjacent MIs. The two disparity maps are encoded using HEVC video coder. Layer 3 uses the reconstruction of layer 1 and 2 as a reference in the displacement Intra and Inter coder

6.5.3.1 Sparse Set of Micro-lens Images

The decimation of the full LF image into a sparse set of MIs is done by selecting every s MI. The input LF image $C(x, y, r, t)$ is described by $N \times M$ MIs with coordinates (x, y), each containing $N_t \times M_t$ pixels with coordinates (r, t). So, the sparse set of MIs is described by $C_s(x_s, y_s, r, t) = C(x \cdot s, y \cdot s, r, t)$, such that $x_s \in [1, N/s]$ and $y_s \in [1, M/s]$. See Fig. 6.17a.

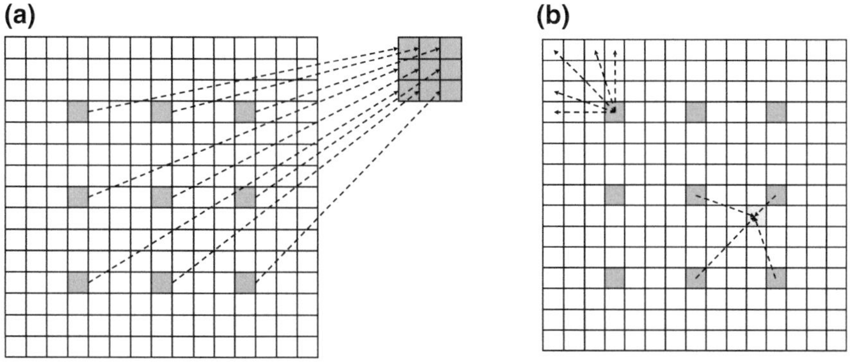

Fig. 6.17 Decimation and reconstruction of sparse set of MIs. **a** The sparse set is constructed by selecting every s MI and combing them into a new LF image of fewer MIs; here $s = 4$. **b** LF image of high resolution is reconstructed by first placing the MIs of the low-resolution LF image into their original positions. The remaining MIs are recovered through predictions that use estimated disparities, see Sect. 6.5.3.2. MIs between those MIs part of the sparse set averaged using all surrounding MIs (bottom right arrows)

The sparse set is itself an LF image with fewer MIs than the original, and therefore has a reduced resolution. The scheme is developed for LF images from focused LF cameras, which has a combined distribution of angular and spatial information throughout the MIs, which means that the decimation implies a reduction in both angular and spatial resolution.

Another consequence of the sparse set being an LF image is that it can be compressed with any codec developed for this kind of data. HEVC was employed in [41] but displacement Intra has a better compression performance for LF image data and is employed in the sequel of this section. The compressed sparse set of MIs constitutes the first layer of the scalable codec. The encoder further includes a decoding of the first sparse set to assure that the other layers receive the same data as the decoder on the receiver side.

6.5.3.2 Disparity Maps

Disparities between adjacent MIs are estimated on the original LF image. They are later used to calculate the disparities from the sparse set of MIs onto the MIs removed from the original LF image. In fact, these disparities can give an estimated disparity between any two MIs in the LF image. The disparities are gathered in two maps, one for horizontal disparity and one for vertical disparity.

The disparity map is computed by finding a best disparity for the whole MI and its neighbor. The total pixel square error between the MI displaced by the disparity and its neighbor is minimized to obtain the disparity value,

$$
\begin{aligned}
D_h(x,y) &= \arg\min_{D_h} \|C(x,y,r+D_h,t) - C(x+1,y,r,t)\|_F \\
D_v(x,y) &= \arg\min_{D_v} \|C(x,y,r,t+D_h) - C(x,y+1,r,t)\|_F
\end{aligned}
\tag{6.2}
$$

where subscript F denotes the Frobenius norm, in which the summation is done over all r and t. Note that the disparity maps are of sizes $(N-1) \times M$ and $N \times (M-1)$, respectively.

These disparity maps can be compressed in many different ways. It is very important that the decoded values are very accurate in order to retain a good prediction result when reconstructing the full LF image. (See Sect. 6.5.4 for the reconstruction process.) In [41], HEVC lossless Intra codec was used to assure the quality of the disparity maps. However, it turns out that HEVC lossy coding of high quality (low QP-value) give higher compression ratio with sufficient quality. HEVC was selected to encode the disparity maps as a sequence of two frames, i.e., the first map is encoded by HEVC lossy Intra coding and the second is Inter-frame predicted. The compressed disparity maps constitute the second layer of the scalable codec.

The encoder further includes a decoding of disparity maps and a reconstruction of the full LF image starting from the sparse set of MIs and calculated disparities. As in conventional encoding settings, this is done to assure the same data as in the decoder for the process of encoding the third layer of the scalable codec.

6.5.3.3 Refinement by Inter and Intra Prediction

The sparse set of MIs and the disparity maps can be used to predict the other MIs of the full resolution LF image. This LF image reconstruction is as other predictions of lower quality than the original. A straightforward way to have a final LF image of high quality would be to compute the residuals with respect to the original LF image, and compress the residuals by common arithmetic coding. In [41], the LF image reconstruction was instead used as a reference frame in the displacement Intra and Inter prediction compression as described in Sect. 6.5.2 and the HEVC RDO chooses the block that gives the least error. The residuals of this prediction are compressed as in the original HEVC scheme. The compression by Inter and Intra prediction constitutes the third layer of the scalable codec that produces the full resolution LF image of highest quality.

6.5.4 *Decoding and Reconstruction*

The encoded LF image data are decoded and reconstructed in three steps that correspond to the three scalability layers, as schematically depicted in Fig. 6.18.

The first layer is decoded according to the disparity Intra prediction scheme defined in [33]. The result of the decoding is the sparse set of MIs, which itself is a

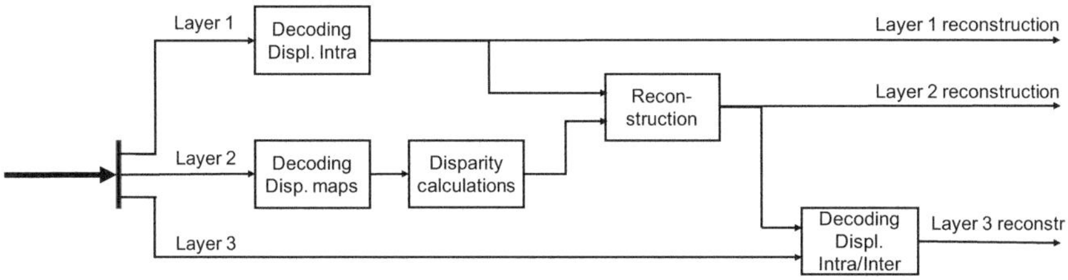

Fig. 6.18 Schematic overview of decoding for sparse set of MIs and disparities. The decoding is divided into three parts, one for each scalability layer. Each layer decoding results in LF images of low resolution (layer 1), high resolution of low quality (layer 2), and high resolution and high quality (layer 3)

low-resolution LF image of the original LF image. It can be used to render images of low spatial resolution and refocusing with limited depth resolution and depth of field as the data contain low angular resolution. This is sufficient for thumbnails and small displays.

The second layer is decoded using HEVC and results in the horizontal and vertical disparity maps, $D_h(x,y)$ and $D_v(x,y)$. These maps are then used to calculate the disparities $D_{hs}(x,y,x_s,y_s)$ and $D_{vs}(x,y,x_s,y_s)$ in (6.3), from the MIs at position (x_s,y_s) in the sparse set to each MI position (x,y) of the original LF image, i.e., to those not being part of the sparse set. See Fig. 6.18.

$$\begin{aligned} D_{hs}(x,y,x_s,y_s) &= \left\{ \begin{array}{ll} \sum_{k=x}^{x_s-1} D_h(k,y) & x<x_s \\ -\sum_{k=x_s}^{x-1} D_h(k,y) & x>x_s \end{array} \right\} \\ D_{vs}(x,y,x_s,y_s) &= \left\{ \begin{array}{ll} \sum_{l=y}^{y_s-1} D_v(x,l) & y<y_s \\ -\sum_{l=y_s}^{y-1} D_v(x,l) & y>y_s \end{array} \right\} \end{aligned} \qquad (6.3)$$

The full LF image of the second layer is reconstructed by shifting the sparse set MIs using the disparities $D_{hs}(x,y,x_s,y_s)$ and $D_{vs}(x,y,x_s,y_s)$, and placing the predicted MI in the corresponding position. If the predicted MI has more than one prediction candidate of MIs in the sparse set, the predicted MIs are averaged. See Fig. 6.17b. In case there are still missing areas in a predicted MI, i.e., pixels have not been assigned a value in the disparity-based prediction, these areas need to be filled with plausible information. For this reason, a dynamic inpainting approach [47] was employed in [40] to obtain the final LF image reconstruction of the second layer. This second layer LF image reconstruction has full spatial and angular resolution but has a lower image quality than when also utilizing the third layer data.

The third layer data is fed into the decoder of the disparity Intra and Inter prediction scheme, along with the output from the second layer. The second layer LF image is put into the reference picture list and is used for Inter prediction along the displacement Intra prediction and the original HEVC Intra prediction in the third layer. Thereby, the final, third layer, LF image is reconstructed that has full resolution and is of high quality.

6.5.5 Evaluation

The coding scheme using a sparse set of MIs and disparities was evaluated in [40, 41]. The lowest bit rate is obtained for a decimation factor of $s = 2$, which led to a bit rate reduction of 50–60% for different input LF images. Larger decimation factors imply a small increase in bit rate. Although it improves the compression efficiency by only a small margin relative to the displacement Intra prediction, it provides a scalable structure for coding and rendering. Compared to HEVC-BC being a single-hypothesis Intra prediction scheme, the sparse set and disparity scheme has a reduction of 20% in bit rate. See Fig. 6.19a.

The third scalability layer contains a fair amount of the bit budget, especially when a very high quality is required for the final LFI. The compression of disparity maps in the second layer results in a very low number of bits when the lossy HEVC coding is employed, whereas the lossless coding use more than 30 times more bits. Yet, the second layer is the smallest component among the three. See Fig. 6.19b.

The objective quality of the second layer reconstruction is much lower than that of the third layer. See Fig. 6.20a. However, the visual quality of the second layer reconstruction is fairly good for the central view rendering, even if improvements can be seen for the full LFI reconstruction. See Fig. 6.20b.

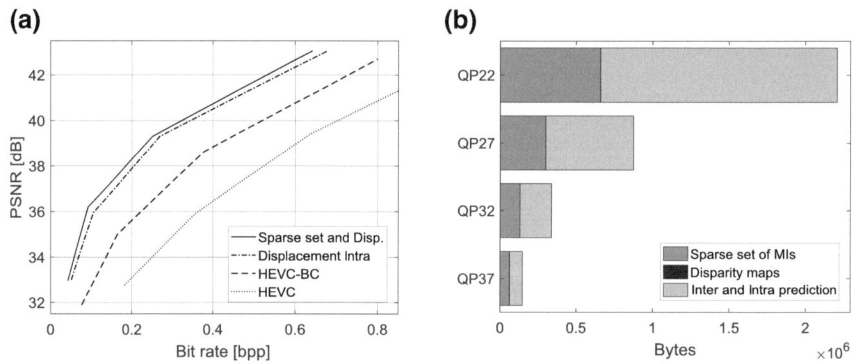

Fig. 6.19 Efficiency of compression schemes. **a** Rate-distortion graphs for evaluated compression schemes. The scheme using sparse set and disparities performs slightly better than the displacement Intra scheme and much better than the single-hypothesis prediction method HEVC-BC and standard HEVC. **b** Distributions of data in the three scalability layers. The third layer contains most data, whereas the second layer (disparities) is constant independent of QP-value, and contains less than 0.5% of the total for QP37. Figures produced from data in [40]

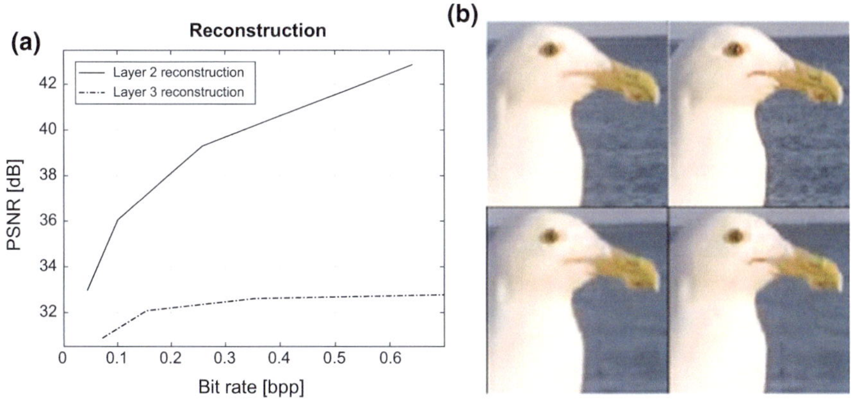

Fig. 6.20 Reconstructed images using decimation factor $s = 2$. **a** Objective quality (PSNR) for reconstructed central view using layer 1–2 and layer 1–3, respectively. **b** Central view reconstruction. Upper images were compressed using QP22, lower used QP37. Left images are reconstructed using layer 1–2, and right images using layer 1–3. Figures produced from data in [40]

6.5.6 Remarks

A compression scheme for LF images from focused LF cameras was presented in this section. It uses a sparse set of micro-lens images (MIs) and disparities between these MIs. The scheme exhibits large compression improvements over both HEVC Intra and HEVC-BC, and moderate improvements over the multi-hypothesis prediction scheme displacement Intra. The computational complexity is increased. Instead, the scheme introduces scalability in both resolution and quality, and so provides a flexible reconstruction of images.

6.6 Conclusions

This chapter covered recent advances in LF coding, based on different approaches. After a brief description of LF representation formats, the coding efficiency of unmodified standard codecs using various LF image data structures was evaluated and discussed for different coding configurations. In general, it was shown that LFs in pseudo-video format provides higher compression efficiency still image formats. Then, a scalable LF coding solution, capable of providing compatible substreams to 2D and 3D decoders, is described and evaluated. A display scalable architecture was presented, using a three-layered hierarchical approach, which allows to support a wide range of end-user displays, from conventional 2D to advanced immersive LF applications (e.g., augmented and immersive virtual reality). Furthermore, another recent approach to encode LFs, which exploits spatial correlation based on the disparity maps and multi-hypothesis prediction, is also presented and discussed. A sparse set of MIs is used as a first layer, which provides a small resolution representation of the visual content. Then, the second and third layers provide higher spatial resolution and the full resolution of the LF, respectively. Such scalable coding schemes also enable seamless interoperability with legacy video systems and smooth transition to emerging applications and services where LFs are increasingly gaining importance and user acceptance.

References

1. Lippmann, G.: Épreuves Réversibles Donnant la Sensation du Relief. J. Phys. Théorique Appliquée **7**, 821–825 (1908)
2. Levoy, M.: Light fields and computational imaging. Computer (Long Beach Calif) **39**, 46–55 (2006). https://doi.org/10.1109/MC.2006.270
3. Levoy, M., Hanrahan, P.: Light field rendering. In: Proceedings of the 23rd Annual Conference on Computer Graphics and Interactive Techniques. SIGGRAPH '96, New Orleans, LA, US, pp. 31–42 (1996)
4. Aggoun, A.: A 3D DCT compression algorithm for omnidirectional integral images. In: 2006 IEEE International Conference on Acoustics, Speech and Signal Processing. Proceedings, Toulouse, France, pp. II-517–II-520 (2006)
5. Olsson, R., Sjöström, M., Xu, Y.: Evaluation of a combined pre-processing and H.264-compression scheme for 3D integral images. In: Proceedings. SPIE 6508, Visual Communications Image Processing. San Jose, CA, US, p 65082C (2007)
6. Olsson, R., Sjöström, M., Xu, Y.: A combined pre-processing and H.264-compression scheme for 3D integral images. In: 2006 International Conference Image Processing. Atlanta, GA, US, pp. 513–516 (2006)
7. Aggoun, A.: Compression of 3D integral images using 3D wavelet transform. J. Disp. Technol. **7**, 586–592 (2011). https://doi.org/10.1109/JDT.2011.2159359
8. Olsson, R., Empirical rate-distortion analysis of JPEG 2000 3D and H. 264/AVC coded integral imaging based 3D-images. In: 2008 3DTV Conference: The True Vision-Capture, Transmission and Display of 3D Video. Istanbul, Turkey, pp. 113–116 (2008)
9. Perra, C., Assuncao, P.: High efficiency coding of light field images based on tiling and pseudo-temporal data arrangement. In: 2016 IEEE International Conference on Multimedia and Expo Work, Seattle, WA, US, pp. 1–4 (2016)
10. Perra, C.: Lossless plenoptic image compression using adaptive block differential prediction. In: 2015 IEEE International Conference on Acoustics, Speech and Signal Processing. IEEE, pp. 1231–1234 (2015)
11. Helin, P., Astola, P., Rao, B., Tabus, I.: Sparse modelling and predictive coding of subaperture images for lossless plenoptic image compression. In: 2016 3DTV-Conference True Vision–Capture, Transmission and Display of 3D Video (3DTV-CON), Hamburg, Germany, pp. 1–4 (2016)
12. Santos, J.M., Assuncao, P.A.A., da Silva Cruz L.A., et al.: Performance evaluation of light field pre-processing methods for lossless standard coding. In: IEEE COMSOC MMTC Communications—Frontiers, vol. 12, pp. 44–49 (2017)
13. Vieira, A., Duarte, H., Perra, C., et al.: Data formats for high efficiency coding of lytro-illum light fields. In: 2015 International Conference on Image Processing Theory, Tools and Applications (IPTA), Orleans, France, pp. 494–497 (2015)

14. Dansereau, D.G.D.G., Pizarro, O., Williams, S.B.S.B.: Decoding, calibration and rectification for lenselet-based plenoptic cameras. In: 2013 IEEE Conference on Computer Vision and Pattern Recognition, Portland, OR, US, pp. 1027–1034 (2013)
15. Draft test conditions for HEVC still picture coding performance evaluation. ISO/IEC JTC1/SC29/WG11 MPEG2013/N13826, Vienna, Austria (2013)
16. Conti, C., Nunes, P., Soares, L.D.: Inter-layer prediction scheme for scalable 3-D holoscopic video coding. IEEE Signal Process. Lett. **20**, 819–822 (2013). https://doi.org/10.1109/LSP.2013.2267234
17. Aggoun, A., Tsekleves, E., Swash, M.R., et al.: Immersive 3D holoscopic video system. IEEE Multimed **20**, 28–37 (2013). https://doi.org/10.1109/MMUL.2012.42
18. Arai, J., Kawakita, M., Yamashita, T., et al.: Integral three-dimensional television with video system using pixel-offset method. Opt. Express **21**, 3474–3485 (2013). https://doi.org/10.1364/OE.21.003474
19. Arai, J.: Integral three-dimensional television. In: 2015 14th Workshop on Information Optic, Kyoto, Japan, pp. 1–3 (2015).
20. NHK STRL Science & Technology Research Laboratories. https://www.nhk.or.jp/strl/index-e.html. Accessed 10 July 2016
21. Wang, J., Xiao, X., Hua, H., Javidi, B.: Augmented reality 3D displays with micro integral imaging. J. Disp. Technol. **11**, 889–893 (2015). https://doi.org/10.1109/JDT.2014.2361147
22. Lanman, D., Luebke, D.: Near-eye light field displays. In: ACM SIGGRAPH 2013 Emerging Technologies—SIGGRAPH '13 1–1 (2013). https://doi.org/10.1145/2503368.2503379
23. Conti, C., Nunes, P., Soares, L.D.: Using self-similarity compensation for improving inter-layer prediction in scalable 3D holoscopic video coding. In: Proceedings of SPIE 8856 Applications of Digital Image Processing, vol. XXXVI. San Diego, CA, US, p. 88561K (2013)
24. Sullivan, G.J., Ohm, J.-R., Han, W.-J., Wiegand, T.: Overview of the high efficiency video coding (HEVC) standard. IEEE Trans. Circuits Syst. Video Technol. **22**, 1649–1668 (2012)
25. Vetro, A., Wiegand, T., Sullivan, G.J.: Overview of the stereo and multiview video coding extensions of the H.264/MPEG-4 AVC standard. Proc. IEEE **99**, 626–642 (2011). https://doi.org/10.1109/JPROC.2010.2098830
26. White Paper on State of the Art in compression and transmission of 3D Video. ISO/IEC JTC1/SC29/WG11 N13364, Geneva, Switzerland (2013)
27. Vetro, A., Müller, K.: Depth-based 3D video formats and coding technology. In: Dufaux, F., Pesquet-Popescu, B., Cagnazzo, M. (eds) Emerging Technologies for 3D Video. Wiley, Chichester, pp. 139–161 (2013)
28. Tech, G., Chen, Y., Muller, K., et al.: Overview of the multiview and 3D extensions of high efficiency video coding. IEEE Trans. Circuits Syst. Video Technol. **26**, 35–49 (2016). https://doi.org/10.1109/TCSVT.2015.2477935
29. Georgiev, T., Lumsdaine, A.: Focused plenoptic camera and rendering. J. Electron. Imaging **19**, 021106 (2010). https://doi.org/10.1117/1.3442712
30. Conti, C., Nunes, P., Soares, L.D.: New HEVC prediction modes for 3D holoscopic video coding. In: 2012 19th IEEE International Conference on Image Processing, Orlando, FL, US, pp. 1325–1328 (2012)
31. Conti, C., Soares, L.D., Nunes, P.: HEVC-based 3d holoscopic video coding using self-similarity compensated prediction. Signal Process. Image Commun. **42**, 59–78 (2016). https://doi.org/10.1016/j.image.2016.01.008
32. Conti, C., Nunes, P., Soares, L.D.: HEVC-based light field image coding with bi-predicted self-similarity compensation. In: 2016 IEEE International Conference on Multimedia and Expo Work, Seattle, WA, US, pp. 1–4 (2016)
33. Li, Y., Sjöström, M., Olsson, R., Jennehag, U.: Coding of focused plenoptic contents by displacement intra prediction. IEEE Trans. Circuits Syst. Video Technol. **26**, 1308–1319 (2016). https://doi.org/10.1109/TCSVT.2015.2450333
34. Sullivan, G.J., Wiegand, T.: Rate-distortion optimization for video compression. IEEE Signal Process. Mag. **15**, 74–90 (1998). https://doi.org/10.1109/79.733497

35. Geogiev T Todor Georgiev Gallery of Light Field Data. http://www.tgeorgiev.net/Gallery/. Accessed 17 Sept 2016
36. 3D Holoscopic Sequences (Download Link). http://3dholoscopicsequences.4shared.com/. Accessed 30 Oct 2016
37. MV-HEVC Reference Software HTM-12.0. https://hevc.hhi.fraunhofer.de/svn/svn_3DVCSoftware/tags/HTM-12.0/. Accessed 22 Dec 2014
38. Bjøntegaard, G.: Calculation of average PSNR differences between RD curves. VCEG-M33, Austin, TX, US (2001)
39. Bossen, F.: Common HM test conditions and software reference configurations. JCTVC-L1100, Geneva, Switzerland (2013)
40. Li, Y., Sjöström, M., Olsson, R., Jennehag, U.: Scalable coding of plenoptic images by using a sparse set and disparities. IEEE Trans. Image Process. **25**, 80–91 (2016). https://doi.org/10.1109/TIP.2015.2498406
41. Li, Y., Sjöström, M., Olsson, R.: Coding of plenoptic images by using a sparse set and disparities. In: 2015 IEEE International Conference on Multimedia and Expo. IEEE, pp. 1–6 (2015)
42. Li, Y., Sjöström, M., Olsson, R., Jennehag, U.: Efficient intra prediction scheme for light field image compression. In: 2014 IEEE International Conference on Acoustics, Speech and Signal Processing. IEEE, pp. 539–543 (2014)
43. Li, Y., Olsson, R., Sjöström, M.: Compression of unfocused plenoptic images using a displacement intra prediction. In: 2016 IEEE International Conference on Multimedia and Expo Work. IEEE, pp. 1–4 (2016)
44. Magnor, M., Girod, B.: Data compression for light-field rendering. IEEE Trans. Circuits Syst. Video Technol. **10**, 338–343 (2000). https://doi.org/10.1109/76.836278
45. Kundu, S.: Light field compression using homography and 2D warping. In: 2012 IEEE International Conference on Acoustics, Speech and Signal Processing. IEEE, pp. 1349–1352 (2012)
46. Rosewarne, C., Sharman, K., Naccari, M., Sullivan, G.: HEVC range extensions test model 6 encoder description. JCTVC-P1013, San Jose, CA, US (2014)
47. Bertalmio, M., Bertozzi, A.L., Sapiro, G.: Navier-stokes, fluid dynamics, and image and video inpainting. In: Proceedings of 2001 IEEE Computer Society Conference on Computer Vision and Pattern Recognition. CVPR 2001. Kauai, HI, US, pp. I-355–I-362 (2001)

Chapter 7
Impact of Packet Losses in Scalable Light Field Video Coding

Caroline Conti, Paulo Nunes and Luís Ducla Soares

Abstract Light field imaging technology has been recently attracting the attention of the research community and the industry. However, to effectively transmit light field content to the end-user over error-prone networks—e.g., wireless networks or the Internet—error resilience techniques are required to mitigate the impact of data impairments in the user quality perception. In this context, this chapter analyzes the impact of packet losses when using a three-layer display scalable light field video coding architecture, which has been presented in Chap. 6. For this, a simple error concealment algorithm is used, which makes use of inter-layer redundancy between multiview and light field content and the inherent correlation of the light field content to estimate lost data. Furthermore, a study of the influence of 2D views generation parameters used in lower layers on the performance of the used error concealment algorithm is also presented.

7.1 Introduction

Light field is an imaging technology that has been attracting the attention of the research community and the industry for providing richer two-dimensional (2D) image capturing systems [1–3], single-camera 3D imaging, and more immersive 3D viewing systems [4–6].

C. Conti (✉) · P. Nunes · L. Ducla Soares
Instituto Universitário de Lisboa (ISCTE-IUL), Av. das Forças Armadas, 1649-026 Lisbon, Portugal
e-mail: caroline.conti@lx.it.pt

P. Nunes
e-mail: paulo.nunes@lx.it.pt

L. Ducla Soares
e-mail: lds@lx.it.pt

C. Conti · P. Nunes · L. Ducla Soares
Instituto de Telecomunicações, Av. Rovisco Pais 1, 1049-001 Lisbon, Portugal

P. A. Assunção and A. Gotchev (eds.), *3D Visual Content Creation, Coding and Delivery*, Signals and Communication Technology,
https://doi.org/10.1007/978-3-319-77842-6_7

However, to progressively introduce this technology into the consumer market and to efficiently deliver light field content to end-users, a crucial requirement is backward compatibility with legacy 2D and 3D devices. Hence, to enable light field content to be delivered and presented on legacy displays, a scalable light field coding approach is required, where by decoding only the adequate subsets of the scalable bitstream, 2D or 3D compatible video decoders can present an appropriate version of the light field content.

Moreover, following the current forecasts indicating that three-quarters of the world's mobile data traffic will be video by 2020 [7], it should be envisaged to efficiently provide light field video services in such type of error-prone environments. To guarantee this, error resilience techniques in the encoding and decoding side are needed to mitigate the impact of data impairments in the user quality perception. The design of an appropriate error resilience technique typically considers the type of network (i.e., error characteristics of the network being used for transmission) and also the type of content (i.e., inherent characteristics of the content) being transmitted. In this sense, due to the different nature of acquisition system and, consequently, the different type of correlation in the light field content, when compared to the conventional 2D and 3D multiview contents, the set of factors which could affect the performance of an error control algorithm may also differ. Hence, it is essential to deeply understand the impact of packet losses in terms of decoding video quality for the specific case of light field content, notably when a scalable approach is used.

To the best of the authors' knowledge, the proposal of error resilience techniques suitable for light field content has been only addressed by the authors in [8]. In this context, this chapter aims to contribute to the discussion of this issue and presents a study of the influence of packet losses in scalable light field content coding. For this, the three-layer scalable light field video coding architecture presented in Chap. 6 (Sect. 6.4.1) is considered. Based on this coding architecture, a simple error concealment algorithm is proposed, which derives from the previously proposed inter-layer prediction method (Sect. 6.4.3.2) to estimate the lost data. Finally, an analysis of the influence of some meaningful parameters in the proposed coding architecture (e.g., 2D view generation parameters used in lower layers) on the performance of the used error concealment algorithm is also presented.

The remainder of this chapter is organized as follows: Sect. 7.2 briefly reviews the used scalable architecture for light field video coding, as well as the inter-layer prediction scheme, in order to better understand the proposed error concealment algorithm; Sect. 7.3 discusses some meaningful factors which affect the inter-layer prediction accuracy, and presents the proposed error concealment algorithm; Sect. 7.4 presents the considered test conditions and studies the influence of packet losses on the accuracy of inter-layer prediction; and finally, Sect. 7.5 concludes the chapter.

7.2 Scalable Light Field Coding

The display scalable architecture for light field coding that has been presented in Sect. 6.4.1 is here considered (see Fig. 7.1). In this case, each layer of this scalable coding architecture represents a different level of display scalability:

- **Base Layer (2D)**—The base layer represents a single 2D view, which can be used to deliver a 2D version of the light field content to 2D displays devices.
- **First Enhancement Layer (Stereo or Multiview)**—This layer represents the necessary information to obtain an additional view (representing a stereo pair) or various additional views (representing multiview content). This is to allow stereo and autostereoscopic devices to play versions of the same light field content.
- **Second Enhancement Layer (Light Field)**—This layer represents the additional data needed to support full light field display.

For generating 2D views from the light field content to compose the content in the base and first enhancement layers, two rendering algorithms, proposed in [9] and referred to as Basic Rendering and Weighted Blending, are adopted here. Essentially, there are two parameters that control these algorithms: (i) the patch size that controls the plane of focus in the generated view; and (ii) the patch position that controls the viewing angle (i.e., the scene perspective).

High compression efficiency is still an important requirement for the scalable coding architecture adopted in this section. Therefore, an inter-layer prediction mode (see Fig. 7.1) is used to further improve the second enhancement layer coding efficiency by removing the redundancy between the light field content and its multiview version from the enhancement layer underneath. For this, an inter-layer

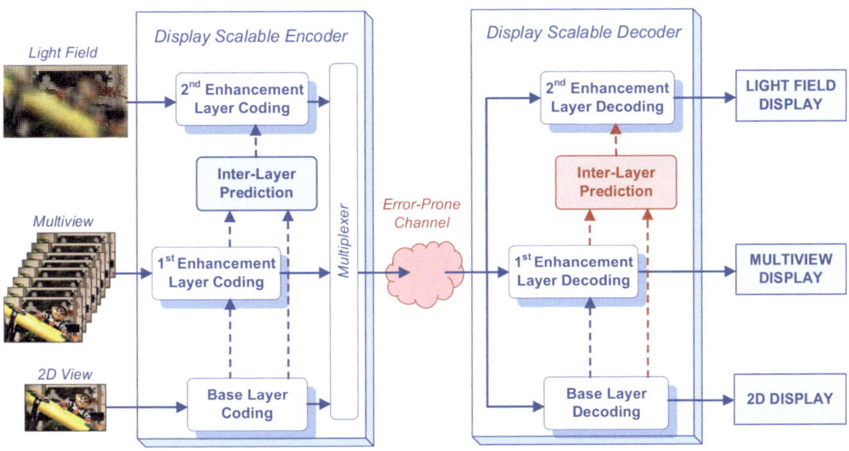

Fig. 7.1 Display scalable light field coding architecture considered in this chapter for analyzing the impact of packet losses when transmitting in error-prone channels

(IL) reference picture is constructed by using the set of reconstructed 2D views obtained by decoding the bitstream in the lower layers. This IL reference picture can be then used as new a reference frame for employing an inter-layer compensated prediction when encoding the light field image. The process for constructing the IL reference picture can be basically divided into the following two steps (which are explained in detail in Sect. 6.4.1):

- **Patch Remapping**—The purpose of this step is to reorganize (remap) the texture samples (patches) from the reconstructed 2D views into its original positions in the light field image.
- **Micro-Image Refilling**—Since most of the light field information is discarded when rendering the 2D views in the lower layers, this step aims at emulating the significant cross-correlation existing in light field content between neighboring micro-images so as to synthesize the missing texture samples as much as possible to complete the IL reference picture.

7.3 Mitigation of Packet Loss Impact on Scalable Light Field Coding

Guaranteeing successful light field video transmission in the presence of channel errors is a challenging issue that requires reliable error resilience mechanisms for fighting the transmission errors and mitigating their impact in the user quality perception.

State-of-the-art error resilience techniques for 2D and 3D multiview video can be typically categorized in three main groups [10]: (i) error resilient encoding techniques, which are introduced into the video encoding process to make the bitstream more robust to errors; (ii) error concealment techniques, which are employed at the decoding process to conceal the effect of errors; and (iii) those that require interactions between encoder and decoder to adaptively consider the network characteristics in terms of information loss.

Since there is a lack of error resilience techniques specific for light field content, a simple error concealment technique is proposed to estimate the lost data making use of the inherent correlation existing in the light field content. In this section, a discussion about some of the relevant factors which affect the inter-layer prediction accuracy is first presented (in Sect. 7.3.1) and, then, the proposed error concealment method is defined (in Sect. 7.3.2).

7.3.1 Relevant Factors for the Inter-layer Prediction Accuracy

Besides the aforementioned advantages of using a light field imaging system, it is important to notice that for representing the full light field in this type of content, there is a massive increase in the amount of information that need to be captured, encoded, and transmitted when compared to legacy technologies. As opposing the MVC approach where each enhancement layer represents an additional 2D view image, there is a considerable jump in the coding information amount between first and second enhancement layers of the proposed scalable coding architecture.

To illustrate the relation between amounts of information in the lower hierarchical layers and the second enhancement layer, consider one frame from the light field test image *Plane and Toy* (frame 123, in Fig. 7.2a), with a resolution of 1920 × 1088 and micro-image resolution of around 28 × 28 pixels. From this light field content, nine views are generated for the first two scalability layers—one for the base layer and eight for the first enhancement layer. These views are generated using the Basic Rendering algorithm with patch size of 4 × 4 and varying the

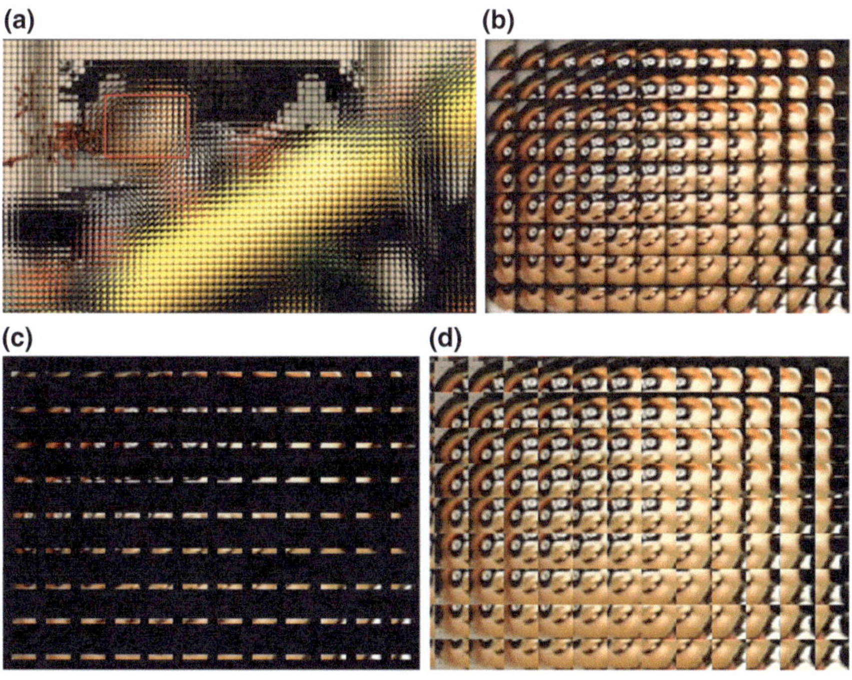

Fig. 7.2 Light field image and corresponding IL picture prediction: **a** light field image Plane and Toy (frame 123); **b** magnified section of 336 × 246 pixels from original image; **c** magnified section of 336 × 246 pixels from sparse IL prediction picture; and **c** Magnified section of 336 × 246 pixels from full IL prediction picture

position of the patch in relation to the center of the micro-image in {−8, −6, −4, −2, 0, 2, 4, 6, 8} pixels. Notice that, from this set of patch positions, adjacent patches contain overlapping areas of the micro-image. Consequently, approximately 12% of the information inside each micro-image is used to build these nine 2D views and the remainder data is discarded. The nine views are then coded independently with the high-efficiency video coding (HEVC) using the "Intra, main" configuration [11].

Afterwar, the nine-coded and reconstructed 2D views are processed to build an IL reference (see Fig. 7.1). In the Patch Remapping step, since there are overlapping areas between adjacent patches, this redundant information is used to refine the pixel values. The resulting sparse light field image is shown in Fig. 7.2b by the enlargement, to illustrate the amount of information that need to be estimated in the Micro-Image Refilling process. After this, by applying the Micro-Image Refilling process, the IL picture prediction is completed, as shown Fig. 7.2c. This IL picture prediction is then used as a new reference picture to efficiently encode the light field content in the second enhancement layer (Sect. 7.3).

Finally, in Fig. 7.3, the used bitrate for encoding all the nine 2D view images independently is compared to bitrate used to encode the light field content with the scalable coding scheme for four different quantization parameter (QP) values. From this, it can be seen that the base layer and the first enhancement layer represent only a small percentage of the scalable bitstream (in this case, about 16% of the scalable bitstream). Therefore, it is expected that losses in the lower hierarchical layers will considerably affect the accuracy of the built IL picture prediction and, consequently, degrade the performance of the proposed scalable coding scheme.

Moreover, it should be also noticed that, as was shown in [12], the performance of the inter-layer prediction scheme is improved when increasing the patches sizes.

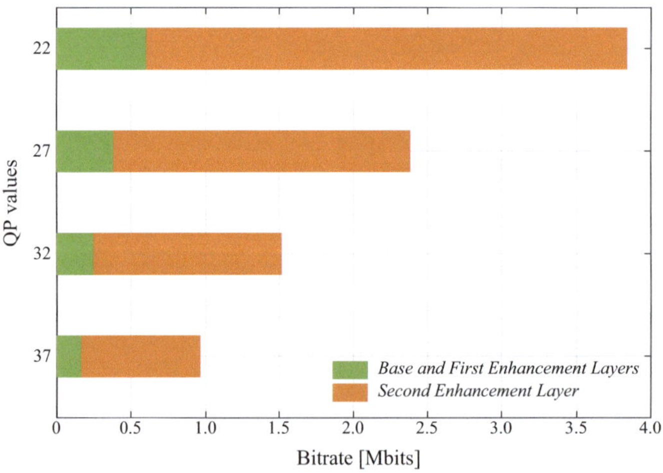

Fig. 7.3 Relation between the amount of data in the bitstream for the base layer and first enhancement layer, compared to the second enhancement layer

This fact is related again to the amount of data from a light field content that is discarded when generating a 2D view image and that need to be estimated in the Micro-Image Refilling process. As the amount of discarded information is a consequence of the chosen patch size and number of views, this means that the parameters which are freely chosen when generating the content in the lower hierarchical layers will also affect the accuracy of the build IL picture prediction.

Considering the Basic Rendering and Weighted Blending algorithms, these parameters are:

- **Patch Size**—During the creative postproduction process, a proper patch size will be selected and will be limited to the used optical depth of field. As mentioned earlier, the quality of the IL picture prediction will improve as relative larger patch sizes are used.
- **Number and position of 2D views**—The choice of number of views and their corresponding positions is based on the type of display that will be used. In this case, as the number of 2D view images increases, less information from the light field content will be discarded and, consequently, the quality of the IL prediction may improve. However, if these 2D views are generated by overlapping patches positions, the amount of relevant information to build the IL prediction picture is smaller, and its performance may decrease.

In other words, there is a large degree of freedom when defining how the light field content will be presented. Therefore, the error resilience problem needs to be analyzed considering the parameters that control the generation of content for the lower hierarchical layers, since the quality of the inter-layer prediction is also dependent on them and may also affect the effectiveness of a resilience error technique.

7.3.2 Proposed Error Concealment Algorithm

Typically, an error concealment algorithm makes use of spatial, temporal, and spectral redundancy of the content to estimate the missing data and mask the effect of channel errors at the decoder.

Although the conventional error concealment tools for the lower layers in the hierarchical scalable architecture can be applied to the second enhancement layer, these methods do not consider the inter-layer correlation between the multiview and light field content, and neither the inherent spatial correlation of the light field content.

When generating the IL picture prediction, the Micro-Image Refilling process is already able to estimate nonexisting data to fill the holes in the IL picture prediction, by making use of the significant cross-correlation existing between neighboring micro-images. Therefore, considering that a 2D view image is lost (see Fig. 7.1), the only difference when building the IL picture prediction is that there will be more holes to be fulfilled in the Micro-Image Refilling process. This means that it is possible to simply derive the error concealment algorithm from the inter-layer prediction method.

Therefore, upon the detection of a lost picture, the following steps are considered by the proposed error concealment algorithm to build the IL picture prediction:

(i) The Patch Remapping process is employed considering only the set of 2D view images that are available (without loss). To illustrate the consequence of a lost 2D view in this step, five nonoverlapping patch positions ({−8, −4, 0, 4, 8}) are used to generate five corresponding 2D view images from the light field image *Plane and Toy (frame 123)*. Then, considering that the central 2D view image (with patch position "0") is not available at the decoder side, the sparse IL picture prediction will contain extra holes where the patches of the lost 2D view were supposed to be placed, as illustrated in Fig. 7.4a.

(ii) The Micro-Image Refilling algorithm can estimate most of the holes by using information from available patch positions, including the set of lost patches in the position "0". This is illustrated in one of the steps of the algorithm (for the first 2D view) in Fig. 7.4b. Finally, it is possible to recall the algorithm also for the lost patch position to fulfill the IL prediction picture, as shown in Fig. 7.4c.

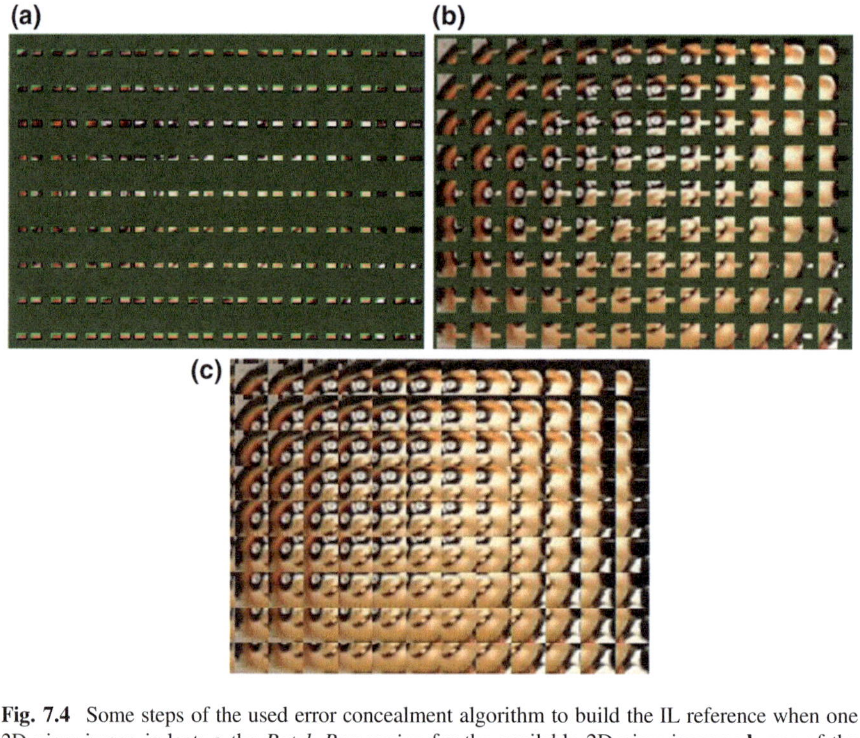

Fig. 7.4 Some steps of the used error concealment algorithm to build the IL reference when one 2D view image is lost: **a** the *Patch Remapping* for the available 2D view images; **b** one of the iterations of the *Micro-Image Refilling* to illustrate the recovery of the lost patches; and **c** the built IL picture prediction

7.4 Experimental Results

To properly analyze the influence of packet losses on the accuracy of the inter-layer prediction, the following test conditions were considered:

- **Light field test images**—Four light field images with different spatial and micro-image resolutions (MI_{resol}) are considered to achieve a set of representative results. These are (see Fig. 7.5): *Plane and Toy* (frames 23 and 123 from a sequence with identical name); *Robot 3D*; and *Laura*. The first three images are available in [13] and the last light field image in [14].
- **Hierarchical Content Generation**—To generate the content for the first two scalability layers, the four test images were processed with the Basic Rendering and Weighted Blending algorithms, proposed in [9]. In this process, nine 2D view images were generated—one for the base layer and eight for the first enhancement layer. Since the resolution of the micro-images varies from one image to another, the patch positions to generated 2D view images were chosen to have nine regularly spaced views within the micro-image limits. Additionally, three different patch sizes were chosen for each test image, which correspond to cases where adjacent patches are taken with and without overlapping areas. One of these patch sizes represents the case where the main object of the scene is in focus. Due to the small size of micro-images in *Plane and Toy* and *Robot 3D* images, an additional set of patch positions is needed to be considered so as to have the case where the patches are taken without overlap areas. In this case, five regularly spaced 2D view images were generated.
- **Network Conditions**—It should be noticed that due to the large number of possible combinations of test conditions (number of views, patch size and patch positions) and since this chapter mainly focuses on analyzing the influence of these parameters on the performance of the error concealment algorithm, it will not yet cover an extensive set of network conditions, which will be, however, considered in future work. To simulate the network conditions, it is considered that an entire 2D view image is coded into only one packet. Hence, loss of a packet implies that the entire 2D view image must be recovered by the error

(a) *Plane and Toy (frame 23)* 1904×1064 $MI_{resol} = 28 \times 28$

(b) *Plane and Toy (frame 123)* 1904×1064 $MI_{resol} = 28 \times 28$

(c) *Robot 3D* 1904×1064 $MI_{resol} = 28 \times 28$

(d) *Laura* 7104×5328 $MI_{resol} = 74 \times 74$

Fig. 7.5 Example of a central view rendered from each light field test image (with the corresponding characteristics below each image)

concealment algorithm. Three different packet loss conditions were considered, where one, two, and three packets are lost. Additionally, packet losses were assumed independent and identically distributed for all 2D view images. For this, a case is considered where the two lower layers are independently encoded, since an enhancement layer would not be decodable if the 2D view image in the base layer was lost.

- **Results Analysis**—The results are presented in terms of the average mean squared error (MSE) (for all the combination of lost 2D view images) of the IL picture prediction built by the error concealment algorithm, compared with the IL prediction picture when there is no packet loss. Alternatively, the average MSE is also shown discarding the cases where the first or the last pictures are lost, since when this happens, a portion of the information cannot be recovered by the *Micro-Image Refilling* algorithm in the border of the IL picture reference.

The experimental results for each tested light field image can be seen in Figs. 7.6, 7.7, 7.8, 7.9, 7.10, 7.11 and 7.12. In each Figs. 7.6, 7.7, 7.8, 7.9, 7.10, 7.11, and 7.12, these results are split in different charts for each used rendering algorithm (Basic Rendering and Weight Blending algorithms) and for each patch size. Finally, each chart shows the corresponding average MSE value for all the possible combinations of lost 2D views (referred to as All Views) as well as the average MSE value when discarding the cases where the first or the last pictures are lost (referred to as Without Border Views). Additionally, the maximum and minimum MSE values in each case are also presented by the error bars.

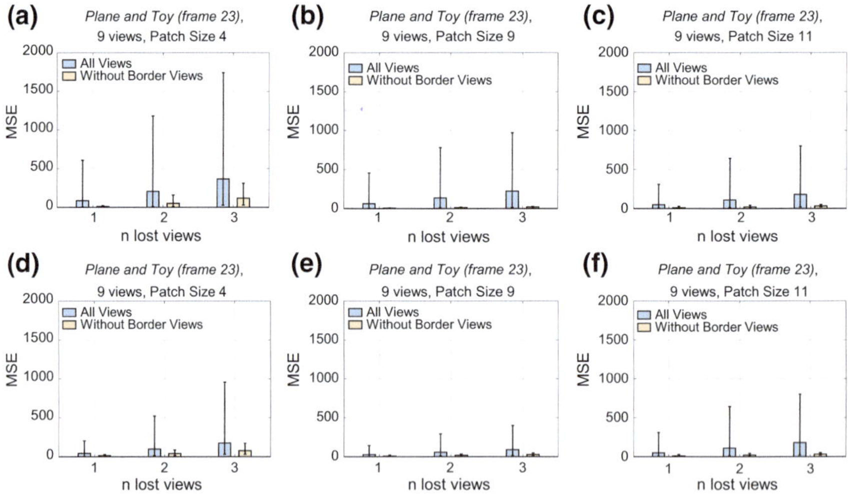

Fig. 7.6 Comparison between qualities of the IL reference when there are lost views. In this case, nine views were generated with three different patches from the tested light field image *Plane and Toy (frame 23)* using: **a** Basic Rendering algorithm; and **b** Weighted Blending algorithm

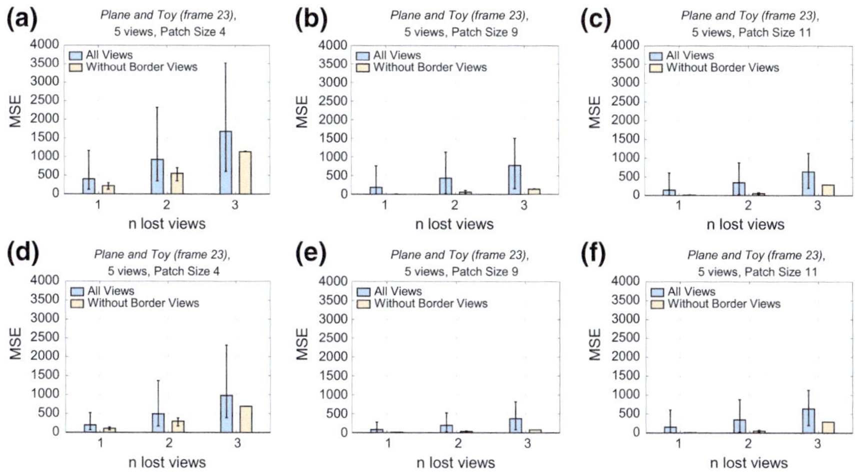

Fig. 7.7 Comparison between qualities of the IL reference when there are lost views. In this case, five views were generated with three different patches from the tested light field image *Plane and Toy (frame 23)* using: **a** Basic Rendering algorithm; and **b** Weighted Blending algorithm

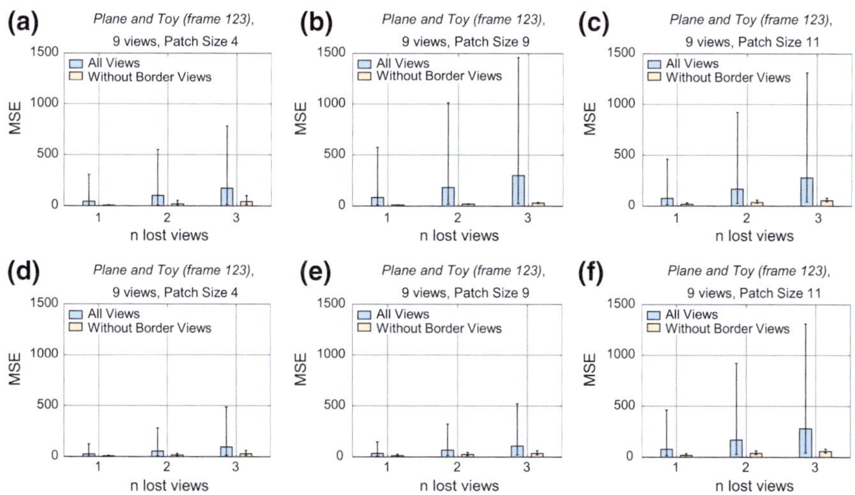

Fig. 7.8 Comparison between qualities of the IL reference when there are lost views. In this case, nine views were generated with three different patches from the tested light field image *Plane and Toy (frame 123)* using: **a** Basic Rendering algorithm; and **b** Weighted Blending algorithm

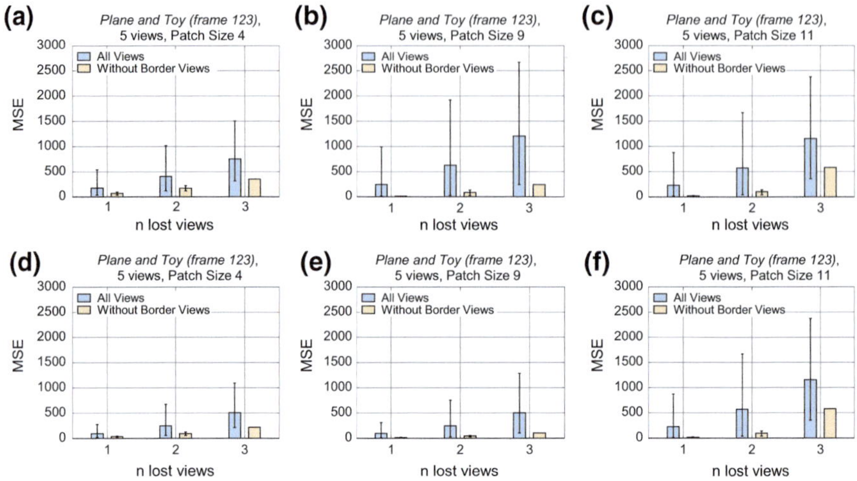

Fig. 7.9 Comparison between qualities of the IL reference when there are lost views. In this case, five views were generated with three different patches from the tested light field image *Plane and Toy (frame 123)* using: **a** Basic Rendering algorithm; and **b** Weighted Blending algorithm

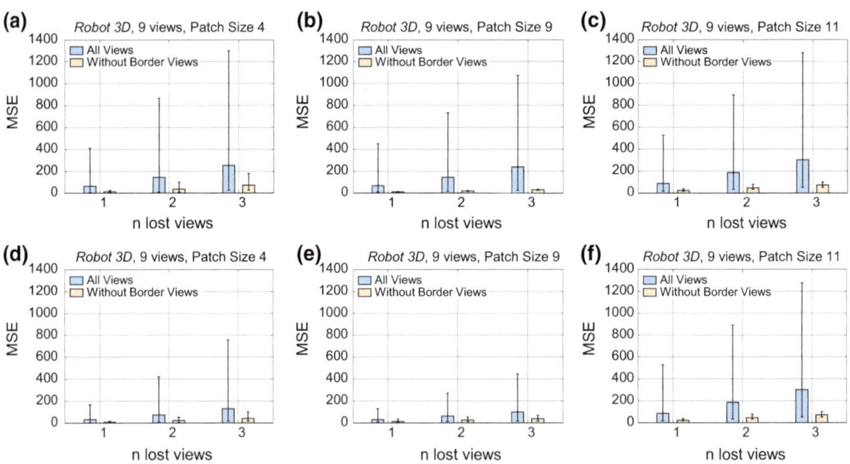

Fig. 7.10 Comparison between qualities of the IL reference when there are lost views. In this case, nine views were generated with three different patches from the tested light field image *Robot 3D* using: **a** Basic Rendering algorithm; and **b** Weighted Blending algorithm

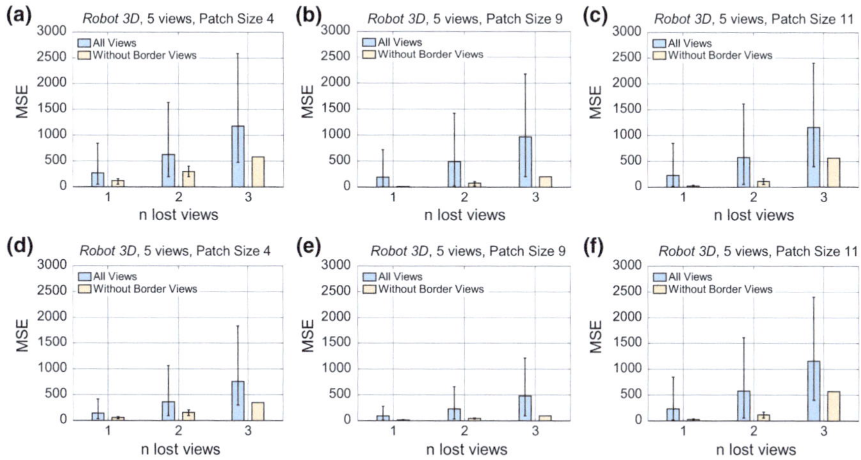

Fig. 7.11 Comparison between qualities of the IL reference when there are lost views. In this case, five views were generated with three different patches from the tested light field image *Robot 3D* using: **a** Basic Rendering algorithm; and **b** Weighted Blending algorithm

Based on these charts, the following conclusions can be drawn in terms of:

- **Number of lost 2D view images**—This analysis compares how the accuracy of the built IL picture prediction varies when different numbers of 2D view images are lost. As expected, in all test conditions, the accuracy of the inter-layer prediction degrades as the number of lost views increases. For instance, considering the tested image *Laura* when using the larger patch size (14) to generate the nine views with the Basic Rendering algorithm (in Fig. 7.12a), the average MSE value goes from 65.23, when only one view is lost, to 253.75, when three views are lost. Moreover, it can be seen that the influence of lost views is stronger when the first or the last 2D views are lost. For example, for the same abovementioned test condition, the corresponding average MSE values for the Without Border Views case are considerably smaller (respectively, 5.9 and 61.63 when one or three views are lost).
- **Different Patch Sizes**—This analysis compares the results when using different patch sizes, for each tested image with the same patch positions and rendering algorithms. Surprisingly, for all results, the patch size corresponding to the case where the main object is in focus was shown to be the less affected by lost 2D views, even when it is the smaller one. For instance, consider the results shown in Fig. 7.8 for *Plane and Toy (frame 123)*, where nine views were generated with the Basic Rendering algorithm. The patch size 4, where the main object is in focus, presented smaller average MSE values than the presented when using larger patch sizes. However, it is known that in this case (patch size 4), more information from the light field image was discarded from the original light field image (when generating the 2D views) and need to be estimated in the *Micro-Image Refilling* process. From this, it can be concluded that in a sequence where

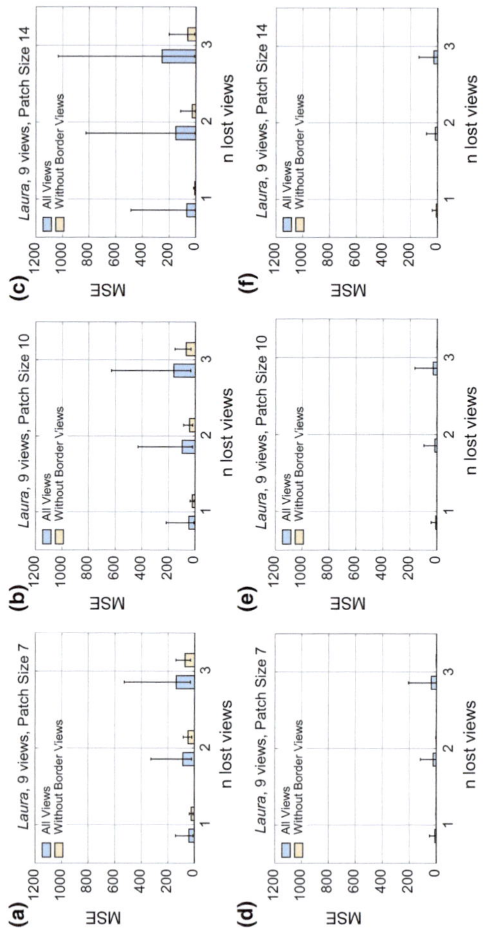

Fig. 7.12 Comparison between qualities of the IL reference when there are lost views. In this case, nine views were generated with three different patches from the tested light field image *Laura* using: **a** Basic Rendering algorithm; and **b** Weighted Blending algorithm

there is interest in varying the patch sizes from one frame to another (e.g., the *Plane and Toy* sequence), the impact of losses will be considerably lower, since the main object is maintained in focus (which is, most of the times, the case).

- **Different Rendering Algorithms**—This analysis compares the results when using one of the rendering algorithms, Basic Rendering (in Figs. 7.6a, 7.7a, 7.8a, 7.9a, 7.10a, 7.11a and 7.12a) and Weighted Blending (in Figs. 7.6b, 7.7b, 7.8b, 7.9b, 7.10b, 7.11b and 7.12b), for each tested image and test conditions shown in Table 7.1. From this, it can be seen that the accuracy of the inter-layer prediction using the Weighted Blending algorithm is generally better than using the Basic Rendering algorithm when one or more views are lost. This can be explained by the high level of blur which is introduced by the weighted average in the Weighted Blending algorithm. Hence, the differences in these blurred images will be less evident than differences in images generated by the Basic Rendering algorithm.
- **Different Number of 2D View Images in Lower Layers**—This analysis compares the results when different numbers of 2D views are generated to the lower layers, using the same patch sizes and rendering algorithms. For this, the results using five and nine views for *Plane and Toy (frame 23)* (in Figs. 7.6 and 7.7), *Plane and Toy (frame 123)* (in Figs. 7.8 and 7.9), and *Robot 3D* (in Figs. 7.10 and 7.11) are compared. As expected, by using less 2D view images in the lower layer, a loss of 2D view images will affect more drastically the accuracy of the built IL picture prediction. This can be understood since, when considering less views: (i) more information from the original light field image is discarded to generate the 2D view images; and (ii) a loss of a 2D view image represents a higher packet loss rate.

It is important to notice that although for these tests it was considered that an entire 2D view image is coded into only one packet, the *Patch Remapping* and *Micro-Image Refilling* processes could easily be adapted to the case where the lost packets represent slices of 2D view images. Moreover, from the presented analysis, it was shown that although the parameters of the scalable coding architecture somehow interfere on the performance of the error concealment algorithm, in some

Table 7.1 Test conditions—patch sizes and positions (in pixels) for generating content for the lower hierarchical layers (for each light field test image in Fig. 7.5)

Test image	Patch sizes	Patch positions (horizontal positions)
Plane and Toy (frame 23)	{4 (in focus), 9, 11}	9 views: {−8, −6, −4, −2, 0, 2, 4, 6, 8} 5 views: {−8, −4, 0, 4, 8}
Plane and Toy (frame 123)	{4, 9(in focus), 11}	9 views: {−8, −6, −4, −2, 0, 2, 4, 6, 8} 5 views: {−8, −4, 0, 4, 8}
Robot 3D	{4 (in focus), 9, 11}	9 views: {−8, −6, −4, −2, 0, 2, 4, 6, 8} 5 views: {−8, −4, 0, 4, 8}
Laura	{7, 10 (in focus), 14}	9 views: {−28, −21, −14, −7, 0, 7, 14, 21, 28}

cases, the used error concealment algorithm is able to recover the IL picture prediction with negligible MSE value (e.g., when the *Weighted Blending* algorithm is used). However, it is also important to consider cases where the losses happen in the *second enhancement layer* of the proposed scalable coding solution, since the information in this layer represents the largest percentage of the scalable bitstream (as shown in Sect. 7.3.1). Therefore, these cases will be considered in future work.

7.5 Conclusions

This chapter proposed to start the discussion about error resilience techniques for light field content and presents a study of the influence of packet losses in display scalable light field content coding. For this, an error concealment algorithm was adopted to estimate the lost data, which was derived from the inter-layer prediction scheme previously proposed by the authors. From the presented study, it could be seen that although the parameters of the scalable coding architecture somehow interfere on the performance of the error concealment algorithm, in some cases, it is possible to recover the inter-layer prediction with negligible differences compared to the prediction when there are no losses. However, it is also important to consider cases where the losses happen in the second enhancement layer of the proposed scalable coding solution. Therefore, future work includes the proposal of error resilience techniques for dealing with transmission errors in this layer.

Acknowledgements This work was supported by FCT (*Fundação para a Ciência e a Tecnologia*, Portugal), under the project UID/EEA/50008/2013.

References

1. Raytrix: Raytrix Website (2012). http://www.raytrix.de/. Accessed 7 July 2014
2. Lytro Inc. (2012). https://www.lytro.com/. Accessed 7 July 2016
3. Georgiev, T., Yu, Z., Lumsdaine, A., Goma, S.: Lytro camera technology: theory, algorithms, performance analysis. In: Proceedings of SPIE 8667, Multimedia Content and Mobile Devices. Burlingame, CA, US, p 86671J (2013)
4. Wang, J., Xiao, X., Hua, H., Javidi, B.: Augmented reality 3D displays with micro integral imaging. J. Disp. Technol. **11**, 889–893 (2015). https://doi.org/10.1109/JDT.2014.2361147
5. Lanman, D., Luebke, D.: Near-eye light field displays. In: ACM SIGGRAPH 2013 Emerging Technologies—SIGGRAPH '13, pp. 1–1 (2013). https://doi.org/10.1145/2503368.2503379
6. Aggoun, A., Tsekleves, E., Swash, M.R., et al.: Immersive 3D holoscopic video system. IEEE Multimed. **20**, 28–37 (2013). https://doi.org/10.1109/MMUL.2012.42
7. Cisco Visual Networking Index: Global mobile data traffic forecast update. 2015–2020 Cisco White Paper (2016)
8. Conti, C., Nunes, P., Soares, L.D.: Impact of packet losses in scalable 3D holoscopic video coding. In: Proceedings of SPIE Optics, Photonics and Digital Technologies for Multimedia Applications, III. Brussels, Belgium, p 91380E (2014)

9. Georgiev, T., Lumsdaine, A.: Focused plenoptic camera and rendering. J. Electron. Imaging **19**, 021106 (2010). https://doi.org/10.1117/1.3442712
10. Wang, Yao, Wenger, S., Wen, Jiantao, Katsaggelos, A.K.: Error resilient video coding techniques. IEEE Signal Process. Mag. **17**, 61–82 (2000). https://doi.org/10.1109/79.855913
11. Bossen, F.: Common HM test conditions and software reference configurations. JCTVC-L1100, Geneva, Switzerland (2013)
12. Conti, C., Nunes, P., Soares, L.D.: Inter-layer prediction scheme for scalable 3-D holoscopic video coding. IEEE Signal Process. Lett. **20**, 819–822 (2013). https://doi.org/10.1109/LSP.2013.2267234
13. Geogiev T Todor Georgiev Gallery of Light Field Data. http://www.tgeorgiev.net/Gallery/. Accessed 17 Sep 2016
14. 3D Holoscopic Sequences (Download Link) (2013). http://3dholoscopicsequences.4shared.com/ (2016). Accessed 30 Oct 2016

Chapter 8
Transmission of 3D Video Content

Emil Dumic, Anamaria Bjelopera, Khaled Boussetta, Luis A. da Silva Cruz, Yuansong Qiao, A. Murat Tekalp and Yuhang Ye

Abstract This chapter describes different video transport technologies that support the existing 3D video formats, such as frame-compatible side-by-side and multi-view video plus depth. Particular emphasis is given to the DVB systems (terrestrial, satellite, and cable) and IP transport, focusing HTTP/TCP streaming, adaptive HTTP streaming, RTP/UDP streaming, P2P Networks, and Information-Centric Networking-ICN. Hybrid transport technologies, combining broadcast and broadband networks for video delivery are also addressed. The chapter highlights important aspects of 3D video transmission over wireless

E. Dumic (✉)
Department of Electrical Engineering, University North, 42000 Varaždin, Croatia
e-mail: emil.dumic@gmail.com

A. Bjelopera
Department of Electrical Engineering and Computing, University of Dubrovnik, 20000 Dubrovnik, Croatia
e-mail: anamaria.bjelopera@unidu.hr

K. Boussetta
L2TI, University of Paris, 13, 99, Avenue J-B Clement, 93430 Villetaneuse, France
e-mail: Khaled.Boussetta@univ-paris13.fr

L. A. da Silva Cruz
Department of Electrical and Computer Engineering, Instituto de Telecomunicações, University of Coimbra, Pólo II, 3030-290 Coimbra, Portugal
e-mail: lcruz@deec.uc.pt

Y. Qiao · Y. Ye
Software Research Institute, Athlone Institute of Technology, Athlone, Ireland
e-mail: ysqiao@research.ait.ie

Y. Ye
e-mail: yye@research.ait.ie

A. Murat Tekalp
Koç University, Istanbul, Turkey
e-mail: mtekalp@ku.edu.tr

P. A. Assunção and A. Gotchev (eds.), *3D Visual Content Creation, Coding and Delivery*, Signals and Communication Technology,
https://doi.org/10.1007/978-3-319-77842-6_8

networks, together with their benefits and limitations in the delivery of this type of content. Recent research results are summarized for different delivery systems and transport technologies.

8.1 Introduction

With the rapid development of different 3D video content, its transmission and delivery to the end user become a challenging problem to the existing network systems due to the ever-increasing capacity needs, different errors that may arise in the transmission chain, used format of 3D content, etc. This chapter describes different video transport technologies that support most of the existing 3D video formats (e.g., side-by-side, multi-view + depth). The DVB systems (terrestrial, satellite, and cable) are addressed and also IP transport (HTTP/TCP streaming, adaptive HTTP streaming, RTP/UDP streaming, P2P Networks, and Information-Centric Networking-ICN), hybrid transport technologies (a combination of broadcast and broadband video delivery), and 3D video transmission over wireless networks, together with their benefits and limitations in 3D video delivery content.

First, in Sect. 8.2 the main processing blocks of DVB systems (DVB-T/T2, DVB-S/S2, and DVB-C/C2) are presented, along with transport of 3D video (side-by-side, multi-view + depth), from compressed video, audio, and ancillary information that form elementary streams, packetized elementary streams. The most important specifications of Program Stream (PS) or Transport Stream (TS), as defined in MPEG-2 Systems are also addressed. Hybrid transport technologies to pool existing DVB broadcast of stereo 3D TV and IP streaming for multi-view 3DTV broadcast applications are highlighted in Sect. 8.3. In Sect. 8.4, IP transport technologies will be discussed in more detail, like multi-casting, content-distribution networks (CDN), peer-to-peer (P2P) streaming. In this section, HTTP/TCP streaming, adaptive HTTP streaming, and RTP/UDP streaming are also explained. Section 8.5 describes Information-Centric Networking (ICN) concept. Section 8.6 describes support of 3D stereo and multi-views video over wireless networks and finally, Sect. 8.7 gives conclusions.

8.2 DVB-T/T2, C/C2, and S/S2 Systems

Terrestrial, cable or satellite broadcast have been the most commonly used delivery method for bringing 3D TV content to home and in the technological context the Digital Video Broadcasting–Terrestrial (DVB-T) standard is of utmost importance, as defined by the European Telecommunications Standards Institute (ETSI) standard EN 300 744 [1], DVB-T2 in ETSI EN 302 755 [2], Digital Video Broadcasting-Satellite (DVB-S) in ETSI EN 300 421 [3], DVB-S2 in ETSI EN 302

307 [4], Digital Video Broadcasting-Cable (DVB-C) in ETSI EN 300 429 [5], and DVB-C2 in ETSI EN 302 769 [6]. DVB systems today usually carry stereo video in one of the frame-compatible formats that combines the left and right video sequences in one high-definition (HD) stream.

- SbS (side-by-side)—left and right images are one next to the other in an HD image;
- TaB (top-and-bottom)—put left and right images one above the other in a HD image.

A frame-compatible stereoscopic video format combines the left-eye and right eye images in a spatial multiplex arrangement which results in a composite image that can be treated like a conventional high-definition television (HDTV) image by the receiver demodulator and compression decoder. SbS and TaB formats are compatible with actual HD systems and can be transmitted using current DVB standards. Of course, at the receiver side, one needs to have 3D monitor to properly show 3D video. Regarding how a decoded signal is sent to a stereo display, current stereoscopic systems usually use a frame-sequential 3D signal. Left and right frames are alternately sent to the display and by diverse systems like active shuttered glasses or polarized glasses are then shown to each eye. This means that the real frame frequency is half the video frame frequency. New standard for DVB-3DTV that is currently being developed, ETSI TS 101 547, in Part 2 describes frame-compatible stereoscopic 3DTV formats (SbS, TaB), ETSI TS 101 547-2 [7].

8.2.1 DVB-T

DVB-T is the DVB European-based consortium standard for the broadcast transmission of digital terrestrial television [1]. This system is used for transport of compressed digital audio, video, and other data, multiplexed into a Moving Picture Experts Group (MPEG) TS [8, 9] using Coded Orthogonal Frequency Division Multiplexing (COFDM) modulation [10]. Rather than carrying the data on a single radio frequency (RF) carrier, Orthogonal Frequency Division Multiplexing (OFDM) works by splitting the digital data stream into a large number of slower digital streams, each of which digitally modulates a set of closely spaced adjacent carrier frequencies. In the case of DVB-T, there are two choices for the number of carriers known as 2K-mode or 8K-mode (4K-mode is rarely used) [1].

A DVB-T transmitter shown in Fig. 8.1 (taken from ETSI EN 300 744 [1]), consists of the following signal processing blocks, explained in more detail in [11].

- *Source coding and MPEG-2 multiplexing*: Compressed video, audio, and private data streams form elementary streams (different video and audio compression). Elementary streams are first cut into packetized elementary streams (PES) and afterward multiplexed into MPEG-2 transport stream (MPEG-2 TS) [12]. Each TS packet is 188 bytes long and can contain data from only one PES

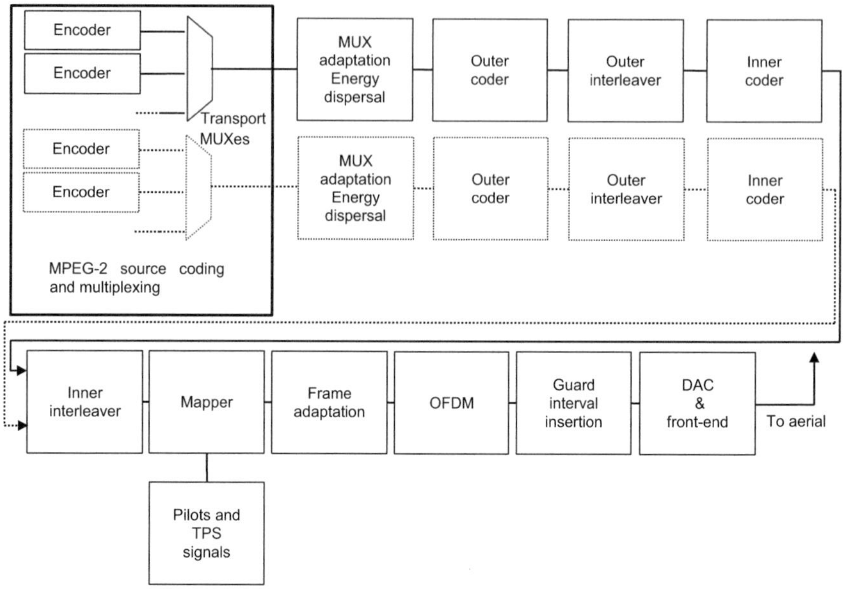

Fig. 8.1 DVB-T transmitter block diagram

packet. The system also allows two-level hierarchical channel coding and modulation, including uniform and multi-resolution constellation. In this case, the functional block diagram of the system shall be expanded to include the modules shown dashed in Fig. 8.1.

- *External encoder (RS encoder)*: Reed–Solomon coding or RS (204, 188, $T = 8$) code is used which is a shortened version of the code RS (255, 239, $T = 8$). Reed–Solomon code RS (204,188) uses 16 parity bytes and it can correct up to eight erroneous bytes per packet.
- *External interleaver (convolutional interleaver, $I = 12$)*: Convolutional interleaver rearranges the transmitted packets with the purpose to increase the efficiency of the Reed–Solomon decoding by spreading the burst errors introduced by the channel over a longer time.
- *Internal encoder (Punctured Convolutional Code)*: Uses convolutional coding and is an efficient complement to the Reed–Solomon coder and external interleaver. Possible Forward Error Correction (FEC) codes: 1/2, 2/3, 3/4, 5/6, and 7/8.
- *Internal interleaver*: Two separate interleaving processes are used to reduce the influence of burst errors, one operating on bits (bit interleaver) and another on groups of bits (symbol interleaver).
- *Mapper (+ pilots and Transmission Parameter Signaling (TPS) carriers)*: All data carriers in one OFDM frame are modulated using Quadrature Phase-Shift Keying (QPSK), 16QAM (Quadrature Amplitude Modulation), 64QAM, nonuniform 16QAM or nonuniform 64QAM constellations.

- *OFDM Transmitter and Guard Interval Insertion*: In DVB-T, OFDM usually uses 2048 or 8192 carriers (2K and 8K mode). Insertion of the guard interval extends symbol duration by 1/4, 1/8, 1/16 or 1/32 to give the total symbol duration. Cyclic prefix serves as a guard interval and eliminates the intersymbol interference from the previous symbol.
- *DAC (digital to analogue converter) and front-end*: Digital signal is transformed into an analogue signal with a DAC and then modulated to radio frequency (very high frequency (VHF), ultra high frequency (UHF)) by the RF front-end. The occupied bandwidth is designed to accommodate DVB-T signal into 5, 6, 7, or 8 MHz channels.

DVB-T receiver consists of below mentioned signal processing blocks, which are as follows:

- Front-end and ADC (analogue to digital converter);
- Time and frequency synchronization;
- Guard interval disposal and OFDM Receiver;
- Channel Estimator and Channel Compensation;
- Demapper;
- Inner Deinterleaver;
- Internal Decoding (Viterbi Decoder);
- External Deinterleaving (Convolutional Deinterleaver, $I = 12$);
- External decoding (RS Decoder);
- MUX (multiplexer) adaptation;
- MPEG-2 demultiplexing and source decoding.

The receiving STB (Set-Top Box) adopts techniques which are dual to those used in the transmission. Its practical performance depends on hardware construction (it is not standardized like encoder). Details of an example simulation of DVB-T transmitter and receiver can be found in [11], while the simulation itself can be downloaded from [13].

8.2.2 DVB-T2

Because of the inefficient frequency capacity usage in the terrestrial television platform, defined in DVB-T standard [1], a more efficient transmission system was developed to fulfill the market demands and allow launching new services, such as 3D and multi-view video delivery. To maximize spectrum efficiency, the DVB Project developed the second-generation digital terrestrial television standard, the DVB-T2 standard [2]. This new specification includes a newer coding scheme, interleaving and modulation techniques which provide increased capacity and robustness in the terrestrial transmission environment, compared to DVB-T. Transmission can be adapted to the characteristics of the actual channel conditions, thanks to all configurable parameters of the new standard.

Similarly to the DVB-T, the DVB-T2 uses COFDM, but new modulation and coding techniques are introduced. The possibility of using the 256QAM mode allows higher number of bits to be carried per data cell, which increases the spectral efficiency and bitrate. This increase is possible due to the better coding scheme Low-density parity-check + Bose–Chaudhuri–Hocquenghem (LDPC + BCH). The support for the 16K and 32K transmission modes allowed increase of the guard interval length without decreasing the spectral efficiency of the system. It is possible to choose between normal or extended carrier modes. The extended carrier mode gives the possibility to use more carriers per symbol, resulting in increased data capacity. Comparison of available modes in DVB-T and DVB-T2 specifications is shown in Table 8.1 [14]. Bolded values in DVB-T2 are newly added in standard.

The DVB-T2 transmitter, shown in Fig. 8.2, consists of several signal processing blocks. First novelty in the DVB-T2 standard is LDPC code [15] in combination with BCH, used as a protection against interference and noise. LDPC and BCH codes offer excellent performance resulting in a robust signal reception in different signal transmission condition. An important new feature is also bit, cell, time, and frequency interleaver. Additionally, a new technique called rotated constellations [16] resulted in improved robustness against loss of data cells. Similarly to the DVB-T, the DVB-T2 uses COFDM, but new modulation and coding techniques are introduced. 256QAM mode allows higher number of bits to be carried per data cell, which increases the spectral efficiency and bitrate. Higher number of bits per symbol, compared with DVB-T, is possible due to the better protection coding scheme in DVB-T2 (LDPC + BCH). The support for the 16K and 32K transmission modes increases the guard interval length without decreasing the spectral efficiency of the system. It is also possible to choose between normal or extended carrier modes. Extended carrier modes achieve even better spectral efficiency, comparing to normal carrier modes. DVB-T2 standard uses eight scattered pilot pattern modes (depending on Fast Fourier Transform (FFT) size and guard

Table 8.1 Comparison of parameters in DVB-T and DVB-T2 standard

	DVB-T	DVB-T2
FEC	Convolutional coding (1/2, 2/3, 3/4, 5/6, 7/8) + Reed–Solomon	LPDC (1/2, **3/5**, 2/3, 3/4, **4/5**, 5/6) + BCH
Modes	QPSK, 16QAM, 64QAM	QPSK, 16QAM, 64QAM, **256QAM**
Guard interval	1/4, 1/8, 1/16, 1/32	1/4, **19/256**, 1/8, **19/128**, 1/16, 1/32, **1/128**
FFT size	2K, 8K	**1K**, 2K, **4K**, 8K, **16K**, **32K**
Scattered Pilots	8% of total	**1%**, **2%**, **4%**, 8% of total
Continual Pilots	2.6% of total	**0.35%** of total
Spectrum	5, 6, 7, or 8 MHz	**1.7**, 5, 6, 7, 8, **10** MHz

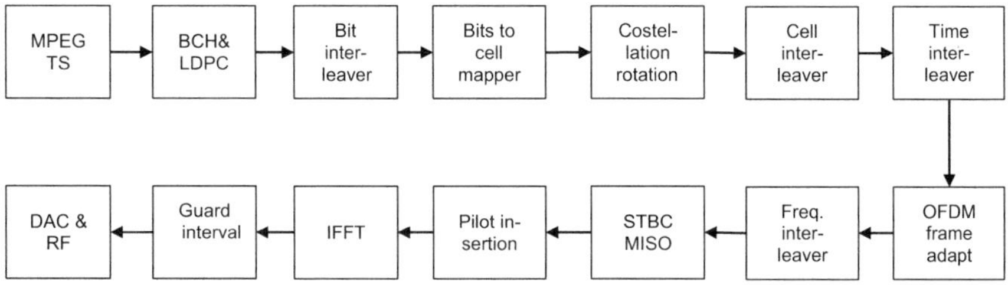

Fig. 8.2 DVB-T2 transmitter block diagram

interval, only some combinations are allowed), in order to maximize the data payload. The DVB-T2 standard also allows the transmission of single or multiple PLPs (Physical Layer Pipes) simultaneously. Each PLP carries one or more logical data streams (DVB-T2 services and signaling data) and can have different physical parameters, like coding rate or constellation. Optionally, the DVB-T2 standard supports Multiple Input Single Output (MISO) systems. Two techniques for Peak to Average Power Ratio (PAPR) reduction can be used: Active Constellation Extension (ACE) for lower order constellations and Tone Reservation method for higher order constellations.

An example of net bitrate calculation application in DVB-T2 network can be downloaded from [13] (written in MATLAB). Net bitrate for specific symbols/ frame can be also calculated (if exists). Supported modes are given as follows (but only with allowed combinations of parameters):

- scattered pilots: PP1–PP8;
- bandwidth: 1.7, 5, 6, 7, 8, 10 MHz;
- FFT: 1k, 2k, 4k, 8k, 16k, 16k-ext, 32k, 32k-ext;
- guard interval: 1/4, 19/128, 1/8, 19/256, 1/16, 1/32, 1/128;
- modulation: QPSK, 16QAM, 64QAM, 256QAM;
- L1 post-modulation: BPSK (Binary Phase-shift keying), QPSK, 16QAM, 64QAM;
- FEC: 1/2, 3/5, 2/3, 3/4, 4/5, 5/6;
- PAPR (Peak to Average Power Reduction): no tr-PAPR, tr-PAPR;
- efficiency mode: High-efficiency mode—HEM (no Deletion of Null Packets), normal (no Input Stream Synchronization, no Deletion of Null Packets).

8.2.3 DVB-S/S2

DVB-S standard is defined in ETSI EN 300 421 [3] for digital video broadcasting over satellite. Functional block diagram is shown in Fig. 8.3. Basically, its structure is similar with DVB-T up to Inner encoder: Outer coder is Reed–Solomon (204,188); Outer interleaver is convolutional interleaver; Inner coder can have FEC

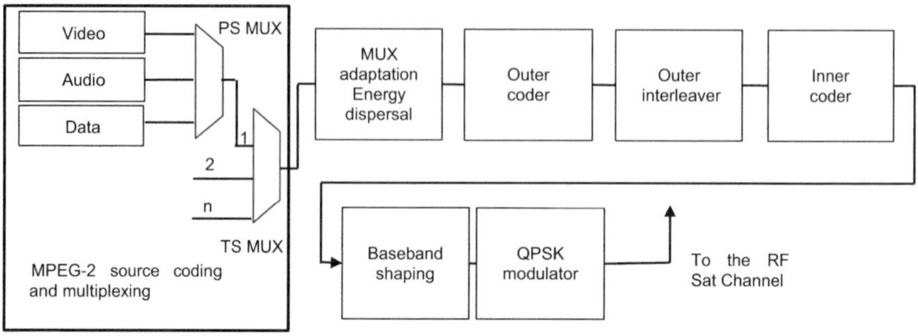

Fig. 8.3 Functional block diagram of DVB-S system

Table 8.2 Comparison of parameters in DVB-S and DVB-S2 standard

	DVB-S	DVB-S2
Input interface	Single transport stream (TS)	Multiple transport stream and generic stream encapsulation
Modes	Constant coding and modulation	Variable coding and modulation and adaptive coding and modulation
FEC	Reed–Solomon (RS) 1/2, 2/3, 3/4, 5/6, 7/8	LDPC + BCH 1/4, 1/3, 2/5, 1/2, 3/5, 2/3, 3/4, 4/5, 5/6, 8/9, 9/10
Modulation	QPSK	QPSK, 8PSK, 16APSK, 32APSK
Roll-off	0.35	0.2, 0.25, 0.35
Pilots	N/A	Pilot symbols

1/2, 2/3, 3/4, 5/6, 7/8. After it, signal is shaped in baseband (square root cosine filter with roll-off factor 0.35) and modulated using QPSK modulation. QPSK modulation, although with lower spectral efficiency, is chosen because satellite services are particularly affected by power limitations in transmission channel.

The second-generation digital satellite television standard, DVB-S2, is explained in detail in ETSI EN 302 307 [4]. Main differences between DVB-S and DVB-S2 are shown in Table 8.2.

In March 2014, an optional extension of DVB-S2 standard was proposed as DVB-S2X [16].

8.2.4 DVB-C/C2

The DVB-C standard, described in detail in ETSI EN 300 429 [5], is used for cable delivery systems A DVB-C transmitter, Fig. 8.4, consists of several signal processing blocks, which are given as follows:

- MPEG-2 source coding and multiplexing;
- MUX adaptation and energy dispersal;

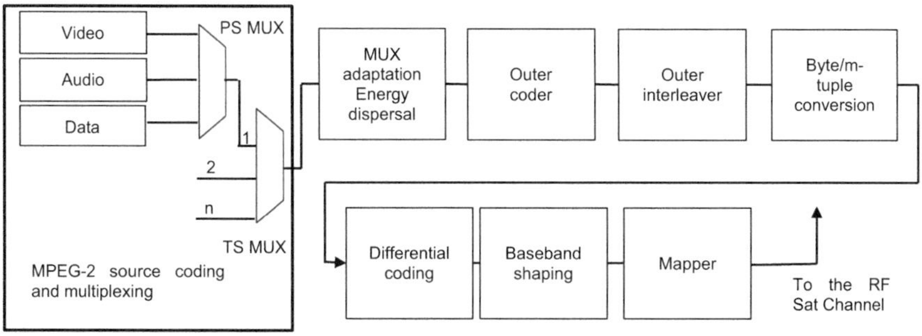

Fig. 8.4 Functional block diagram of DVB-C system

- *Outer encoder*: Reed–Solomon RS (204, 188) code;
- Outer interleaver;
- *Byte/m-tuple conversion*: This unit shall perform a conversion of the bytes generated by the interleaver into QAM symbols (m = 4, 5, 6, 7 or 8).
- *Differential coding*: In order to get a rotation-invariant constellation, this unit shall apply a differential encoding of the two Most Significant Bits (MSBs) of each symbol.
- Baseband shaping
- *QAM Mapper and RF front-end*: Five allowed modulation modes are 16QAM, 32QAM, 64QAM, 128QAM, and 256QAM.

DVB-C STB has techniques which are dual to those used in the transmission. DVB-C2, second-generation digital cable television standard is explained in detail in ETSI EN 302 769 [6]. Main differences between DVB-C and DVB-C2 are shown in Table 8.3.

Table 8.3 Comparison of parameters in DVB-C and DVB-C2 standard

	DVB-C	DVB-C2
Input interface	Single transport stream (TS)	Multiple transport stream and generic stream encapsulation
Modes	Constant coding and modulation	Variable coding and modulation and adaptive coding and modulation
FEC	Reed–Solomon (RS)	LDPC + BCH 1/2, 2/3, 3/4, 4/5, 5/6, 8/9, 9/10
OFDM	–	4K IFFT
Modulation schemes	16QAM to 256QAM	16QAM to 4096QAM
Guard interval	–	1/64 or 1/128
Interleaving	Bit-interleaving	Bit, time, and frequency interleaving
Pilots	–	Scattered and continual pilots

8.2.5 *Transport of 3D Video in DVB Systems*

Compressed video, audio, and ancillary information can be transported over DVB physical layers after multiplexing and packetization using the formats defined in MPEG-2 Systems specification, as laid out in ISO/IEC (International Organization for Standardization/International Electrotechnical Commission) 138181-1 [8] and ETSI technical standard 101 154 [9]. MPEG-2 Systems describes two packetizing formats, one at program stream (PS) level adequate for use in storage or transmission over error-free communication systems, and another one at transport layer level (TS—Transport Stream) with provisions for multiplexing of multiple program streams and better suited for transmission over error-prone channels such as those typical of the DVB systems described in Sects. 8.2.1–8.2.4. In both cases, the source encoded media information is first formatted as elementary video, audio, and other data streams that are packetized and prepended with headers containing timing information to form elementary stream packets, as illustrated in Fig. 8.5. The figure shows packets with variable lengths, reflecting the fact that the media information can be sourced at different data rates. The PES associated with a given program usually share a common time base and are multiplexed, defining a program stream. The packet headers carry important timing information such as presentation time stamps (PTS) and decode time stamps (DTS) which are used by the audio and video decoders to decide in which order the packet contents should be decoded and rendered and also allows synchronization of the audio and video streams at play time.

The program stream packets can be stored or transmitted directly over error-free transmission channels. Alternatively, PES packets belonging to different programs possibly with different time bases can segment into smaller packets, to form TS packets which can be passed on to the lower layers of the DVB system for transmission after addition of FEC bits. The segmentation of PES packets into TS packets is shown in Fig. 8.6 and the multiplexing of TS packets from different programs into a final transport stream is sketched in Fig. 8.7.

The transport layer also fulfills very important functions of synchronization, both end-to-end and between elementary streams, relying on a reference signal provided

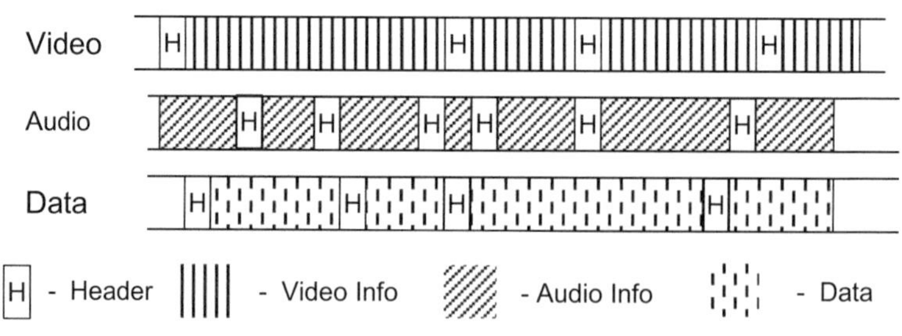

Fig. 8.5 Packetized elementary streams

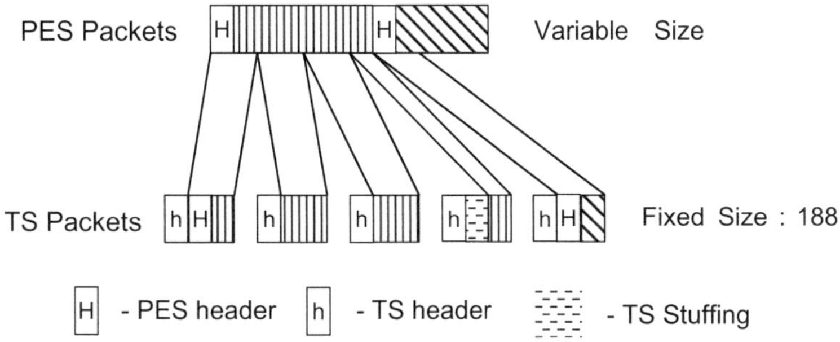

Fig. 8.6 PES packetization into TS packets

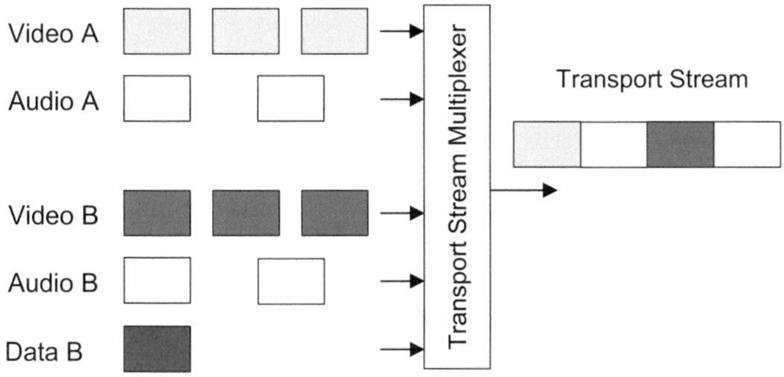

Fig. 8.7 Multiplexing of TS packets

by the system time clock (STC) whose value is periodically inserted in the TS packets.

Originally, MPEG-2 Systems was designed for use in the transmission of 2D video but it has since suffered amendments to support the transmission of 3D video in stereo, multi-view video, and in texture plus depth formats. Encoded 3D stereo video can be transported in several different ways over MPEG-2 TS systems, some choices depending on how the left and right views are encoded. The simplest solution is applicable to video which has been encoded in “simulcast”, i.e., both views have been encoded separately. In this case, the two streams can be transported as two elementary streams of a given program stream or two program streams of a transport stream and the decoder receives information about the 3D nature of the content through specific signaling. A slightly more efficient method involves using layered video coding, such as in MPEG-2 with the Multi-view Profile, to encode one of the views as a base view and encode the other with reference to the base view. The two streams can then be multiplexed into a transport stream. These two transport methods are very convenient as they are easily compatible with 2D decoders which just have to ignore one of the elementary video streams and decode the other.

Instead of encoding both views and separate video streams, frame-compatible encoders assemble a composite frame which is the encoded using a regular 2D video encoder. The composite frames can be constructed either by joining the left and right view frames side-by-side, or in a top–bottom arrangement or using some other spatial multiplexing scheme possibly with some previous spatial sub-sampling operations. The encoded video is then delivered using a normal MPEG-2 TS complemented with information to signal the decoder that the stream contains a frame-compatible-encoded stereo video. Several variants of these schemes are documented in ETSI technical standard series 101 547 [17–20].

Both H.264/AVC (Advanced Video Coding) and HEVC (High-Efficiency Video Coding) have extensions for encoding multi-view video, according to which view is independently encoded (base view) while the others are encoded using the frames of the base view as reference for inter-view prediction. Several streams result from the encoding procedure, which are interleaved on a frame basis to produce a multi-view elementary stream that can be delivered using an MPEG-2 transport stream together with signaling information.

Video plus depth emerged as 3D video format with several advantages over stereo video, as it supports compatibility with 2D decoders in a natural way and results in a relatively small additional amount of bits when compared to 2D video. Due to these advantages and the foreseeable wide adoption of the format, the MPEG group standardized a method to encode auxiliary video information such as the depth component of this 3D video format (the standard also supports representing parallax information). The standard is informally known as MPEG-C Part 3, or ISO/IEC 23002-3 and also specifies how to signal the receiver that a stream contains 3D video in video plus depth format. Both the video part as well as the depth component can be transported over MPEG-2 TS systems, together with supplemental information which can specify the encoding standard used to encode the depth information. Further details can be found in [21].

In [22], the authors describe how to deliver layered media (such as MVC and SVC) over DVB-T2 using multiple PLPs. The key is to use the common PLP to deliver one representation of the layered media stream and transmit the second representations of the layered media stream in a data PLP. The combination of layered media codecs with multiple PLPs in DVB-T2 can enable flexible and cost-efficient delivery of high-quality HDTV and 3DTV services. Figure 8.8 shows flexible implementation of 3D services using MVC and 2D services using SVC.

A review of state of the art in 3D video formats, coding methods for different transport options and video formats, IP streaming protocols, and streaming architectures is presented in [23]. The authors also describe asymmetric stereoscopic video streaming, adaptive, and peer-to-peer (P2P) streaming of multi-view video, view-selective streaming, and future directions in broadcast of 3D media over IP and jointly over DVB and IP. In [24], the authors present a complete framework of an end-to-end error resilient transmission of 3D video over DVB-H and provides an analysis of transmission parameters. Figures 8.9 and 8.10 show basic principle of proposed 3D video transmitter and receiver over DVB-H networks.

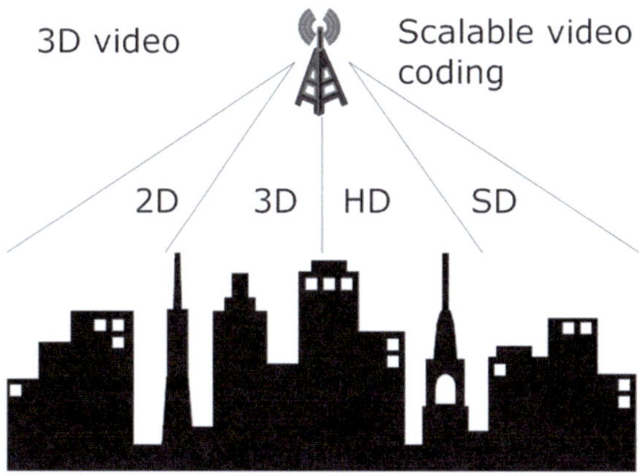

Fig. 8.8 3D services with MVC and 2D services with SVC

The effect of different slice modes and protection methods on the error performance of video + depth-based 3D video broadcast over DVB-H was studied in [25]. In [26], the authors propose a complete framework for end-to-end transmission of stereo video for regular services using a digital video broadcasting-terrestrial version 2 (DVB-T2) system. The proposed system incorporates existing services (such as fixed and mobile) to deliver stereoscopic 3D content in order to maintain backward compatibility without requiring any additional bandwidth. Figure 8.11 shows transmitter side of Hybrid DVB-T2 3DTV and Fig. 8.12 shows receiver side of the proposed system.

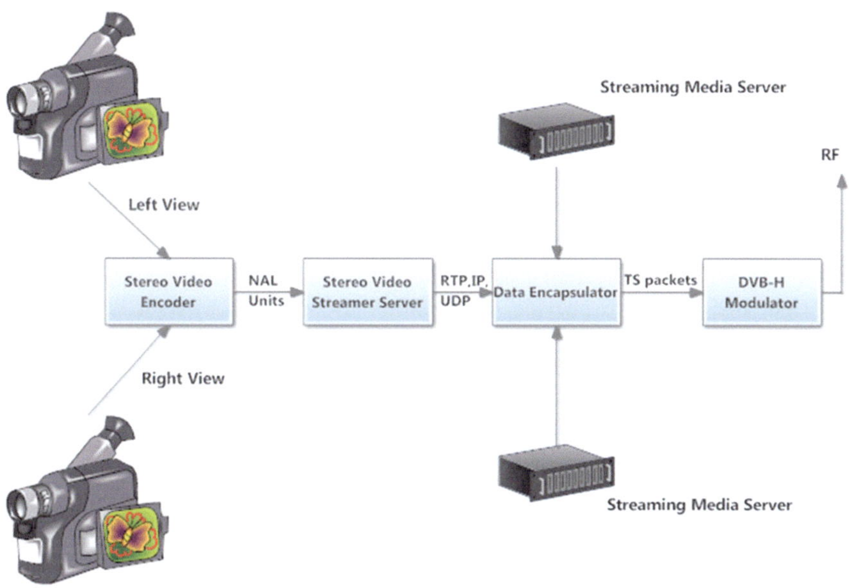

Fig. 8.9 3D video transmitter over DVB-H

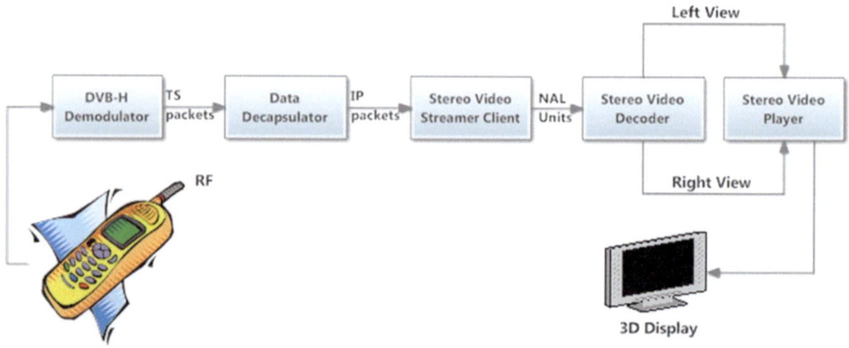

Fig. 8.10 3D video receiver over DVB-H

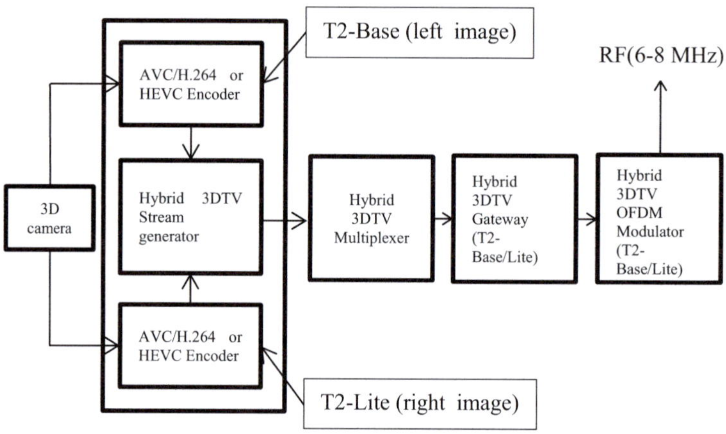

Fig. 8.11 Hybrid DVB-T2 3DTV system overview (transmitter side)

The work described in [27] presents DVB-T2 and T2- Lite/NGH hybrid 3DTV system design and implementation for DVB-T2 terrestrial 3DTV broadcasting services. The proposed system uses the video streams from two different PLP as left and right eye videos to provide high-quality 3D services, while it is backward compatible with DVB-T2/T2-Lite and NGH standard.

8.3 Hybrid Broadcast/Broadband 3DTV

The broadcast transport and existing IP networks do not sufficiently scale up to carry multi-view video with depth maps. Hence, it may be of interest to pool existing DVB broadcast of stereo 3D TV and IP streaming for multi-view 3DTV broadcast applications. A particular system, where frame-compatible stereo

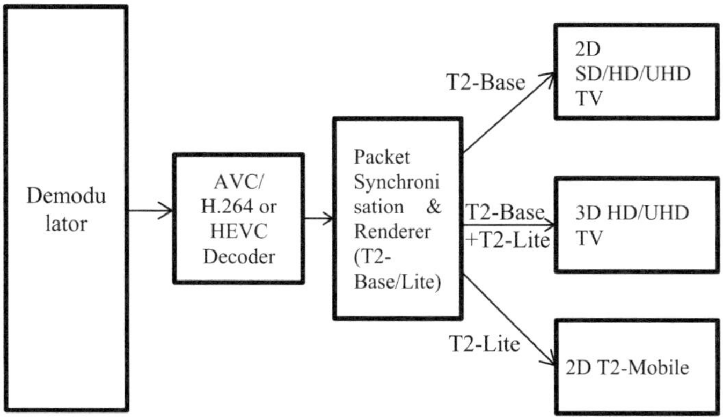

Fig. 8.12 Hybrid DVB-T2 3DTV system overview (receiver side)

broadcast is supplemented by several synchronized auxiliary views and associated depth maps over IP has been proposed in [28]. A hybrid system is constructed in the proposed work, where the DVB network and a P2P overlay network transport multi-view media together. The DVB network has been used to deliver part of the 3D service owing to its robustness and wide availability, as a mechanism to guarantee the minimum 3D Quality of Experience (QoE). P2P transport technology is used for real-time (overlay) multimedia delivery, together with its benefits that include a user preference-aware adaptation mechanism, adaptive redundant chunk scheduling for robustness, incentives to decrease the load on the content server for improved system scalability, and resynchronization capability with the DVB transmission.

A converged broadcast and broadband platform in order to deliver 3D media to both mobile and fixed users with guaranteed minimum quality of experience (QoE) is described in [29]. It offers an ideal business model for operators having both digital video broadcast and Internet Protocol (IP)-based media services. In [30], the authors propose a new synchronization and transport system target decoder (T-STD) model of 3D video distribution based on heterogeneous transmission protocol in a hybrid network environment, where a broadcasting network and broadband network are combined. Proposed technology has been proved to be successfully used as a core element for synchronization and T-STD model in a hybrid 3D broadcasting.

In [31], a general reference model was devised to allow the convergence of 3DTV and 3D Web by defining a general architecture and some extensions of current Smart TV concepts as well as the related standards, like HbbTv (Hybrid Broadcast Broadband TV) [32] and HTML-5 [33]. The authors propose two scenarios: the first scenario is designed for TV sets which include 3D rendering resources and where the apps that are executed on the device have access both to Web and broadcast content. The second scenario is less restrictive and only needs capability to display the remotely rendered 3D content overlaid on top of the broadcasting signal.

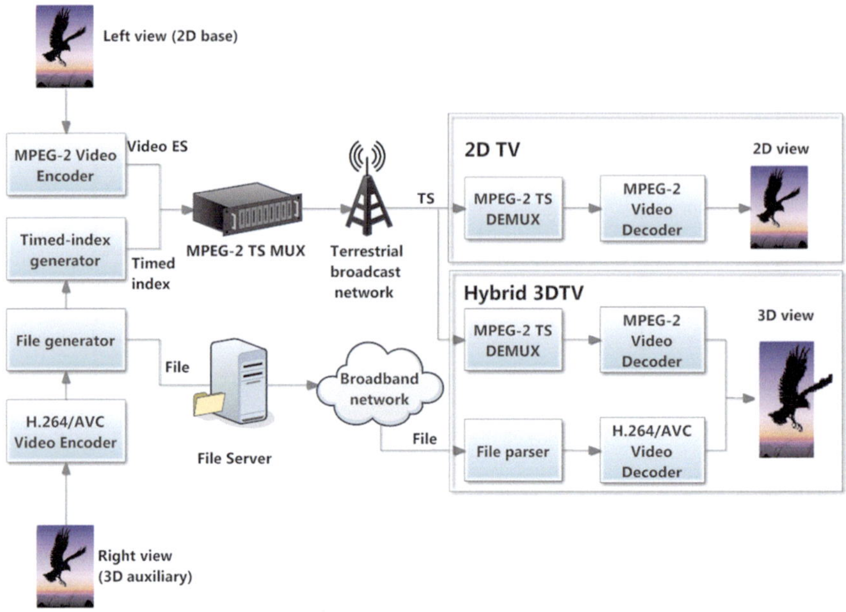

Fig. 8.13 Hybrid broadcast/broadband system overview for hybrid 3DTV service

In [34], the authors propose a hybrid 3DTV broadcasting system, which utilizes both a terrestrial broadcast network and a broadband network. In the proposed system, two elementary streams of left and right views for a stereoscopic video service are transmitted over a terrestrial broadcasting network and a broadband network, respectively. In addition, the proposed system suggests a new mechanism for synchronization between these two elementary streams. Basic diagram of the proposed hybrid 3DTV broadcasting service is shown in Fig. 8.13.

8.4 3D Video Delivery Over IP

In this section, different IP transport technologies that can be used to deliver 3D video content are discussed: HTTP/TCP streaming, adaptive HTTP streaming, RTP/UDP streaming, multi-casting, content-distribution networks (CDN), peer-to-peer streaming (P2P).

8.4.1 HTTP and RTP-Based 3D/Multi-view Streaming

Delivery of media over the Internet Protocol (IP), where the client player can start playback before the entire file has been sent is called streaming. Streaming over IP is a flexible transport system for multi-view video since the transmission bitrate can be configured according to requirements of video format and user equipment.

Streaming systems can be classified as server-client or peer-to-peer (P2P) systems. A server-client streaming system consists of a streaming server and a client that communicate using a set of standard protocols. The client may be a stand-alone player or a Web browser. In the server-client model, typically a different stream is sent to each client. This model is not scalable since server traffic increases linearly with the number of stream requests. Different solutions have been proposed to solve this problem including multi-casting, content-distribution networks (CDN) and P2P streaming. Multi-casting is a one-to-many delivery system, where the server sends each packet only once and the nodes in the network replicate packets only when necessary to reach multiple clients. It can be implemented at the network (IP) or application level. In P2P streaming, clients (peers) forward packets to other peers (as opposed to network nodes) to minimize the load on the source server. Some P2P technologies employ the multi-cast concept when distributing content to multiple recipients, known as peer-casting.

Streaming applications can be classified as one-way video-on-demand, broadcast, or live streaming sessions and two-way real-time communication (RTC) sessions. In a video-on-demand session, the server streams from a pre-encoded and stored file. Live streaming refers to live content delivered in real-time over the Internet, which requires a live camera and a real-time encoder on the server side [35].

A streaming service can be a pull-application (client-driven stream request) or a push-application (server driven packet transmission). In order to avoid the burden of processing individual client status information on the server side, one-way streaming applications are often configured as pull-applications using HTTP/TCP (Hypertext Transfer Protocol/Transmission Control Protocol), where the server is a simple HTTP server and all intelligence reside on the client side. Streaming over HTTP works by breaking a stream into a sequence of small HTTP-based file downloads, each download loading one short *chunk* of the whole stream. Delay critical applications, such as real-time communication sessions, are set up as push applications using Real-Time Streaming Protocol (RTSP). RTSP is an open standard published by the Internet Engineering Task Force (IETF) in 1998. RTSP servers use the Real-time Transport protocol (RTP) over User Datagram Protocol (UDP) for media stream delivery, which supports a range of media formats. Smartphone platforms also include support for RTSP as part of the Third-Generation Partnership Project (3GPP) standard. The main problem with UDP-based streaming is that streams are frequently blocked by firewalls since they are not being sent over HTTP (port 80). In order to alleviate this problem, protocols have been extended to allow for stream encapsulation within HTTP requests, called tunneling, as a fallback solution. Streaming protocols also have secure variants that use encryption to protect the stream.

Since the Internet is a best-effort channel, packets may be delayed or dropped by the routers and the effective end-to-end bitrates fluctuate in time. Adaptive streaming aims to adapt the video encoding rate according to estimated available end-to-end network rate. One possible solution is stream switching, where the server encodes video at multiple preselected bitrates and the client requests switching to the stream encoded at the rate that is closest to its network access rate. An alternative solution is based on scalable video coding, where one or more enhancement layers of video may be dropped to reduce the bitrate as needed.

HTTP streaming solutions include support for adaptive streaming (bitrate switching) to allow clients dynamically switch between streams of varying quality and chunk size during playback, in order to adapt to changing network conditions and available central processing unit (CPU) resources. HTTP streaming allows chunks to be cached within Internet service providers (ISP) or corporations, which would reduce the bandwidth required to deliver HTTP streams. Different vendors have implemented different HTTP-based streaming solutions, such as HTTP Live Streaming (HLS) by Apple, Smooth Streaming by Microsoft, HTTP Dynamic Streaming (HDS) by Adobe, which use similar mechanisms but are incompatible. MPEG-DASH (Dynamic Adaptive Streaming over HTTP) is an international standard, published in April 2012, for adaptive bitrate HTTP-based streaming that is audio/video codec agnostic in order to address this fragmentation problem.

The work done in [36] addresses multi-view video streaming over HTTP for emerging multi-view displays. In this research, the effect of various adaptations of decision strategies is evaluated and, as a result, a new quality-aware adaptation method is designed. The proposed method is benefiting from layer-based video coding in such a way that high Quality of Experience (QoE) is maintained in a cost-effective manner. A different approach was followed in [37], where the authors proposed a method of efficient 3D adaptive streaming service based on the DASH, which covers service-compatible stereoscopic content in a single segment sequence. The 3D adaptive HTTP streaming method introduced in this chapter is able to provide seamless streaming service for various kinds of stereoscopic contents.

In [38], the authors propose a dynamic adaptive rate control system and its associated rate-distortion model for multi-view 3D video transmission, which improves the user's quality of experience in the face of varying network bandwidth. Rate control system has been built on top of two state-of-the-art key technologies: HEVC and MPEG-DASH.

In [39], the authors investigated the possibility of greatly improving the average quality and also attenuating the quality variations for a better user experience, by leveraging multipath communication requiring no extra bandwidth. Algorithm exploits concurrently multiple paths using a Content Delivery Network (CDN) compliant distributed streaming infrastructure and standard HTTP range requests.

8.4.2 3D Video Distribution Over P2P Networks

Due to the large bandwidth consumptions to deliver 3D video content (e.g., stereoscopic and multi-view video), it is a challenge to achieve service scalability and satisfy increasing numbers of recipients [40]. Therefore, the throughput provided by the traditional server-client model may not fulfill the requirements of high-quality 3D video delivery. On the contrary, the peer-to-peer model enables video streaming clients to share content with others, which enhances the utilization of spare bandwidth of the users, such that the overall throughput can be improved and the burden of service providers can be mitigated.

The P2P protocols were originally designed to distribute larger files over the Internet via an overlay solution [41]. In the early stage, the P2P streaming protocols choose a tree-based topology [42] and hierarchical connections among peers. As this approach is difficult to cope with network dynamics, e.g., peer churns, and cannot utilize the outbound bandwidth of the peers at the leaves of the tree, mesh-based topology has become the widely adopted structure in practice [43]. Usually, a tracker server [44] is deployed to provide a list of relevant peers for a content object, which allows any peer to download the missing content chunks from other peers. Later, due to the creation of the distributed hash table (DHT) [45] for discovering peers without centralized storage and inquire, the tracker becomes redundant and is deprecated in some protocols.

A popular P2P file sharing protocol—BitTorrent [46] operates according to a divide-and-conquer principle and divides content into equal size chunks to facilitate delivery. Different successors are proposed for optimizing the video streaming quality via modifying chunk scheduling policies [47, 48] and interleaving chunks [49]. In the original BitTorrent protocol, two scheduling policies of downloading chunks are ***rarest-first*** and ***tit-for-tat***. The rarest-first policy downloads the chunk that is the least distributed within the swarm, which is inferred as the least distributed chunk are most likely to be required by neighbors. The tit-for-tat policy determines the qualities (contributions) of the neighboring peers and rejects the requests from the low-rank peers, which prevents leeches from downloading much more than their uploading [50].

For video streaming applications, the rarest-first policy may encounter some serious issues. For instance, when a peer has started playback of the video, the rarest-first policy may prefer to download the chunks that are far away from the playback time and consequently stall the playback. To solve this problem, a novel method [47] that combines the sequential downloading and rarest-first is proposed. The peer has a priority to retrieve the chunks sequentially within a ***ready-to-play*** buffer that guarantees the least playback time without stalling. If more bandwidth is available, the peer downloads content chunks with the rarest-first policy to improve the content richness of the swarm.

Streaming the 3D video can be either supported by a push-based P2P [45, 51] or a pull-based P2P [52]. In push-based P2P, peers are arranged in a tree structure and the video chunks are pushed from a parent node to its child nodes. It is an efficient solution if connections between peers are stable. However, tree regeneration is required when a parent node leaves. Thus, some solutions make use of redundant parents to prevent tree loss (disconnection) caused by leaving nodes. In pull-based P2P, peers can be arranged in a mesh-based structure and peers pull (request) the needed content from neighbors. In order to achieve so, each peer maintains a bitmap of chunks that are currently available by it or its neighbors and downloads the required video chunks from the neighbor with the highest bandwidth.

In 2D video streaming, video quality adaptation is the main design concern. For example, in delivering of scalable videos (e.g., H.264/SVC (Scalable Video Coding) [53]), the base layer is indispensable and the enhancement layer can be abandoned according to the adaptation logic when bandwidth is insufficient.

In contrast, 3D video content delivery is more complicated than 2D video delivery because of the complexity of the video layer dependency. The scheduling and adaptation logics in 3D video delivery require considering more factors besides the quality of a view, e.g., current viewpoint, viewpoint switching, and the possibility of missing viewing synthesis. Taking multi-view streaming, for example, the viewpoint of the viewer (peer) determines the importance of different views, the farther the less important it is. The adaptation logic should guarantee the chunks of the view in the current field of interest. The loss of the enhancement layer in this view may affect the QoE (Quality of Experience) much greater than the loss of the layers that is outside the field of interest. Furthermore, as the viewer can switch the viewpoint during playback, the views outside the field of interest also need to be pre-fetched. Otherwise, playback stalling or image distortion caused by missing chunks will largely affect QoE. The prediction of viewer's behavior becomes critical, which allows viewers to achieve adequate playback quality by spending less bandwidth. Finally, the coding techniques affect the video delivery design and performance significantly. In the early stage, the different views are coding independently (simulcasting) and symmetrically [54]. The delivery and adaptation algorithms proposed for 2D video can be utilized here without lots of modifications. With the development of asymmetric coding [54] for different views, peers need to distinguish the coding scheme for each view and perform optimal adaptation. Moreover, if the coding scheme supports to recover the lost view from the depth map and the reference views [55], peers can proactively abandon some views by determining the importance of each one, which enhances QoE with a limited bandwidth.

8.5 3D Video Distribution in ICN

The Information-Centric Networking (ICN) [56] concept is an important approach to the future Internet research activities. It is proposed to shift the Internet infrastructure away from the host-centric paradigm to a network architecture based on named data objects. In ICN, content is named and routed directly in the network and it can be retrieved through in-network caching and adaptive routing, which is a promising solution to improve efficiency, scalability, and robustness of the network.

An obvious benefit of ICN is that it enables caching of content in intermediate nodes to reduce congestion and improve delivery speed. For instance, if a large group of users is accessing the same data source (e.g., website or video) on the Internet simultaneously, the bottleneck link can be overloaded in the current TCP/IP architecture. By caching content in network [56] or at edge [57], the requests from multiple clients can retrieve content from intermediate nodes along the forwarding path, which can shorten the delay and offload the burden of bottlenecks. This applies to scenarios like large content delivery such as 3D video delivery. Currently, there are basically no studies on 3D video streaming on ICN. In this section, the challenges and possible solutions are discussed.

For the ICN implementations (e.g., Content-Centric Networking [58]) that support adaptive forwarding, the traffic control is still undergoing work which may affect the deployment of 3D video streaming. As interest packets (requests) of a video can be forwarded to multiple content producers concurrently, the consumer (i.e., a node is requesting the content) can hardly predict the congestion of the network based on its local information (e.g., timeout or duplicate acknowledgment) as used in TCP/IP. To this end, recent works [59, 60] prefer to utilize the explicit congestion notification signals (e.g., ECN—Explicit Congestion Notification or NACK—Negative Acknowledgement) to notify the congestion explicitly and another type of promising solutions is to use path-labeling [61] to detect congestion for each path.

Although it is similar to P2P, that ICN enables a consumer to download different content chunks from multiple producers simultaneously, there are some fundamental differences between ICN and P2P. For instance, P2P allows peers to communicate with other peers directly and routers are not responsible for discovering peers or balancing the load. By removing the host information in ICN, consumers are not designed to communicate to producers directly. Instead, routers are responsible for balancing the traffic to peers or servers to maximize the throughput of peers. Thus, the peer selection in traditional P2P is mapped to content publishing and adaptive forwarding. For the time being, the design of forwarding strategy becomes critical to maximize the downloading performance of peers.

As ICN introduces in-network caching, how and where to cache 3D video chunks becomes a novel topic that must be explored. In layered video streaming, the base layer is more important than the enhancement layers from the decoding point of view, thus differentiated service can be applied to cache chunks with different priority if the cache is limited.

Another possibility is to apply differentiated service to selective forwarding. When the network cannot provide sufficient bandwidth to deliver the content, some ICN solution enables routers to drop interest packets proactively. If the priority of content is considered, routers are able to discard the interest packets that do not have large effects on QoE.

8.6 3D Stereo and Multi-view Video in Wireless Networks

The support of 3D stereo and multi-view video over wireless networks is still an emerging research topic and very few research works have investigated the related challenges. First categories of works have focused on assessing the quality of experience obtained when transmitting 3D contents over wireless networks. In [62], the authors have conducted several experiments with subjective QoE evaluation of a 3D stereoscopic video files transported through TCP connections and over 802.11n WiFi access network. Several metrics were ranked, such as video continuity, 3D visual quality or video/voice synchronization while varying some network QoS (Quality of Service) metrics, such as WiFi channels and available bandwidth. The study showed the strong correlation between QoS degradation, in particular, bandwidth capacity, and the perceived QoE.

In [63], the authors address the support of 3D video streaming over LTE wireless networks and propose a framework that aims to dynamically control the system parameters in order to optimize the perceived 3D QoE. The key idea relies on a proxy that enables, through a Machine learning approach, real-time control and monitoring of the parameters impacting the 3D QoE. In [64], the authors extend this idea of context-aware 3D rendering and streaming over heterogeneous wireless networks to the cloud paradigm. The real-time control and monitoring of the network resources are facilitated thanks to the SDN (Software-defined networking) controller. The dynamic adjustment of the 3D coding accordingly with the available cloud resources and the 3D video visual quality is then enabled thanks to a second module.

Actually, the critical issue associated with 3D/multi-view video streaming in nowadays wireless networks is related to the limited and variable wireless bandwidth capacity and their inability to support the huge data rates associated with 3D contents. Moreover, packet losses and disconnections are quite frequent, either due to user's mobility or are as a consequence of interferences. In addition, expected use case scenarios are those were wireless resources are shared among several users. Dedicated or reserved wireless resources to support 3D streaming is a technically possible approach, but it is not economically viable for large public. To cope with the wireless bandwidth limitations, a first direction is to dynamically adjust the encoding and transmission bitrate depending on the available wireless bandwidth. In [65], authors proposed an encoding and transmission technique over WiFi networks of 3D objects of an MPEG-4 video depending on their perceived importance. The objects are segmented thanks to the depth information of 3D content.

Another recent proposal is the collaborative rendering of 3D contents. This approach can be used for instance to stream 3D video games contents in cloud-based services. In such services, the conventional approach is to stream the video games 3D contents to thin clients (tablets, smartphones, etc.). While this solution offers several advantages, such as the independence between the games and the terminals, it requires the satisfaction of stringent QoS constraints, including short delays and large bandwidth. To accommodate the users with the limited bandwidth of wireless networks, the authors of [66] propose to offload part of the GPU computation. Two strategies are analyzed. The first one assumes that the client renders each frame with reduced details. Therefore, the cloud has to stream only the per-frame differences between the high detail version and the client frame. In the second strategy, the client renders with high details, a subset of frames while the server streams only the remaining ones.

In [67], a rate adaptation method that uses the packet buffer size as an indicator of network load was proposed. During congestion, transmitter proactively drops packets belonging to layers with less impact. On the receiver side, data from the dropped layers are approximated using error concealment strategy, based on synthesizing the missing texture and depth frames when possible. A different strategy was followed in [68], where the authors proposed a multi-view video-aware transmission over MIMO wireless systems. The basic idea is to exploit the channel diversity of multiple antennas and the source coding characteristics so to achieve unequal error protection against channel errors. To achieve this goal, authors developed a nonlinear mixed integer programming framework to perform antenna selection and power allocation and proposed low-complexity algorithms to assign these resources.

A multiple description coding approach for multipath transmission of free-viewpoint video, with joint interview and temporal description recovery capability was investigated in [69]. Even frames of the left view and the odd frames of the right view are separately encoded and transmitted as one description on one path. Remaining frames in the two views are encoded and transmitted over a second path. If the receiver receives only one description due to burst loss in the other path, the missing frames in the other description are partially reconstructed using newly proposed frame recovery procedure from the same authors.

Another strategy was followed in [70], in mobile delivery of 3D content, where the authors propose to leverage depth-image-based rendering (DIBR) in multi-view 3D, which allows each mobile client to synthesize the desired view from nearby left and right views, in order to effectively reduce the bandwidth consumption in wireless networks. The authors developed the Multi-View Group Management Protocol (MVGMP) for multi-view 3D multicast. When a user joins the video multicast group, it can receive the most suitable right and left views, so that the view failure probability is guaranteed to stay below a threshold. When a user leaves the video multicast group, MVGMP carefully selects and withdraws a set of delivered views to reduce the network load, so that the video failure probability for other users will not exceed the threshold. Bandwidth consumption can be effectively reduced since it is not necessary to deliver all the views subscribed by the clients.

Another scheme called the Temporal Synchronization Scheme (TSS) for live 3D video streaming over 802.11 wireless networks was developed in [71]. The TSS scheme delivers video frames for the left and right views in the same frame order with the same transmission priority and compensates for frame damage and loss during the decoding phase. A new metric called the Stereoscopic Temporal Variation Index (STVI) is also proposed to measure the degree of temporal asynchrony in 3D video.

While the main idea in previous approaches is to act on the 3D contents in order to fit with the wireless network capacity, another promising direction is to the increase the wireless bandwidth in the next generations of wireless access technologies. Hence, regarding WLAN, new amendments to the IEEE 802.11 standard are already under development. Precisely, the 802.11ax is the next WiFi access technology that is actually under development with the targeted maximum bitrate around 10 Gbit/s. In complementary, 802.11ay is expected to replace cabled Ethernet LAN. This technology is predicted to provide up to 176 Gbit/s.

Regarding cellular network technologies, the next fifth-generation wireless systems (5G) is also under active development. The standard is expected at the horizon of 2020. Several promising features are announced such as tens of Mb/s per connection for thousands of users and very short latencies.

8.7 Conclusion

This chapter presented different network systems that can be used to deliver different 3D video content. A short description was provided about the DVB broadcast systems (DVB-T/T2, DVB-S/S2 and DVB-C/C2), IP transport technologies (multi-casting, content-distribution networks, peer-to-peer streaming, Information-Centric Networking concept, HTTP/TCP streaming, adaptive HTTP streaming, RTP/UDP streaming), hybrid transport technologies (a combination of broadcast and broadband), and 3D video over wireless networks that can be used to efficiently transmit 3D media content.

References

1. ETSI EN 300 744: Digital video broadcasting (DVB); framing structure, channel coding and modulation for digital terrestrial television. V.1.6.1, Sept 2008
2. ETSI EN 302 755: Digital video broadcasting (DVB); frame structure channel coding and modulation for a second generation digital terrestrial television broadcasting system (DVB-T2). V1.1.1, Sept 2009
3. ETSI EN 300 421: Digital video broadcasting (DVB); framing structure, channel coding and modulation for 11/12 GHz satellite services. V.1.1.2, Aug 1997
4. ETSI EN 302 307: Digital video broadcasting (DVB); second generation framing structure, channel coding and modulation systems for broadcasting, interactive services, news gathering and other broadband satellite applications (DVB-S2). V.1.2.1, Aug 2009
5. EN 300 429: Digital video broadcasting (DVB); framing structure, channel coding and modulation for cable systems. V.1.2.1, Apr 1998
6. ETSI EN 302 769: Digital video broadcasting (DVB); frame structure channel coding and modulation for a second generation digital transmission system for cable systems (DVB-C2). V.1.1.1, Oct 2010
7. ETSI TS 101 547-2: DVB plano-stereoscopic 3DTV; Part 2: frame compatible plano-stereoscopic 3DTV. V1.2.1, Nov 2012
8. ISO/IEC 13818-1: Information technology—generic coding of moving pictures and associated audio—Part 1: Systems, July 2015
9. ETSI TS 101 154 (V1.8.1): Digital video broadcasting (DVB); specification for the use of video and audio coding in broadcasting applications based on the MPEG-2 transport stream, July 2007
10. Stott, J.H.: The how and why of COFDM. EBU Techn. Rev. **278** (1998)
11. Dumic, E., Sisul, G., Grgic, S.: Evaluation of transmission channel models based on simulations and measurements in real channels. Frequenz **66**(1–2), 41–54 (2012)

12. Reimers, U: Digital Video Broadcasting. Springer, Berlin (2001)
13. http://beam.to/datasets
14. DVB fact sheet: DVB-T2—2nd generation terrestrial broadcasting. DVB Project Office, Apr 2012
15. Richardson, T., Urbanke, R.: The renaissance of Gallager's low-density parity-check codes. IEEE Commun. Mag. **41**, 126–131 (2003)
16. ETSI EN 302 307-2: Digital video broadcasting (DVB); second generation framing structure, channel coding and modulation systems for broadcasting, interactive services, news gathering and other broadband satellite applications: Part 2: DVB-S2 extensions (DVB-S2X). V1.1.1, Mar 2014
17. ETSI TS 101 547-1: DVB plano-stereoscopic 3DTV; Part 1: Overview of the multipart, V1.2.1, Dec 2015
18. ETSI TS 101 547-2: DVB plano-stereoscopic 3DTV; Part 2: frame compatible plano-stereoscopic 3DTV, V1.2.1, Nov 2012
19. ETSI TS 101 547-3: DVB plano-stereoscopic 3DTV; Part 3: HDTV service compatible plano-stereoscopic 3DTV, V1.1.1, Nov 2012
20. ETSI TS 101 547-4: DVB plano-stereoscopic 3DTV; Part 4: service frame compatible planostereoscopic 3DTV for HEVC coded services. V1.1.1, June 2016
21. ISO/IEC 23002-3: Information technology—MPEG video technologies—Part 3: Representation of auxiliary video and supplemental information (2007)
22. Hellge, C., Wiegand, T., Torre, E.G., Gomez-Barquero, D., Schierl, T.: Efficient HDTV and 3DTV services over DVB-T2 using multiple PLPs with layered media. IEEE Commun. Mag. **51**(10), 76–82 (2013). https://doi.org/10.1109/MCOM.2013.6619569
23. Gurler, C.G., Gorkemli, B., Saygili, G., Tekalp, A.M.: Flexible transport of 3-D video over networks. Proc. IEEE **99**(4), 694–707 (2011). https://doi.org/10.1109/JPROC.2010.2100010
24. Bugdayci, D., Akar, G.B., Gotchev, A.: Optimized transmission of 3D video over DVB-H channel. In: IEEE Consumer Communications and Networking Conference (CCNC), Las Vegas, NV, pp. 20–24 (2012). https://doi.org/10.1109/ccnc.2012.6181056
25. Buğdaya, D., Bid, M.O., Aksay, A., Demirtaş, M., Akar, G.B.: Video + depth based 3D video broadcast over DVB-H. In: 2010 IEEE 18th Signal Processing and Communications Applications Conference, Diyarbakir, 2010, pp. 902–905. https://doi.org/10.1109/siu.2010.5652310
26. Hossen, M.S., Kim, S.H., Kim, K.D.: Stereoscopic video transmission over DVB-T2 system using future extension frame. IEEE Trans. Broadcast. **62**(4), 817–825 (2016). https://doi.org/10.1109/TBC.2016.2590831
27. Kim, S.-H., Lee, J., Jeong, S., Choi, J., Kim, J.: Development of fixed and mobile hybrid 3DTV for next generation terrestrial DTV. In: 2013 3DTV Vision Beyond Depth (3DTV-CON), Aberdeen, pp. 1–2 (2013)
28. Ekmekcioglu, E., Gurler, G., Kondoz, A., Tekalp, A.M.: Adaptive multi-view video delivery using hybrid networking. IEEE Trans. Circ. Syst. Video Technol. **27**(6), 1313–1325 (2017)
29. Lykourgiotis, A., et al.: Hybrid broadcast and broadband networks convergence for immersive TV applications. IEEE Wirel. Commun. **21**(3), 62–69 (2014). https://doi.org/10.1109/MWC.2014.6845050
30. Yun, K., Cheong, W., Kim, K.: A synchronization and T-STD model for 3D video distribution and consumption over hybrid network. IEICE Trans. Inf. Syst. **98**(10), 1884–1887 (2015), Oct 2015. https://doi.org/10.1587/transinf.2014edl8242
31. Olaizola, I.G., Pérez, J., Zorrilla, M., Martín, Á., Laka, M.: Reference model for hybrid broadcast web 3D TV. In: Proceedings of the 18th International Conference on 3D Web Technology (Web3D '13). ACM, New York, NY, USA, pp. 177–180, June 2013. https://doi.org/10.1145/2466533.2466560
32. ETSI TS 102 796 v1.4.1: Hybrid broadcast broadband TV, Aug 2016
33. https://www.w3.org/TR/2016/REC-html51–20161101/. Accessed 30 Sept 2017

34. Lee, J., Yun, K., Kim, K.: A 3DTV broadcasting scheme for high-quality stereoscopic content over a hybrid network. IEEE Trans. Broadcast. **59**(2), 281–289 (2013). https://doi.org/10.1109/TBC.2013.2256678
35. Tekalp, A.M.: Digital Video Processing. Prentice Hall (2015)
36. Ozcinar, C., Ekmekcioglu, E., Kondoz, A.: Quality-aware adaptive delivery of multi-view video. In: IEEE International Conference on Acoustics, Speech and Signal Processing (ICASSP), pp. 1–6, Mar 2016
37. Park, G., Lee, J., Lee, G., Kim, K.: Efficient 3D adaptive HTTP streaming scheme over internet TV. In: IEEE International Symposium on Broadband Multimedia Systems and Broadcasting (BMSB), pp. 1–6, June 2012
38. Su, T., Javadtalab, A., Yassine, A., Shirmohammadi, S.: A DASH-based 3D multi-view video rate control system. In: 2014 8th International Conference on Signal Processing and Communication Systems (ICSPCS), pp. 1–6, Dec 2014
39. Gouache, S., Bichot, G., Bsila, A., Howson, C.: Distributed & adaptive HTTP streaming. In: IEEE International Conference on Multimedia and Expo (ICME), pp. 1–6, July 2011
40. Gürler, C.G., Tekalp, M.: Peer-to-peer system design for adaptive 3D video streaming. IEEE Commun. Mag. **51**(5), 108–114 (2013)
41. Lua, E.K., Crowcroft, J., Pias, M., Sharma, R., Lim, S.: A survey and comparison of peer-to-peer overlay network schemes. IEEE Commun. Surv. Tutor. **7**(2), 72–93 (2005)
42. Hudzia, B., Kechadi, M.-T., Ottewill, A.: Treep: a tree based p2p network architecture. In: Cluster Computing, 2005. IEEE International, pp. 1–15 (2005)
43. Magharei, N., Rejaie, R., Guo, Y.: Mesh or multiple-tree: a comparative study of live p2p streaming approaches. In: INFOCOM 2007. 26th IEEE International Conference on Computer Communications. IEEE, pp. 1424–1432 (2007)
44. Sen, S., Spatscheck, O., Wang, D.: Accurate, scalable in-network identification of p2p traffic using application signatures. In: Proceedings of the 13th International Conference on World Wide Web, pp. 512–521 (2004)
45. Galuba, W., Girdzijauskas, S.: Distributed hash table. In: Encyclopedia of Database Systems. Springer, Berlin, pp. 903–904 (2009)
46. B. Inc; BitTorrent. Available: http://www.bittorrent.com/. Accessed 11 June 2017
47. Vlavianos, A., Iliofotou, M., Faloutsos, M.: BiToS: enhancing BitTorrent for supporting streaming applications. In: INFOCOM 2006. 25th IEEE International Conference on Computer Communications. Proceedings, pp. 1–6 (2006)
48. Paris, J.-F., Shah, P.: Peer-to-peer multimedia streaming using BitTorrent. In: Performance, Computing, and Communications Conference, 2007. IPCCC 2007. IEEE International, pp. 340–347 (2007)
49. Liu, Z., et al.: H. 264/MVC interleaving for real-time multiview video streaming. J. Real-Time Image Process **10**(3), 501–511 (2015)
50. Wang, P., Wu, L., Aslam, B., Zou, C.C.: A systematic study on peer-to-peer botnets. In: Proceedings of 18th International Conference on Computer Communications and Networks, 2009. ICCCN 2009, pp. 1–8 (2009)
51. Bracciale, L., Piccolo, F.L., Luzzi, D., Salsano, S., Bianchi, G., Blefari-Melazzi, N.: A push-based scheduling algorithm for large scale P2P live streaming. In: Telecommunication Networking Workshop on QoS in Multiservice IP Networks, 2008. IT-NEWS 2008. 4th International, pp. 1–7 (2008)
52. Zhang, M., Zhang, Q., Sun, L., Yang, S.: Understanding the power of pull-based streaming protocol: can we do better? IEEE J. Sel. Areas Commun. **25**(9) (2007)
53. SVC extension of H.264/AVC—Fraunhofer Heinrich Hertz Institute. Available: https://www.hhi.fraunhofer.de/en/departments/vca/research-groups/image-video-coding/research-topics/svc-extension-of-h264avc.html. Accessed 11 June 2017
54. Gurler, C.G., Savas, S.S., Tekalp, A.M.: Quality of experience aware adaptation strategies for multi-view video over p2p networks. In: 19th IEEE International Conference on Image Processing (ICIP), pp. 2289–2292 (2012)

55. Fehn, C.: Depth-image-based rendering (DIBR), compression, and transmission for a new approach on 3D-TV. In: Electronic Imaging 2004, pp. 93–104 (2004)
56. Ahlgren A, Dannewitz C, Imbrenda C, Kutscher D, Ohlman B, "A survey of information-centric networking", *IEEE Commun. Mag.*, vol. 50, no. 7, 2012
57. Fayazbakhsh, S.K., et al.: Less pain, most of the gain: incrementally deployable icn. ACM SIGCOMM Comput. Commun. Rev. **43**, 147–158 (2013)
58. Jacobson, V., Smetters, D.K., Thornton, J.D., Plass, M.F., Briggs, N.H, Braynard, R.L.: Networking named content. In: Proceedings of the 5th International Conference on Emerging Networking Experiments and Technologies pp. 1–12 (2009)
59. Zhang, F., Zhang, Y., Reznik, A., Liu, H., Qian, C., Xu, C.: A transport protocol for content-centric networking with explicit congestion control. In: 2014 23rd International Conference on Computer Communication and Networks (ICCCN), pp. 1–8 (2014)
60. Zhou, J., Wu, Q., Li, Z., Kaafar, M.A., Xie, G.: A proactive transport mechanism with explicit congestion notification for NDN. In: IEEE International Conference on Communications (ICC), pp. 5242–5247 (2015)
61. Carofiglio, G., Gallo, M., Muscariello, L., Papalini, M., Wang, S.: Optimal multipath congestion control and request forwarding in information-centric networks. In: 21st IEEE International Conference on Network Protocols (ICNP), pp. 1–10 (2013)
62. Kulik, I., Trinh, T.: Investigation of quality of experience for 3D video in wireless network environment. In: 18th European Conference on Information and Communications Technologies (EUNICE), Aug 2012, Budapest, Hungary. Lecture Notes in Computer Science, LNCS, vol. 7479, pp. 340–349, Information and Communication Technologies. Springer, Berlin (2012)
63. Politis, I., Lykourgiotis, A., Dagiuklas, T.: A framework for QoE-aware 3D video streaming optimisation over wireless networks. Mobile Inf. Syst. **2016**, Article ID 4913216, 18 p (2016
64. Ho, D., Kim, H., Kim, W., Park, Y., Chang, K., Lee, H., Song, H.: Mobile cloud-based interactive 3D rendering and streaming system over heterogeneous wireless networks. IEEE Trans. Cir. Sys. Video Technol. **27**(1), 95–109 (2017)
65. Nasir, S., Hewage, C.T.E.R., Mrak, M., Worrall, S., Kondoz, A.M.: Depth based object prioritisation for 3D video communication over Wireless LAN. In: 2009 16th IEEE International Conference on Image Processing (ICIP), Cairo, pp. 4269–4272 (2009)
66. Cuervo, E., Wolman, A., Cox, L.P., Lebeck, K., Razeen, A., Saroiu, S., Musuvathi, M.: Kahawai: high-quality mobile gaming using GPU offload. In: Proceedings of the 13th Annual International Conference on Mobile Systems, Applications, and Services (MobiSys '15). ACM, New York, NY, USA, pp. 121–135 (2015)
67. Oztas, B., Pourazad, M.T., Nasiopoulos, P., Leung, V.C.M.: Adaptive 3D-HEVC video streaming over congested networks through layer prioritization. In: 24th International Conference on Telecommunications (ICT), Limassol, pp. 1–5 (2017). https://doi.org/10.1109/ict.2017.7998258
68. Chen, Z., Zhang, X., Xu, Y., Xiong, J., Zhu, Y., Wang, X.: MuVi: multi-view video aware transmission over MIMO wireless systems. IEEE Trans. Multim **99**, 1–1. https://doi.org/10.1109/tmm.2017.2713414
69. Liu, Z., Cheung, G., Chakareski, J., Ji, Y.: Multiple description coding and recovery of free viewpoint video for wireless multi-path streaming. IEEE J. Sel. Top. Signal Process. **9**(1), 151–164 (2015). https://doi.org/10.1109/JSTSP.2014.2330332
70. Lin, C.H., Yang, D.N., Lee, J.T., Liao, W.: Efficient error-resilient multicasting for multi-view 3D videos in wireless network. In: IEEE Global Communications Conference (GLOBECOM), Washington, DC, pp. 1–7 (2016). https://doi.org/10.1109/glocom.2016.7841779
71. Yoon, Y., Kim, M., Lee, B., Go, K.: Temporal synchronization scheme in live 3D video streaming over IEEE 802.11 wireless networks. In: Proceeding of IEEE International Symposium on a World of Wireless, Mobile and Multimedia Networks 2014, Sydney, NSW, pp. 1–7 (2014). https://doi.org/10.1109/wowmom.2014.6918944

Chapter 9
3D Video Tools

Emil Dumic, Khaled Boussetta, Luis A. da Silva Cruz,
Tasos Dagiuklas, Antonio Liotta, Ilias Politis, Yuansong Qiao,
A. Murat Tekalp, Maria Torres Vega and Yuhang Ye

Abstract This chapter presents an overview of different tools used in research and engineering of 3D video delivery systems. These include software tools for 3D video compression and streaming, 3D video players, and their interfaces. Other types of tools widely used in research studies and development of new networking solutions, such as network simulators, emulators, testbeds, and network analysis tools are also covered. In addition, several 3D video evaluation tools, which have been specifically designed for testing and evaluation of 3D video sequences subject to network impairments, are further described. The chapter also presents several examples of recent works that have been carried out based on one or more

E. Dumic (✉)
Department of Electrical Engineering, University North, Varaždin 42000, Croatia
e-mail: emil.dumic@gmail.com

K. Boussetta
L2TI, University of Paris 13, 99, Avenue J-B Clement, 93430 Villetaneuse, France
e-mail: Khaled.Boussetta@univ-paris13.fr

L. A. da Silva Cruz
Department of Electrical and Computer Engineering, Instituto de Telecomunicações,
University of Coimbra, Pólo II, 3030-290 Coimbra, Portugal
e-mail: lcruz@deec.uc.pt

T. Dagiuklas
Department of Computer Science, London South Bank University, London, UK
e-mail: tdagiuklas@lsbu.ac.uk

A. Liotta
Data Science Research Centre, University of Derby, Quaker Way,
Derby DE1 3HD, UK
e-mail: a.liotta@derby.ac.uk

I. Politis
Department of Electrical and Computer Engineering,
University of Patras, Patras, Greece
e-mail: ipolitis@ece.upatras.gr

Y. Qiao · Y. Ye
Software Research Institute, Athlone Institute of Technology, Athlone, Ireland
e-mail: ysqiao@research.ait.ie

P. A. Assunção and A. Gotchev (eds.), *3D Visual Content Creation, Coding and Delivery*, Signals and Communication Technology,
https://doi.org/10.1007/978-3-319-77842-6_9

simulation, emulation, test, and/or evaluation tools in research studies or innovative solutions for relevant problems affecting 3D multimedia delivery.

9.1 Introduction

Advances in 3D video capturing, compression, transmission, and display technologies have enabled the use of 3D media services in different sectors. Although rapid improvements in IP-based networks have led to capacity increases, it is critical that network operators and content providers optimize the user experience of 3D video content. By using and combining several simulation/emulation tools, it can be observed how different factors will influence subjective quality. The delivery and distribution of 3D content to large numbers of subscribers include several key factors ranging from 3D multimedia creation parameters (3D video representation format, used coding technology), networking conditions (transmission standards, quality of service) to end-user's device processing (autostereoscopic, display with shutter/passive glasses, etc.). This chapter examines different aspects of 3D video tools: Software tools for 3D video compression; Streamers and 3D video players; Network simulators, emulators, testbeds, and network analysis tools; 3D video evaluation tools. Overview of existing simulation tools that can be used for this purpose will be given. Also, some of the newly developed tools will be explained in more detail.

In Sect. 9.2, available software tools for video compression are presented. After a brief introduction to the H.264/AVC and HEVC video coding standards emphasizing their 3D video encoding extensions, the FFmpeg multimedia framework is described in the final subsection. Section 9.3 addresses streamers and 3D video players by describing some of the existing Open-Source video streamers and players (OpenSVC Decoder, 3D Open SVC Decoder Extensions, and VLC player). In Sect. 9.4, several network simulators, emulators, testbeds, and network analysis tools are presented. These include simulation tools (DVB-T, DVB-T2, NS-2, NS-3, and ndnSIM), emulators (Network Link Emulators, Virtual Network Emulators, Omnet++ and MANE), testbeds (PlanetLab and Network-Impairing Multimedia Testbed), and network analysis tools (Wireshark). Section 9.5 describes several 3D video evaluation tools, which have been specifically designed for testing and evaluation of 3D

Y. Ye
e-mail: yye@research.ait.ie

A. M. Tekalp
Koç University, Istanbul, Turkey
e-mail: mtekalp@ku.edu.tr

M. Torres Vega
IDLab - imec, Ghent University, Zwijnaarde-technologiepark 15, 9052 Ghent, Belgium
e-mail: maria.torresvega@ugent.be

video sequences, but they can also be easily adapted for 2D video. Such tools comprise the generator of degradations in 3D SBS video sequences, Crowd3D and 3D MOS using DSCQS method. Finally, Sect. 9.6 concludes the chapter.

9.2 Software Tools for 3D Video Compression

This section provides a brief introduction to the H.264/AVC and the HEVC video coding standards emphasizing their 3D video encoding extensions. The FFmpeg multimedia framework is introduced in the final subsection. In all the cases, useful references are provided for interested readers to get more detailed information about these standards and available media processing tools.

9.2.1 *H.264 and 3D Extensions*

In May 2003, the joint video team (JVT), an initiative of the ITU-T Video Coding Experts Group (VCEG) and of the ISO/IEC JTC1 Moving Picture Experts Group (MPEG), the final version of the H.264/AVC video coding standard was approved. The standard was developed mostly during the years from 2001 through 2003, but the founding ideas and call for key technologies go back as far as 1998, to the H.261L proposals. After the final version of the H.264/AVC standard was approved, additional work was continued or started on the development of several extensions to address the need for H.264/AVC-compatible scalable and 3D video coding, among others. The core coding engine of the H.264/AVC and extensions differs from previous generations of coders mostly on the following aspects:

- adoption of two-level structure with coding and transmission functionalities separated into a video coding layer (VCL) and a network abstraction layer (NAL),
- adoption of smaller sized integer transforms that approximate the discrete cosine transform (DCT)
- adoption of smaller size and asymmetrical inter-frame prediction partitions for motion compensation and prediction residue computation,
- adoption of multi-frame inter prediction with the possibility of forward and backward references,
- use of quarter-pel accuracy interpolation in motion-compensated prediction step,
- introduction of an intra-frame prediction mechanism with support for directional prediction,
- adoption of an in-loop deblocking filter to reduce the visibility of block-edge artificial contours and improve coding efficiency,

- introduction of adaptive variable length coding (CAVLC) and binary arithmetic (CABAC) entropy encoders designed for the syntactical elements defined for use in the coder,
- use of several features like slices and other frame partitioning arrangements designed to increase the robustness of encoded bitstream to transmission impairments

This short list is not exhaustive and leaves out other also important innovations such as improved higher level signaling through supplemental enhancement information (SEI) and frame numbering schemes. More detailed information can be found in [1, 2].

In terms of encoding performance, the H.264/AVC standard represented a big leap over the major predecessor, MPEG-2, with a reduction in bitrate of about 50%, depending on the type and characteristics of the source signal.

Of great relevance to the tasks involving evaluations of 3D video transmission are, of course, the H.264/AVC extensions for coding multi-view video as well as stereoscopic video [3]. The multi-view video coding (MVC) extension of H.264/AVC handles the multiple video views by using one of them (indicated via encoder configuration information) as a base view, which is then encoded independently of the remaining views. The frames of the other views are encoded supplementing the usual time-domain prediction with inter-view prediction, where the reconstructed frames from the base view and (depending on the coding structure) the other views are used as targets for the view prediction procedure. The base-view bitstream is formatted according to the H.264/AVC syntax for single-view video and segmented into base-view-specific NAL units. The bits representing other views are packed into NAL units specific for this use. It is important to note that the MVC extension does not allow predicting a frame based on frames of other views occurring at different time instants. Besides MVC-specific NAL formats, this extension introduces new SEI messages which together with the MVC encoding functionalities allow random and view-specific access to encoded data as well as easy view switching and base view only decoding for compatibility with single-view decoders. Concerning encoding performance, for the case of two views, published results indicate that using MVC enables bitrate reductions of from 20 to 50% when compared to simulcast encoding [4].

Coding of stereoscopic video using H.264/AVC can be done using MVC or using the stereo-specific frame-compatible mode, where the left and right views are composited into a larger frame which is then encoded using regular H.264/AVC single-view encoding tools and an enhanced SEI message set. The new SEI messages indicate which frame-compatible format from the allowed options, «Top-bottom», «Side-by-side», «Row interleaved», «Column interleaved», or «Checkerboard», is in use. Other than the different SEI message, all the stereo video encoding proceeds as if it were a single-view video with twice the number of pixels.

9.2.2 HEVC and 3D Extensions

The successor of H.264/AVC, the HEVC [5] video encoding standard was approved almost 10 years to the day after H.264/AVC, on April 2013 after roughly 3 years of development under the auspices of the joint coding team on video coding (JCT-VC) of the VCEG and MPEG standardization bodies. HEVC had as goals achieving higher coding efficiency for video contents of very high resolution where H.264/AVC was shown to perform below expectations. Reflecting the need for a standard able to handle ultra-high-definition resolutions (UHD) equal and higher than 3840 $\times$ 2560 (4K), a coding structure was adopted which is based on larger coding blocks up to 64 $\times$ 64 pixels in size together with quadtree partitioning structures. Other noteworthy differences relative to H.264/AVC are

- an increased gamut of directional prediction orientations for intra coding,
- advanced motion compensation through the use of motion vector prediction based on candidate lists and motion information merging,
- larger choice of partitions possibilities for inter-mode prediction,
- use of sample-adaptive-offset filtering to reduce banding effects,
- improved entropy coding of transformed and quantized residues and syntactical elements,
- introduction of tiling scheme to partition frames into large units for independent processing and transmission,
- inclusion of parallel-processing-oriented tools like wavefront parallel processing (WPP),
- support for larger bit depths.

In terms of encoding performance, HEVC achieved the notable result of reducing the bitrate necessary by H.264/AVC for the same quality encoding by about 40%. However, this improvement was met with a significant increase of computational complexity, of up to 500% as reported in [6]. Also, in the case of HEVC, the basic codec was and still is being complemented with several extensions [7] for range extension, scalable coding, screen content coding, high dynamic range content coding, multi-view video coding and 3D video coding, and others. Concerning multi-view and 3D video coding, apart from the differences in the basic encoder technology, HEVC differs from H.264/AVC in that depth information coding is supported with the possibility of prediction from texture and vice versa.

The HEVC extensions most relevant in the context of this text are surely the MV-HEVC for coding of multi-view video and the 3D-HEVC for coding of multi-view video plus depth (MVD). MV-HEVC is similar in concept to the H.264/AVC extension MVC but uses new tools such as Neighboring Block-Based Disparity Vector Derivation (NBDV), Inter-View Motion Prediction, Inter-View Residual Prediction, and Illumination Compensation. According to [7] compared to simulcast HEVC, i.e., independent view encoding, MV-HEVC achieves close to 30% bitrate savings for two-view encoding and nearly 40% for three-view encoding where in the latter case, in general, each dependent view is encoded with 60% less bits than if they were encoded independently.

The 3D-HEVC extension shares some encoding concepts with MV-HEVC but adds depth to the data processed. These depth data are not only encoded but also used to improve the compression of the video component through mechanisms like view synthesis prediction (VSP) and inheritance of motion information between video and depth.

Compared to simulcast encoding of video and depth using HEVC, 3D-HEVC achieves a bitrate reduction of about 41% for encoding 3 views. When compared to MV-HEVC a smaller advantage is observed for 3D-HEVC with a reduction of about 16%. Additionally, the encoding of depth in 3D-HEVC can be done by optimizing the tradeoff between synthesized quality and bitrate, a process called view synthesis optimization (VSO), bringing a performance improvement relative to MV-HEVC to about 30% bitrate reduction. As for the other side of the coin 3D-HEVC is around 18% more complex (run-time complexity) than MV-HEVC and 11% more complex that simulcast HEVC [7].

9.2.3 FFmpeg

FFmpeg is a software package for multimedia content processing, which was the brain child of Fabrice Bellard who started its development in 2000. The project was later continued by Michael Niedermayer. FFmpeg comprises a set of tools for manipulating multimedia that supports most existing video and audio formats and that can be used as standalone programs or integrated into other programs through the use of functions from the several libraries provided. The set of tools and functionalities is quite vast and its use can be customized through numerous options and interconnection schemes that allow the construction of complex filtering/processing chains [8]. A short not very detailed list of major characteristics and components of FFmpeg is as follows:

- Audio and Video support for many formats such as MPEG-2, H.264/AVC, HEVC, MP3, AAC, HE-AAC, JPEG, JPEG 2000, JPEG-LS, PNG.

  Several types of functions/tools:

- Built-in Encoders and Decoders and wrappers for common external encoders like x265.
- Input and Output Devices to read and write information from and into files and network pipes.
- Filters to operate on audio, video, and images.
- Muxers and Demuxers to combine and split streams.
- Bitstream filters to operate directly on bitstream, supporting direct conversion, bitstream chopping, and other.
- Video converters to, e.g., convert resolution, chroma spaces, and other.

- Audio processing tools.

Individual standalone tools

- Ffmpeg
- Ffplay
- Ffprobe
- Ffserver
- Libraries:
  - libavutil
  - libswscale
  - libswresample
  - libavcodec
  - libavformat
  - libavdevice
  - libavfilter
- Well-developed supporting website.

Among all the tools listed above, the most useful for simulation of transmission of 3D video are quite likely the Muxer and Demuxer tools, and the streaming able tool FFmpeg, and the player FFplay. All these tools are easily configurable through the use of numerous flags and parameters and can be interconnected with the input and output filters quite easily to build complex transmission chains that can include downsampling operations, joining of multiple streams, transcoding, and many other operations. Important parts are, of course, the encoder and decoder tools; FFmpeg has native support for some decoders like HEVC and encoders like MPEG-4 Visual. Many external encoders can be used through wrappers, as is the case with a second MPEG-4 encoder and the x265 encoder.

Besides the easy configurability of the tools included with FFmpeg other positive aspects worth mentioning are its good performance that can allow, e.g., real-time operation of processing chains involving transcoding, and the fact that it is licensed under the GNU Lesser General Public License (LGPL) and so can be a low-cost solution for research projects that require the use of multimedia processing and transmission tools. For more detailed information, interested readers should consult [8].

9.2.3.1 FFmpeg as a Research Tool

In recent years, several works have been carried out and papers have been published using FFmpeg as a research tool. First, FFmpeg can be used as a general encoding/transcoding/decoding tool. In [9], the authors describe the history of video encoding/decoding standards development. Through the software debug (using FFmpeg and sdl2.0 function library [10]), they realized the encoding/decoding and playing H264 and HEVC format video in the paper, and also realized the mutual format conversion between H264 and HEVC format. In [11], the authors discuss the implementation of transcoding queue after video upload on the web service side. They use FFmpeg as the video transcoding tool and also propose a combination of MySQL database transcoding queue approach which ultimately reduces the pressure of the server. The work presented in [12] describes an implementation of a software solution for processing WebVTT (Web Video Text Track) [13] subtitles during playback of HLS (HTTP Live Stream) [14] streams. The software solution is created by using the functionality of the FFmpeg library. The result of using such software solution on HLS streams is an overall transport stream containing video, audio, and subtitle components.

FFmpeg has been used in different papers related to network performance systems. For instance, in [15], the design and implementation of an FFmpeg-based decoding process and stream analysis system are described. The system supports real-time analysis of media data from a different type of transmission protocols, media container formats, and video/audio coding standards. In the work presented in [16], a new method to implement the synchronized transmission of FFmpeg multimedia data on Android platform is proposed. Experimental results have shown that the timestamp-based multimedia audio- and video-synchronized algorithm can effectively guarantee multimedia data synchronization on the embedded Android system. A framework to simulate a real-time video streaming over wireless channels is presented in [17]. The system is divided into several modules and is simulated with different tools. Dummynet (explained in Sect. 9.4.2.1) is used for the network simulator while FFmpeg is for coder/encoder and sender/receiver module. The simulation results show that the simulated system can estimate accurately the quality of the videos transmitted over wireless network. In [18], the authors propose a smart, easy-deployed remote monitor system based on Android system. The system is based on the open-source library live555, FFmpeg, and x264, thus it is extendable. In this system, the video is recorded using V4L2 interface [19] and encoded by x264. The live555 media server [20] then forwards the video stream over RTSP to the client. Finally, the client decodes the video and plays it.

FFmpeg can be also used as a tool to develop different degradation types for 2D/3D quality evaluation. For instance, in [21] a new 2D no-reference video quality measure is proposed to accurately predict the video quality when transmission distortions are introduced due to poor networking conditions. FFmpeg was used as encoder and decoder, while packet loss simulator simulated random packet losses. In the same work, the authors test 3D Video transmission quality with 29 different degradation types (mostly symmetrical), related to compression (H.264/AVC or MVC), resizing, and optional packet losses in transmission. FFmpeg has been used as H.264/AVC encoder. The work described in [22] consists of a research study on the subjective assessment of 3D video quality using a newly constructed 3D video database (3DVCL@FER). Generator of degraded 3D SBS video sequences (described in Sect. 9.5.1) has been used to develop degraded sequences. Some of the degradations were made using FFmpeg as encoder and (or) decoder. Another work [23] proposes a different approach, based on a non-reference (NR) objective model to predict the quality of lost frames in 3D video streams. The model is based only on header information from three different packet-layer levels: Network Abstraction Layer (NAL), Packetized Elementary Streams (PES), and Transport Stream (TS). FFmpeg was used to develop dataset, to encapsulate video stream into a TS stream. Results show that the proposed model is capable of estimating the SSIM quite accurately using only the lost frames estimated sizes. In [24], a comparison study is presented for different consumer 3D display technologies, by means of subjective assessment. Four 55-inch displays have been considered: one autostereoscopic display, one stereoscopic with polarized passive glasses, and two with active shutter glasses. FFmpeg has been used to convert original sequences to avi files containing uncompressed video with 24 fps without audio, to avoid the introduction of possible coding degradations and synchronization errors, and the influence of audio in the observers' QoE. Results show performance improvement of active shutter glasses technology, the high performance of the polarized glasses technology in terms of quality and comfort, and the need of improvement of the autostereoscopic displays to complement the visual comfort to reach a global high-quality visual experience.

In the research work described in [25], the authors focus on light-field 3D displays, outline typical use cases for such displays, analyze processing requirements for display-specific and display-independent light fields, and discuss how these map to MVC as the underlying 3D video compression method. The study also includes an overview of available MVC implementations, and the support these provide for multi-view 3D video. Although there have been several attempts toward integrating MVC into open-source H.264 codecs into FFmpeg, and x264 encoder, none of these patches were successfully integrated into the mainline development branch.

9.3 Streamers and 3D Video Players

Recently, several tools have been developed to stream and playback multi-view video content. This section focuses only on Open-Source video streamers and players.

9.3.1 OpenSVC Decoder

The OpenSVC Decoder [26, 27] is an H.264/Scalable Video Coding (SVC) decoding library created by the IETR/INSA of Rennes. It has been integrated into mplayer and The Core Pocket Media Player (TCPMP). SVC support temporal, spatial, and quality scalabilities. It encapsulates coded video data into Network Abstract Layer Units (NALU). The NALU header contains 3 fields to specify the spatial, temporal, and quality level of a layer representation, i.e., Dependency ID, Temporal ID, and Quality ID. According to the convention, the abbreviations for

Dependency, Temporal, and Quality are D, T, and Q respectively. The Dependency, Temporal, and Quality IDs of the base layer all equal to 0. Higher layers have larger IDs. OpenSVC Decoder combines Dependency ID and Quality ID into one variable *DQId*, i.e., *DQId* = (Dependency ID $\ll$ 4) + Quality ID). For example, when Dependency ID equals to 2 and Quality ID equals to 1, DQId is 33. The example below plays the SVC video *example_video.264* with the specified quality: Dependency ID = 2, Quality ID = 1, and Temporal ID = 2. During video playback, it is also possible to switch layers using hot-keys.

./mplayer -fps 25 -vo sdl -setlayer 33 -settemporalid 2 example_video.264

9.3.1.1 3D Open SVC Decoder Extensions

The "Open SVC Decoder" is an open-source decoder for scalable video streams, making it ideal for real-time decoding [27] of H.264/SVC video, i.e., the scalable extension of H.264/AVC standard. By default, Open SVC is capable of reading and decoding locally stored SVC .264 files to YUV file. The open-source code has been modified to support Stereo Video streaming in real time. Key features for 3D Video Streaming include the following [28]:

1. Real-Time Decoding of RTP packets from both views: Without loss of generality, we assume that the encoding parameter sets (SPS) are transmitted prior the RTP payload session using an out-of-band reliable and asynchronous control transmission protocol (e.g., TCP). The Client application establishes a TCP connection with the Streaming Server. Without loss of generality, a dedicated port is used for the RTP packets of each view. The Client receives RTP packets containing NALU units from two different ports in order to manage the left and the right view separately
2. Adaptive Jitter Buffer Management: An Adaptive Jitter buffer dynamically adapts its size in time interval taking into account the playback performance and the packets delivery behavior in order to maximize the size in case where all or more than the expected data packets are received and minimize it in case where less than the expected data packets are received
3. Stereo Synchronization: In order to synchronize the frames for both views, a Master–Slave technique has been used for the rendering process. Left view has been assigned as the Master guiding the right view as the Slave and the sequence number of the frames for both views remains equal
4. 3D Error Concealment techniques: Two different error concealment techniques have been tested to cope with the Frame Loss or Frame distortion: Frame Copy and *Switching 3D to 2D technique*

Figure 9.1 illustrates the key elements of the 3D Open SVC framework to support 3D Video Streaming. 3D Open SVC decoder has been integrated with a Stereo Player. The Stereo Player was created using "Microsoft DirectX SDK 28 June 2010" and it is used along with a Stereo Nvidia Graphics Card, a Stereo

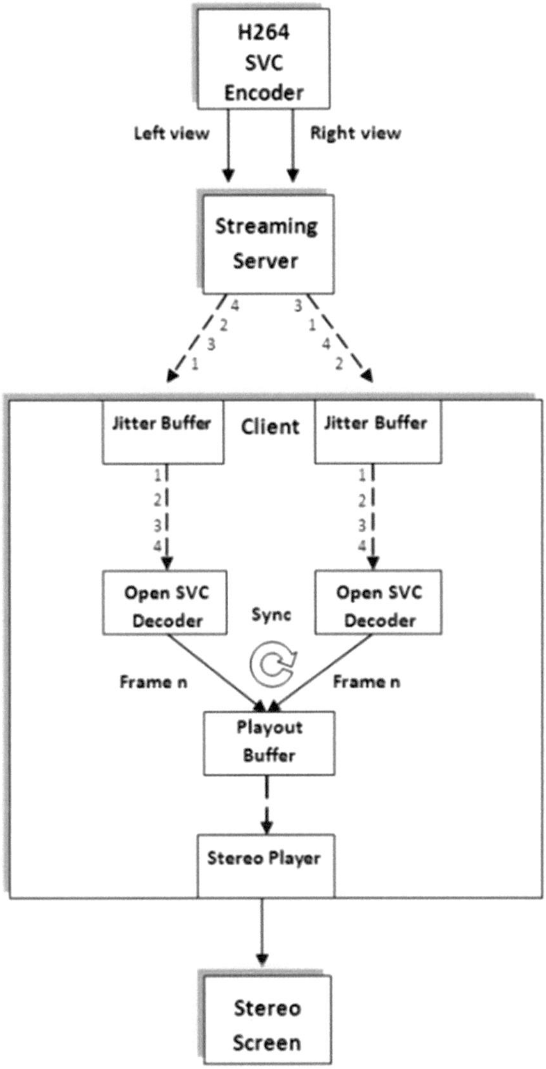

Fig. 9.1 Open SVC extensions to support 3D video streaming

Screen, and a pair of Active Stereo 3D Nvidia Glasses. The 3D Open SVC Decoder is freely available and can be downloaded from the following site: http://cones.eap.gr/?q=node/5.

In [29], the authors assess and evaluate stereo video quality within an end-to-end video communication platform in heterogeneous networking conditions using previously described enhanced Stereo Client. Several experiments have been done in a realistic networking environment with various packet loss rates and delay variation so that the robustness of the stereo client is assessed and evaluated. Through measurements, it is shown that video quality is more sensitive to delay variation than packet loss. In case of the lost frame in either view, switching from 3D to 2D is the best error resilience technique.

9.3.2 VLC Player

VLC is an open-source cross-platform media player that plays most multimedia files, DVDs, audio CDs, and various streaming protocols [30]. It has been developed by the VideoLAN project, which turned into a multi-national non-profit organization with developers from more than 20 nations. It comes with GNU General Public License (GPL), which allows anyone to edit and redistribute modified source code without copyright issues.

VLC supports many video and audio file formats and codecs, including the x264 codec that can encode/decode H.264/MPEG-4 AVC video in real time, since it includes the libavcodec library of the FFmpeg project. The inclusion of libavcodec allows the player to support DVD Video. VLC also supports some codecs that are not included in the FFmpeg's libavcodec. VLC has its own implementation of muxers and demuxers and hence does not include the FFmpeg libavformat library. It also has its own streaming protocol implementations. VLC version 2.1 and later has built-in support for viewing 3D videos using anaglyph filter for side-by-side (SBS) stereo format.

Several papers use VLC player as a tool to show 3D content. In [31], the authors propose the extended No-reference objective Video Quality Metric (eNVQM), an innovative metric for real-time 3D video quality assessment. eNVQM estimates the 3D video quality by taking as the input parameters network packet loss, video transmission bitrate, and frame rate. VLC player has been used at one end for encoding video streams into the H.264/MPEG-4 AVC format for stereoscopic 3D videos. At the other end, the 3D video stream is captured and decoded using VLC into sequence pairs of left and right views. During the transmission over the network, Dummynet [32] (emulator described in Sect. 9.4.2) is used to control the desired packet loss rate in the network. An algorithm for bandwidth-efficient 3D video streaming over a best-effort Internet was proposed in [33]. The authors use VLC video player for decoding, rendering, and display adaptation.

9.4 Network Simulators, Emulators, Testbeds and Network Analysis Tools

A simulator to be used in 3D video delivery systems is a software package implementing one or more models of communications modules for analysis and experimentation of effects that different system parameters may have on the real system based on such models. Sirannon is an example simulator that supports both

H.264/SVC and H.264/MVC, typically over RTP protocol [34]. An emulator is a software that replicates all functionalities of a hardware element or a system, which can be used as a substitute of the actual hardware element or system. For example, Media-Aware Proxy Element (MANE) [35] is an emulator for 3D video delivery. Testbed is a hardware implementation of a real system for test purposes. Network analysis tool can be used to analyze network protocols, examine security problems, learn protocol implementations, etc.

9.4.1 Simulators

This subsection describes different simulators that can be used for 3D video streaming.

9.4.1.1 Sirannon

Sirannon, formerly known as xStreamer, is a modular multimedia streamer and receiver [34]. The modularity is inspired by the Click Modular Router project and Direct Show filters. The user configures Sirannon by combining a collection of components in a workflow with each component performing basic functions such as reading video frames from a file, packetizing frames into smaller packets, or multiplexing video and audio into a transport stream. Figure 9.2 shows such the configuration interface.

Sirannon has been accepted as part of the reference toolchain, defined in the final test plan of the Video Quality Experts Group (VQEG) Hybrid Perceptual/Bitstream project, for streaming video sequences and simulating network impairments. An example use is shown in Fig. 9.2. It can be found in Sect. 9.5.1, Generator of degradations in 3D SBS video, degradation number 15. It simulates packet losses in H.264/AVC stream, generated with Gilbert–Elliot model (specific parameters for Gilbert–Elliot model: $\alpha = 0.01$, $\beta = 0.1$, $\gamma = 0.4$, $\delta = 0.01$). Sirannon scripts can be run from both command line or GUI in Python (from Unix), or alternatively, from Windows command line just as a script with different input and output parameters (such as input and output file name, different transmission, impairment parameters, etc.). However, source files can be also used to compile Sirannon on other platforms.

As another example, the authors in [36] propose error-resilient multi-view video transcoding algorithm. The proposed tool is configured to mitigate the error propagation resulting from channel conditions. Sirannon has been used as a tool to simulate packet losses on a wireless link.

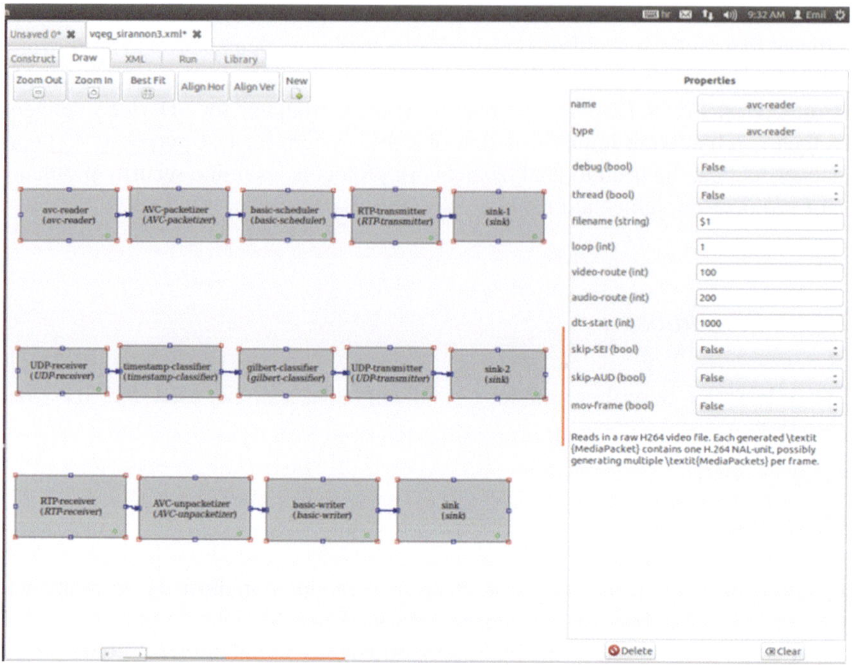

Fig. 9.2 Model of the script for H.264/AVC packet losses impairment using Gilbert–Elliot model

9.4.1.2 DVB-T Simulator in Simulink

This simulator covers main DVB-T processing blocks described in ETSI EN 300 744 [37]. It is made in Simulink, integrated in MATLAB. Simulink is an environment for multidomain simulation and Model-Based Design for dynamic and embedded systems. Simulation input is 188 bytes-long random binary sequence (as it should be similar to 188 transport stream bytes after energy dispersal). Simulation works in 8 k mode (which is used in almost all DVB-T systems). Different parameters can be chosen prior to opening Simulink model:

- modulation type ($\alpha = 1$): QPSK, 16-QAM or 64-QAM;
- FEC: 1/2, 2/3, 3/4, 5/6, 7/8;
- Guard interval: 1/4, 1/8, 1/16, 1/32.
- channel compensation: zero-forcing or minimum mean square error.

More details about the simulation can be found in [38], while simulation can be downloaded from [39]. Channel compensation can be made using channel compensation: "zero-forcing" or "minimum mean square error" methods. As its input is standard transport stream (random 188 bytes-long sequence), it could be adjusted used to degrade any 3D video packet in the transport stream, using different propagation models (e.g., Gaussian, Ricean).

9.4.1.3 DVB-T2 Simulator in MATLAB

DVB-T2 Common Simulation Platform (CSP) simulator [40] implemented in MATLAB covers DVB-T2 processing blocks described in ETSI EN 302 755 [41]. Its input is also 188 bytes-long random binary sequence, so it can be adjusted for any 3D video stream in .ts packets. Simple comparison between DVB-T and DVB-T2 systems are given in Fig. 9.3. Common parameters used in simulations for DVB-T and DVB-T2 networks are as follows:

- DVB-T: 64QAM, guard interval 1/4, FFT mode 8k; code rates: 1/2, 2/3, 3/4, 5/6, and 7/8
- DVB-T2: 256QAM (rotated), guard interval 19/256, FFT mode 32k extended, Pilot Pattern 4; code rates: 1/2, 3/5, 2/3, 3/4, 4/5, 5/6; other parameters are the same as defined in the verification and validation (V&V) reference model VV004-8KFFT

C/N values represent minimum values to obtain quasi-error-free reception, which means BER (Bit Error Rate) < $2 * 10^{-4}$ after internal decoder in DVB-T and BER < 10^{-7} after internal decoder in DVB-T2. Channel compensation was done using "ideal channel estimation" in both DVB-T and DVB-T2 simulators. Details can be found in [42, 43].

9.4.1.4 NS-2

NS-2 [44] (Network Simulator Version 2) is a discrete-event simulator for networking research. It supports a large number of networking protocols across different network layers and different network types, e.g., LAN and Satellite. NS-2

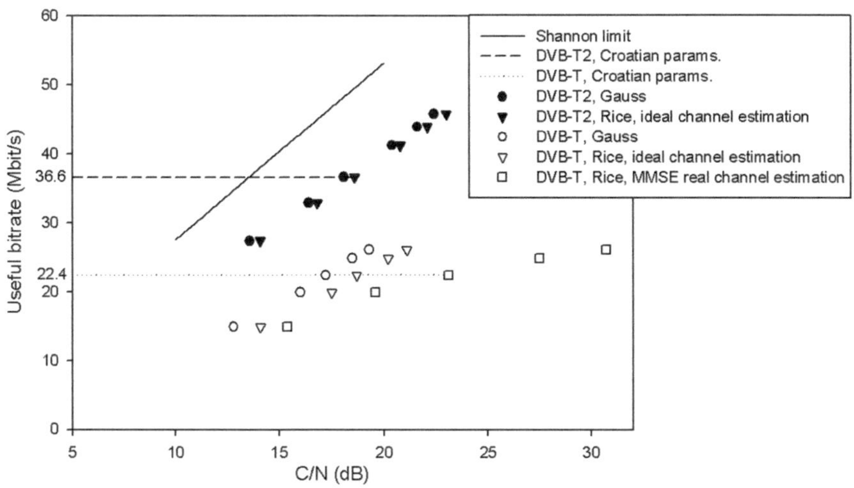

Fig. 9.3 Simulation results for DVB-T2 and DVB-T system with similar parameters

does not support the transmission of real application data. However, it is possible to simulate this feature through extensions. For example, Evalvid [45] connects video streaming data to NS-2.

The NS-2 software is written in both C++ and OTcl languages. There are two class hierarchies in the two languages with one-to-one mapping between the classes. The C++ code is for efficient execution of the simulation and the OTcl code is for easy configurations of the simulation setup The NS-2 software contains many modules. For simplicity, users can download the *allinone* package and run the *install* script under the *ns-allinone-<version>* folder to compile the software. Afterward, the environmental parameters need to be set up in the shell. Table 9.1

Table 9.1 Scripts to install NS-2 on Ubuntu-14.04

```
sudo apt-get update
sudo apt-get install build-essential autoconf automake libxmu-dev
sudo apt-get install gcc-4.4
cd ~/
tar xzvf ns-allinone-2.35

# Change void eraseAll() { erase(baseMap::begin(), baseMap::end()); }
# To
# void eraseAll() { this->erase(baseMap::begin(), baseMap::end()); }
sed -i 's/void eraseAll() { erase/void eraseAll() { this->erase/' ~/ns-allinone-2.35/ns-
2.35/linkstate/ls.h

# Change CC=                @CC@
# To
#                    CC=                gcc-4.4
sed –i 's/CC=\t\t@CC@/CC=\t\tgcc-4.4/' ~/ns-allinone-2.35/otcl-1.14/Makefile.in

cd ~/ns-allinone-2.35/
./install

# Add environmental parameters into .bashrc
cat >> ~/.bashrc <<EOF
export LD_LIBRARY_PATH=~/ns-allinone-<version>/otcl-1.14:/path-to/ns-allinone-2.35/lib
export TCL_LIBRARY=~/ns-allinone-2.35/tcl8.5.10/library
export PATH=~/ns-allinone-2.3.5/bin:$PATH
EOF

. ~/.bashrc
# Start NS-2 to test if the installation is correct.
ns
```

shows an example for setting up the latest NS-2 version (2.35) in Ubuntu-14.04. As the version was released in 2011, some modifications are required to successfully compile the software.

An NS-2 script normally contains the following parts: (1) Initialize the Simulator; (2) Setup the trace files; (3) Setup network topology and various settings for the network components; (4) Schedule the events during the simulation execution; (5) Start the simulation. An example of a NS-2 script is shown in Table 9.2. The script creates the topology in Fig. 9.4. The tracing results are shown in *ns.out*. The simulation process can also be visualized by running *nam* on the trace file *nam.out*.

In [33], an algorithm for bandwidth-efficient 3D video streaming over a best-effort Internet is proposed. The proposed algorithm offers a continuous streaming and achieves a high rendering quality, despite the variations of available bandwidth common to best-effort networks. The performance is evaluated using realistic simulations of Internet transmission conditions, including the impact of competing Internet traffic and real-world protocol implementations, NS-2 simulator.

9.4.1.5 NS-3

NS-3 is a cross-platform open-source network simulator for discrete-event simulations [46] that has been widely used in protocol development and topology optimization. It is necessary to remind that NS-3 is not an upgraded version of the NS-2, but is totally a brand-new simulation software. In NS-3, the simulation scripts are written in C++ or Python. Besides the NetAnim (Network Animator), it supports the Python binding to visualize the network topology and control the simulation.

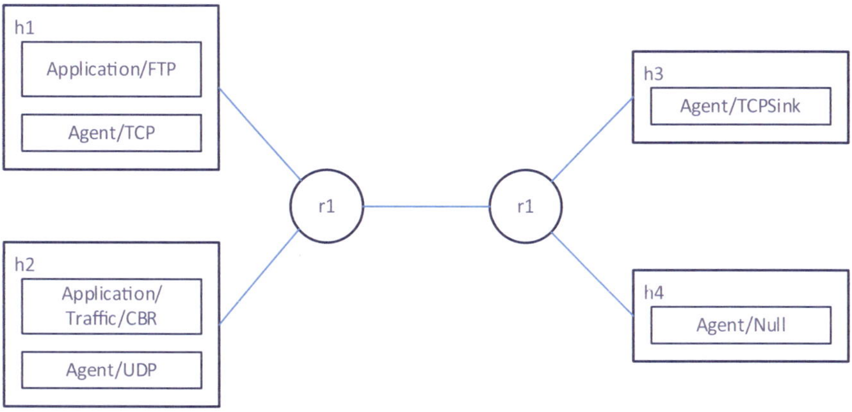

Fig. 9.4 Example topology

Table 9.2 NS-2 example script

```
#Create a simulator object
set ns [new Simulator]

#Open the NAM trace file
set namlog [open nam.out w]
$ns namtrace-all $namlog
#Open the NS trace file
set nslog [open ns.out w]
$ns trace-all $nslog

#Define a 'finish' procedure
proc finish {} {
    global ns namlog
    $ns flush-trace
    #Close the NAM trace file
    close $namlog
    #Run NAM using the trace file
    exec nam out.nam &
    exit 0
}

#Create four nodes
set h1 [$ns node]
set h2 [$ns node]
set r1 [$ns node]
set r2 [$ns node]
set h3 [$ns node]
set h4 [$ns node]

#Create links between the nodes
$ns duplex-link $h1 $r1 2Mb 10ms DropTail
$ns duplex-link $h2 $r1 2Mb 10ms DropTail
$ns duplex-link $r1 $r2 2Mb 100ms DropTail
$ns duplex-link $h3 $r2 2Mb 10ms DropTail
$ns duplex-link $h4 $r2 2Mb 10ms DropTail

#Setup a TCP connection
set tcp [new Agent/TCP]
$ns attach-agent $h1 $tcp
set sink [new Agent/TCPSink]
$ns attach-agent $h3 $sink
$ns connect $tcp $sink
```

Table 9.2 (continued)

```
#Setup a FTP over TCP connection
set ftp [new Application/FTP]
$ftp attach-agent $tcp
$ftp set type_ FTP

#Setup a UDP connection
set udp [new Agent/UDP]
$ns attach-agent $h2 $udp
set null [new Agent/Null]
$ns attach-agent $h4 $null
$ns connect $udp $null

#Setup a CBR over UDP connection
set cbr [new Application/Traffic/CBR]
$cbr attach-agent $udp
$cbr set packet_size_ 1000
$cbr set rate_ 1mb
$cbr set type_ CBR

#Schedule events for the CBR and FTP agents
$ns at 0.1 "$cbr start"
$ns at 1.0 "$ftp start"
$ns at 4.0 "$ftp stop"
$ns at 4.5 "$cbr stop"

#Call the finish procedure after 5 seconds of simulation time
$ns at 5.0 "finish"

#Run the simulation
$ns run
```

NS-3 provides a very friendly script interface to allow users create arbitrary network topology. The real-world network stacks are abstracted into helpers in NS3, which can be installed on the virtual NS-3 devices. NS-3 provides plenty of modules which cover both the IP and some non-IP-based networks. Additionally, NS-3 provide very friendly module interface for users to create their own network modules and simulation applications. Thanks to the C++-based development, NS-3 is able to make the virtual network devices to interact with real-world applications and devices. Table 9.3 shows the available modules in NS-3.

Table 9.3 NS-3 modules

Name	Description
NetAnim	QT4-based software to visualize the network topology and monitor the packets flow between devices
Antenna	Class that provides an interface to model the radiation pattern of different types of antennas
AODV	Model based on the base specification of Ad hoc On-demand Distance Vector (AODV) protocol
BRITE	Representative Internet topology generator to generate the realistic Internet topologies
CSMA	Emulator of the carrier sense multiple access protocol for simple bus network encouraged by Ethernet
DSDV	Proactive, table-driven routing protocol for mobile ad hoc networks
DSR	Reactive routing protocol designed specifically for multi-hop wireless ad hoc networks
Internet models	Collection of popular Internet protocols which include IPv4, ARP, UDP, TCP, IPv6, neighbor discovery, etc.
LR-WPAN	Implementation of the low-rate wireless personal area networks specified by IEEE 802.15.4
LTE	Complete collection of the LTE radio protocols which include RRC, PDCP, RLC, MAC and PHY, and some EPC protocols
MPI	Parallel simulation tool using the standard Message Passing Interface (MPI)
OLSR	Implementation of the optimized link state routing protocol for dynamic MANET unicast
OpenFlow	Implementation of the OpenFlow switches referred as OFSID
PointToPoint	Emulator of a full duplex RS232 or RS422 link with NULL modem and no handshaking
6LoWPAN	Modules which implemented the IPv6 packet compression over IEEE 802.15.4
UAN	Enables to model a variety of underwater network scenarios
WAVE	Indented to build a systematic architecture for wireless-based vehicular communications
Wi-Fi	Collection of the protocols specified by IEEE 802.11 to enable the communication between Wi-Fi devices
Wi-Fi Mesh	Extension module based on the NS-3 Wi-Fi to support the mesh networking specified by IEEE 802.11.s
WiMAX	Collection of the protocols based on IEEE 802.16 for WiMAX communication

A network reputation-based stereoscopic 3D video quality enhancement scheme (NRQ-3D) is devised in [47] for heterogeneous networks. A network reputation module is proposed to report the network quality based on the quality of service-related parameters (i.e., throughput, signal strength, delay, and loss) and price aspects. The performance of the proposed NRQ-3D was evaluated using Network Simulator version 3.17.

9.4.1.6 ndnSIM

The ndnSIM simulator [48] is an extension of NS-3 to research the behavior of Named Data Network (NDN). NDN concept performs content routing based on the content name rather than using location information (like IP address) [49]. Specifically, NDN consumer retrieves content by sending Interests with unique names to the network. Any NDN producer can return the content if it owns it. The simulator has 3 major releases (version 1.0, 2.0, 2.1.). The release of version 1.0 reimplements basic NDN primitives as a network layer protocol to run on top of any link layer protocol. The reimplementation includes the NDN interfaces (e.g., physical face and application face), traffic control, content forwarding/caching strategies, and the content consumer/producer applications. The later releases (version 2.0 and 2.1) use the real code from the NDN forwarding daemon to allow more realistic simulations. However, they no longer support the nice features implemented in the first release. The ndnSIM simulators have implemented the basic functionality of NDN. Similar to NS-3, it allows the users to customize the NDN applications according to the requirements.

The work described in [50] shows an implementation of 360/virtual reality videoconferencing system implemented over NDN, including producing content, formatting into NDN format, transmitting over NDN network, managing the flow of interest/content requests, and displaying in a web browser so as to show 360° rotation and zoom in/out features.

9.4.2 *Emulators*

There are different types of network emulators that are suitable to emulate either network links with transmission impairments or complete virtual networks.

9.4.2.1 Network Link Emulators

Perhaps the most popular network link emulators [51] are NetEm [52] and Dummynet [32], since they are both production quality and are available embedded in operating systems. NetEm is built into Linux Traffic Control subsystem, and Dummynet is integrated with FreeBSD. Dummynet is also used as a traffic shaper in the popular EmuLab testbed.

Both NetEm and Dummynet use the same principle to emulate network link impairments, such as variable delay, loss, duplication, and reordering. They run on either a general-purpose computer (a Linux or FreeBSD bridge) or a dedicated emulation device in a local area network (LAN) to capture incoming or outgoing packets and alter the packet flow in a way that imitates the behavior of application traffic according to some network impairment parameters. The device or system incorporates a variety of network attributes including the round-trip time across the network (latency), the amount of available bandwidth, a given degree or range of packet loss, duplication of packets, reordering packets, and/or the severity of network jitter. Emulation over a LAN keeps the testing environment secure and allows full control over the topology. Dummynet can create multiple links by reinjecting traffic into the emulator multiple times to emulate small networks with multiple hops and routers.

For example, in [31] the authors propose the extended No-reference objective Video Quality Metric (eNVQM), metric for real-time 3D video quality assessment. They use Dummynet to control the desired packet loss rate in the network (simulated packet loss follows a uniform distribution). In [53], a framework for quality of experience-aware delivering of three-dimensional video across heterogeneous wireless networks is proposed. In this work, NetEm was used for emulating diverse networking conditions, variable delay, and loss in accordance with the described experimental scenarios.

9.4.2.2 Virtual Network Emulators

Virtual network emulators use one or more virtual machines (VM) to emulate larger networks with desired topology and hosts. An example of VM-based virtual network emulator is DieCast which supersedes ModelNet. However, VM size and overhead may limit scalability and reduce performance.

Container-based virtualization provides a scalable and efficient network emulation environment [54]. Mininet [55] uses Linux containers that allow running production quality soft switches (e.g., OvS [56]) and actual video streamer and player codes on virtual hosts. Mininet provides functional realism (the same functionality as real hardware switches and hosts), timing realism, and traffic realism (generating real traffic by running video streamers or software tools such as *iperf*). It also provides topology flexibility and easy replication of experimental results.

9.4.2.3 Omnet++

Omnet++ is an open-source discrete-event simulator [57]. A complete simulation model is based on several C++ modules. Models could contain several simple modules, possibly organized in complex components, which interact using message passing mechanism. In order to simplify the creation of realistic networks simulations, different independent contributors have created Omnet++ library, also called frameworks, which implement different network components. Among the most used ones, one can cite INET framework [58], which contains a large number of Internet protocols models. This library is composed of low layers as well as high layer protocols of the communication stack. Precisely, access technologies, like Wi-Fi or Ethernet, IPv4, and IPv6 network layer protocols, UDP, TCP, and different application protocols. Mobility is also supported, using, for instance, the

Veins framework [59]. MixiM [60], Castalia [61], and INETMANET [62] are alternative frameworks, which are specially adapted to simulate wireless networks, ranging from cellular technologies to multi-hop wireless networks or low-power embedded devices like Wireless Sensor Networks.

Using these frameworks, different simulation scenarios can be created for research, test, or validation. The description of the network structure, such as the network infrastructure, protocols, and topology, relies on a specific high-level language, named NED (NEtwork Description). The modules parameters that can be set in the NED language are in the omnetpp.ini file, which uses an XML-type language. A convenient way to set the simulation scenarios is to use an Eclipse-based IDE. The execution of a simulation scenario can be done in command-line or GUI environment. Difference tools are integrated within this GUI, such as 3D topology visualization, event tracing, and parameters monitoring and debugging tools. In addition to discrete-event simulations, Omnet++ has extensions to allow real-time emulations. Since some simulations are time consuming, it is also possible multi-core processor or Grids to run Omnet++ parallel-distributed simulations.

Omnet++ can be used to simulate either the transmission of 3D contents (e.g., stereoscopic images, or 3D audio/video file) as well as the 3D audio/video streaming. In the first case, setting a network simulation scenario (e.g., using wireless technologies) where the 3D content is transmitted using FTP or any application layer protocols, including overlay and P2P networks [63], is quite simple and does not require a lot of effort from the user. When using the real-time Omnet++ emulation mode it is possible to use an external 3D streamer, such as VideoLAN (VLC) that interacts with the Omnet++ simulator. Unfortunately, up to the writing of this text, the discrete-event simulations of 3D streaming is not as simple since the user has yet to develop and integrate into the Omnet++ engine a 3D streamer module to transmit 3D contents generated by any offline 3D encoder.

9.4.2.4 MANE

In general the Media-Aware Proxy Element (MANE) [35] is defined as a transparent user-space module, responsible for low-delay adaptation and filtering of scalable video streams. As such MANE can be utilized for on-the-fly video streaming adaptation by selectively omitting or forwarding video packets to the receiving video users. Specifically, MANE standard can be considered as either a middlebox or an application layer getaway capable of aggregating or thinning RTP streams by selectively dropping packets that have the less significant impact on the user's video experience.

MANE has been proposed as an intermediate system that is capable to receive and de-packetize RTP traffic in order to customize the encapsulated network abstraction layer units, according to client's and access network's requirements. Within the context of 3D video delivery systems, MANE's role is twofold. First, it can act as a central point of decision in order to overcome networking limitations imposed by firewalls and Network Address Translation (NAT) protocol that are extensively used in real life networks. Second, it receives and parses RTP streams and customizes the streaming according to the video client's requirements and network conditions, based on Adaptation Decision Taking Engine (ADTE). From a system's implementation perspective, MANE can be considered as a transparent proxy of the mobile client responsible to parse the packets that are destined for all mobile users over multiple ports. It can be designed to run in Linux kernel level, thus ensuring minimum impact on the end-to-end delay.

In Fig. 9.5, all three transmission modes of H.264/SVC [64] inherited by the H.264/MVC standard [65] are illustrated. It can be observed that MANE is a combination of the Single-Stream Transmission (SST) mode, where all MVC data are carried in a single point-to-point unicast RTP session and the Multi-Stream Transmission (MST) mode, where the MVC bitstream is transmitted over multiple RTP sessions and each one corresponds to one RTP stream.

In [53], the authors propose a framework for quality of experience-aware delivering of three-dimensional video across heterogeneous wireless networks. The proposed architecture combines a Media-Aware Proxy (application layer filter based on the Media-Aware Network Element (MANE) standard), an enhanced version of IEEE 802.21 protocol for monitoring key performance parameters from

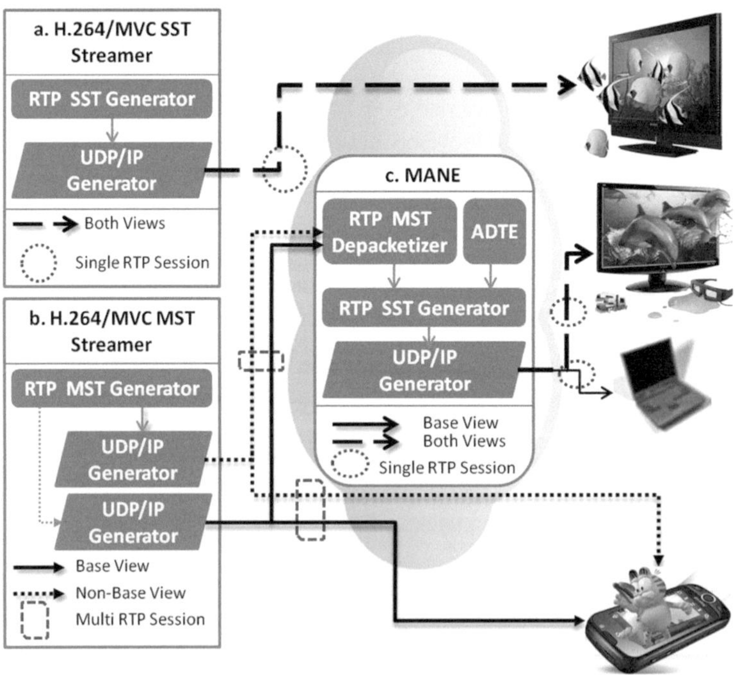

Fig. 9.5 MVC over RTP transmission modes and the role of MANE

different entities and multiple layers, and a QoE controller with a machine learning-based decision engine, capable of modeling the perceived video quality.

9.4.3 *Testbeds*

In this subsection, the PlanetLab testbed and Network-Impairing Multimedia Testbed are described.

9.4.3.1 PlanetLab

PlanetLab is a global experimental network for testing planetary-scale services. It consists of more than one thousand nodes hosted in sites worldwide [66]. A node is a dedicated server running the PlanetLab services. PlanetLab offers different memberships. For academic institutions, it is free to access the testbed. However, two dedicated servers are required as a contribution to the testbed. Each research project runs on a Planet *Slice*. From a user's perspective, a *Slice* is a set of virtual machines running on the nodes chosen by the user. Currently, the virtual machines run Fedora 14 (Linux-2.6.32).

PlanetLab defines three roles for each site:

(1) Principal Investigator (PI). Normally each site has one PI. The PI is responsible for managing users and slices at the site.
(2) Technical Contact (Tech Contact). Each site has at least one Tech Contact. The Tech Contact is responsible for installation, maintenance, and monitoring the nodes at the site.
(3) User. Users run their applications on PlanetLab.

To access PlanetLab services, the following step is required for a PlanetLab user:

(1) The user creates an account on the PlanetLab website (https://planet-lab.eu/db/persons/register.php) and selects a specific site (Fig. 9.6).
(2) The PI of the site activates the user's account, creates a Slice in PlanetLab and then adds the user to the Slice.
(3) The user logs in PlanetLab, uploads his public key, (Fig. 9.7) and then adds nodes into the Slice (Fig. 9.8).
(4) The user accesses the chosen nodes using SSH with the Slice name as the login name and the generated private key as the identity file.

A user manual is provided at https://planet-lab.eu/doc/guides/user.

An example of PlanetLab platform usage can be found in [67], where hybrid multi-view video content delivery solution has been tested. Several test scenarios were used, where each scenario is anticipated to evaluate certain aspects of the proposed content hybrid delivery solution.

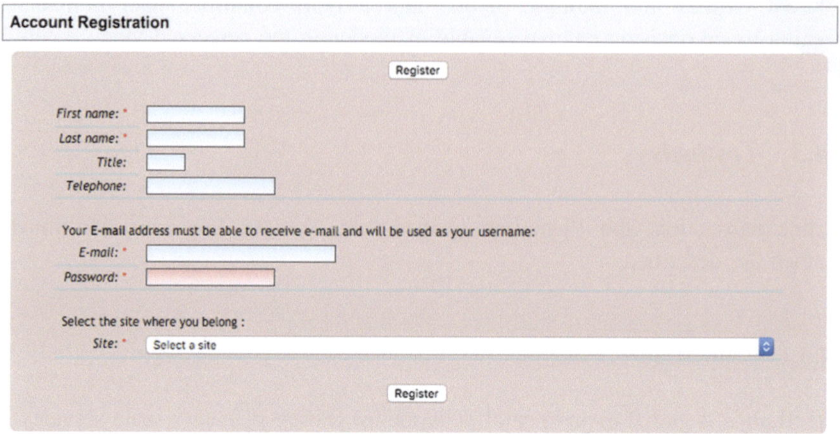

Fig. 9.6 User registration on PlanetLab

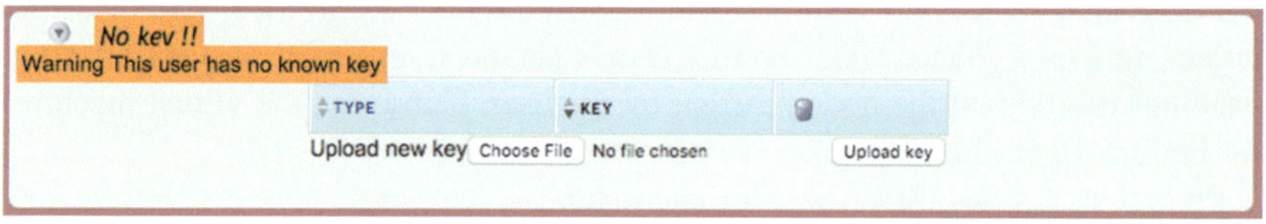

Fig. 9.7 Upload user's public key

9.4.3.2 Network-Impairing Multimedia Testbed

As the number of devices and the complexity of network increase, multimedia systems get affected in unexpected ways. In this case, understanding the effect of real network impairments in multimedia becomes crucial [68, 69] for network designers and service engineering. The purpose of this testbed is to understand the effect of network impairments on multimedia content.

Figure 9.9 shows the structure of the network-impairing multimedia testbed. A multimedia content server is connected to a client in an emulated network [70], in a wired or wireless manner depending on the network to be tested. Between server and client, a Hurricane II PacketStorm emulator [71] is installed. This emulator provides WAN emulation and network simulation. It emulates any network condition in a repeatable and controllable lab setting. To re-create any network condition, the emulator can impair IP and non-IP traffic, provide dynamic, time-varying impairments, and use a fully independent and flexible GUI architecture. Data generation provides network simulation without utilizing network resources or equipment. Application examples can be: Enterprises, Developers, Quality Assurance, Manufacturers, Carriers, etc. The flexibility provided by the network emulator and the content server and client make the Network-Impairing Multimedia

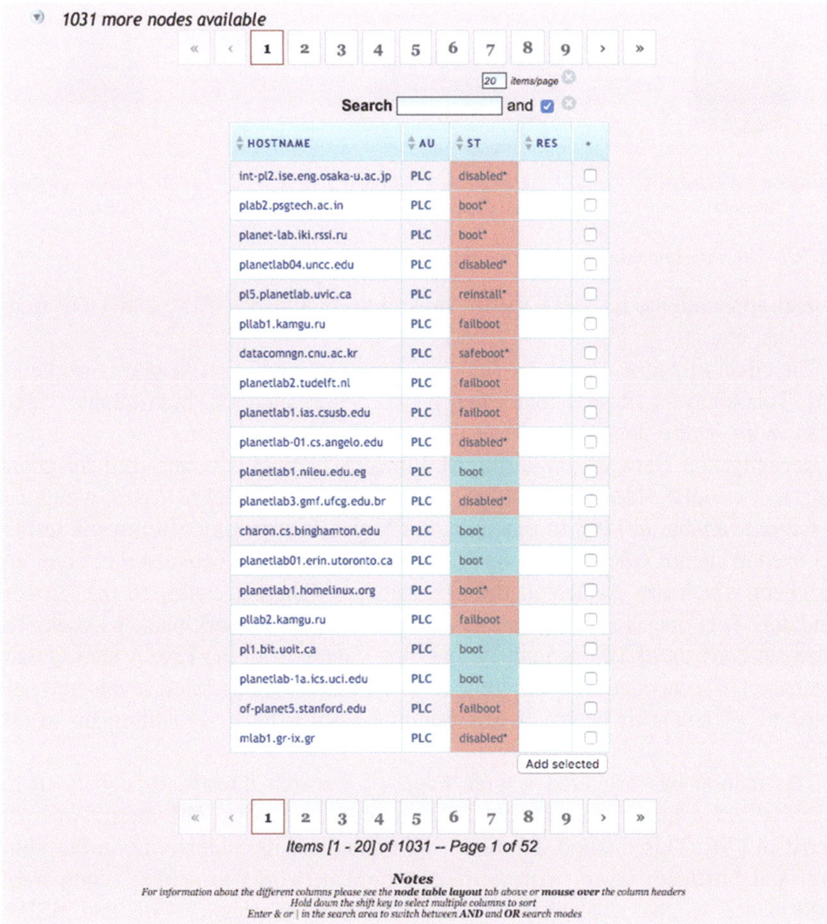

Fig. 9.8 Add nodes to slice

testbed suited to test an ample variety of multimedia applications and network conditions.

Adaptive streaming applications over standard Wi-Fi networks on light-weighted devices (Android mobile devices) were tested in [72]. In this experiment, a new adaptive streaming method based on online learning techniques was developed and its performance is assessed in the presence of wireless networks impaired with bandwidth, delays, or packet loss constraints (impairments generated by the testbed emulator).

In [73], the testbed was used in a Radio-over-Fiber network to study the effect of this type of networks together with impairments generated by the network emulator

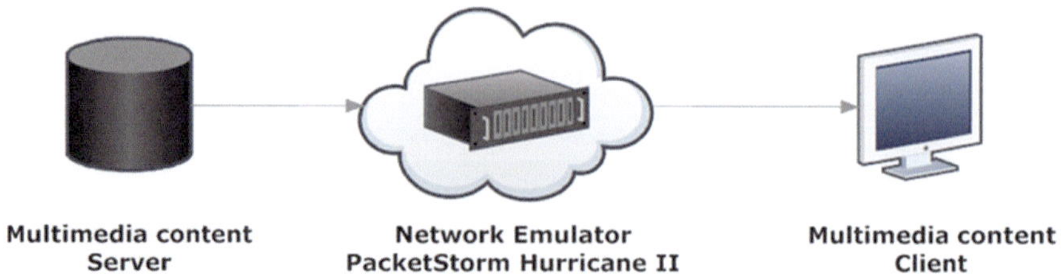

Fig. 9.9 Network-impairing multimedia testbed

on real applications, i.e., HD Video Streaming, high-speed FTP, and UDP transmissions, etc.

The effect of packet losses on video quality in streaming services was studied in [74]. This analysis showed that when packet losses increase, high-quality videos suffer more severe degradation than the lower quality ones.

Recently, the Network-Impairing Multimedia Testbed is being used for generating an extended video dataset (960 videos) affected by packet losses, which can be found available in [75]. In this case, the Network-Impairing Multimedia testbed was used to stream videos using the RTP protocol (FFmpeg) between the server and the client which are connected through standard Ethernet cables to the network emulator. This one is set to generate a wide range of network packet losses. The video set consists of 10 original, 25 fps, raw videos from the Live Video Quality Database [76], encoded at 8 different H.264/MPEG4 compression levels and subjected to 12 levels of packet losses (ranging from 0 to 10%) adding up to 960 videos.

This dataset has triggered a wide range of research threads, especially in the objective video quality assessment field. A predictive novel NR metric was presented in [70]. This method combines light-weighted measurements on the video pixel and bitstream layers with artificial neural networks to achieve comparable performance to the state-of-the-art full reference metrics (PSNR and SSIM). Finally, in [77, 78], by means of the full dataset, a study of the accuracy of an extended range of no-reference metrics was performed.

9.4.4 Network Analysis Tools

In this subsection, the well-known Wireshark network analysis tool is presented.

9.4.4.1 Wireshark

Wireshark is an open-source network packet analyzer [79]. It captures network packets and displays that packet data as detailed as possible. Some example usages can be

- to troubleshoot network problems
- to examine security problems
- to debug protocol implementations
- to learn network protocol internals

An example of the video stream from IPTV platform is shown in Tables 9.4, 9.5, 9.6, 9.7, and 9.8. Table 9.4 shows general information about analyzed frame packet. Table 9.5 shows Ethernet header, e.g., source and destination MAC addresses and type of encapsulated protocol (Internet Protocol, IP in this case). Table 9.6 shows IP header information: version, header length, source and destination IP address, encapsulated protocol—User Datagram Protocol (UDP) in this case, TTL (Time to Live), etc. Table 9.7 shows information for UDP protocol that is used (in this example) to transmit video PES: source and destination port number, length, and checksum. Finally, Table 9.8 shows an example for one packetized elementary stream (PES) data (there can be more PES packets per frame): PES header (PTS, DTS), PES extension, and PES data.

As another example, in paper [31], the authors propose the extended No-reference objective Video Quality Metric (eNVQM), metric for real-time 3D video quality assessment. They use Wireshark at the receiver side to monitor the stream and calculate the packet loss rate.

Table 9.4 Wireshark analysis for video delivery—frame description

Frame 2861: 1358 bytes on wire (10864 bits), 1358 bytes captured (10864 bits)
Encapsulation type: Ethernet (1)
Arrival Time: Feb 15, 2012 20:58:41.932,923,000 Central European Standard Time
Timeshift for this packet: 0.000000000 seconds
Epoch Time: 1329335921.932923000 seconds
Time delta from previously captured frame: 0.002715000 seconds
Time delta from the previous displayed frame: 0.002715000 seconds
Time since reference or first frame: 39.763294000 seconds
Frame Number: 2861
Frame Length: 1358 bytes (10864 bits)
Capture Length: 1358 bytes (10864 bits)
Frame is marked: False
Frame is ignored: False
Protocols in frame: eth:ethertype:ip:udp:mp2t:mpeg-pes:mpeg-pes
Number of per-protocol-data: 2
[ISO/IEC 13818-1, key 0]
[User Datagram Protocol, key 5]
Coloring Rule Name: UDP
Coloring Rule String: udp

Table 9.5 Wireshark analysis for video delivery—Ethernet

Ethernet II, Src: Cisco_e6:f2:00 (00:1b:0d:e6:f2:00), Dst: IPv4mcast_0a:01:04 (01:00:5e:0a:01:04)
Destination: IPv4mcast_0a:01:04 (01:00:5e:0a:01:04)
Source: Cisco_e6:f2:00 (00:1b:0d:e6:f2:00)
Type: IP (0x0800)

Table 9.6 Wireshark analysis for video delivery—IP layer

Internet Protocol Version 4, Src: 95.128.232.187 (95.128.232.187), Dst: 224.10.1.4 (224.10.1.4)
Version: 4
Header Length: 20 bytes
Differentiated Services Field: 0 × 88 (DSCP 0x22: Assured Forwarding 41; ECN: 0x00: Not-ECT (Not ECN-Capable Transport))
Total Length: 1344
Identification: 0x5fa7 (24487)
Flags: 0x00
Fragment offset: 0
Time to live: 60
Protocol: UDP (17)
Header checksum: 0xf033 [validation disabled]
Source: 95.128.232.187 (95.128.232.187)
Destination: 224.10.1.4 (224.10.1.4)
Source GeoIP: Unknown
Destination GeoIP: Unknown

Table 9.7 Wireshark analysis for video delivery—UDP

User Datagram Protocol, Src Port: 10444 (10444), Dst Port: 1234 (1234)
Source Port: 10444 (10444)
Destination Port: 1234 (1234)
Length: 1324
Checksum: 0xe59f [validation disabled]
Stream index: 6

9.5 3D Video Evaluation Tools

This subsection describes several 3D video evaluation tools, which have been specifically designed for 3D video sequences, although they can be easily adapted for 2D video. Those tools include: Generator of degradations in 3D SBS video (generates up to 22 different degradations in 3D side-by-side video), Crowd3D (php/JavaScript platform to subjectively test 3D MOS scores for quality, depth and comfort; either in laboratory environment or remotely), 3D MOS using DSCQS (tool for subjective evaluation of 3D MOS scores using double stimulus continuous quality scale in laboratory environment).

9.5.1 Generator of Degradations in 3D SBS Video Sequences

Generator of degraded 3D SBS video sequences is written in MATLAB, but it also uses many external programs. It is expected to run primarily on Windows x64 platform using a processor with at least 4 cores (for parallel processing), however

Table 9.8 Wireshark analysis for video delivery—packetized elementary stream

ISO/IEC 13818-1 PID=0x701 CC=2 Header: 0x47070112 0100 0111 .... = Sync Byte: Correct (0x00000047) .... 0... .... = Transport Error Indicator: 0 .... .0.. .... = Payload Unit Start Indicator: 0 ....0. .... = Transport Priority: 0 ....0 0111 0000 0001 .... = PID: Unknown (0x00000701) .... 00.. .... = Transport Scrambling Control: Not scrambled (0x00000000) ....01 .... = Adaptation Field Control: Payload only (0x00000001) .... 0010 = Continuity Counter: 2 MPEG2 PCR Analysis 11 Message fragments (1892 bytes): #2859(184), #2859(184), #2859(184), #2860(184), #2860(184), #2860(184), #2860(184), #2860(184), #2860(184), #2861(184), #2861(52) MPEG TS Packet (reassembled) **Packetized Elementary Stream** prefix: 000001 stream: video-stream (0xe0) PES extension length: 1886 1... .... must-be-one: True .0.. .... must-be-zero: False scrambling-control: not-scrambled (0) .... 0... priority: False .... .1.. data-alignment: True .... ..0. copyright: False ....0 original: False 1... .... pts-flag: True .1.. .... dts-flag: True ..0. .... escr-flag: False ...0 .... es-rate-flag: False .... 0... dsm-trick-mode-flag: False .... .0.. additional-copy-info-flag: False .... ..0. crc-flag: False ....0 extension-flag: False header-data-length: 12 PES header data: 370ed5f96b170ed5dd4bffff presentation time stamp (PTS): 36482.789566666 seconds decode time stamp (DTS): 36482.749566666 seconds PES data: 000000010950000001060101148000000 1019e13176a4df4... etc...

this can be adjusted in the program. A basic description can be found in Table 9.9, while the program itself can be downloaded from [39]. Input video sequences should be placed in the same folder, in uncompressed .avi format, left and right view separately. The input format has to be defined and the input frame size should be multiple of 8 (for HEVC-encoded sequences). The output is also in uncompressed .avi format, divided into folders according to the degradation type number in Table 9.9. In the end, the generator produces side-by-side 3D video sequences from previously produced degraded sequences, using x.264 and vp8 encoders from FFmpeg, with constant CRF factor (constant quality). Those sequences can be then played using "Crowd3d" application explained in the next subsection.

9.5.2 Crowd3D

The application "Crowd3D" is easily customizable and can be used with different web browsers. In the study reported in this section, it was set up to be used with Google Chrome and Mozilla Firefox web browsers. The application was programmed using the JavaScript and PHP languages and customized to display 3D video on computers equipped with a 3D monitor. The application collects and saves the subjective scores in a result's database. The application can be found on the links [89, 90], while source files for Crowd3D are in [39].

The start page of the application (shown in facsimile in Fig. 9.10) presents instructions about the testing procedure to the test subject. This GUI is also used to collect some information about the test subject such as age, gender, and e-mail address and some additional information about the monitor type. Several control mechanisms are implemented in the application to ensure the validity of the scores collected. The most important one is that allows the application to switch automatically to full screen during the whole duration of the assessment. If the test subjects exit the full-screen mode, then the test procedure automatically stops and the corresponding result is discarded. Only the results from the subjective assessment that run from start to finish in full-screen mode are flagged as valid, stored in the resulting database, and used in the final results analysis.

The application can be used in a laboratory environment and also used for pure web-based evaluation, which is planned for future developments. The application is developed to be used in two different setups: one is by using Chrome web browser and 3DTV with capability of manual switching to 3D mode [89], and the other is by using Firefox browser and 3D monitor with a Nvidia 3D vision system [90]. 3D content is first preloaded on the local machine, prior running the test. This was done by using Chrome or Firefox cache and it is very useful because this makes evaluation independent on the download speed (of the machine on which the test is being done). However, this makes the overall duration of the test at least twice as long, compared to the case in which preloading would not be used. Details of both setups can be found on the starting web pages [89, 90]. Preloading option could be also skipped, which is useful if 3D content is already stored on the machine from which the test is being done (e.g., in the laboratory), or if server and client machine are both on the high-speed network (preferably 1 Gbit/s).

The assessment of the subjective quality of the 3D videos using the system is based on Absolute Category Rating (ACR) with hidden reference (ACR-HR). In

Table 9.9 Degradation types in the generator of 3D video degradations

Number	Degradation	Detailed description	Tools and settings
1	2D view	Left + right view becomes left + left view	–
2	Resizing	4 × 4 down and up using lanczos3 filter	FFmpeg-x64 downloaded from [80]
3	Frame rate reduction	To 1/3 of the original fps	FFmpeg-x64 [80]
4	Brighten	y value + 15, right view only	FFmpeg-x64 [80]
5	Change gamma	To 0.6, right view only using	FFmpeg-x64 [80]
6	Horizontal disparity	Left view 30 pixels left, right view 30 pixels right	FFmpeg-x64 [80]
7	Horizontal disparity	Left view 30 pixels right, right view 30 pixels left using	FFmpeg-x64 [80]
8	Vertical disparity	Left view 20 pixels down, right view 20 pixels up	FFmpeg-x64 [80]
9	Geometric distortion	Left view only	stirmark [81]
10	2D to 3D conversion	2D to 3D conversion	FFmpeg-x64 [80] + AviSynth 2.6 downloaded from [82]; AviSynth script based on [83]
11	H.264/AVC coding	QP = 32 (both views): specific setting QP = 32	H.264/AVC reference encoder and decoder version 18.6, downloaded from [84]
12	H.264/AVC coding	QP = 44 (both views): specific setting QP = 44	H.264/AVC reference encoder and decoder version 18.6 [84]
13	H.264/AVC coding	Left QP = 32, right QP = 44: specific setting for QP parameter (asymmetric)	H.264/AVC reference encoder and decoder version 18.6 [84]
14	H.264/AVC coding	QP = 32 with edge enhancement, strength 75%	AviSynth 2.6 [82] and toon filter; toon filter can be downloaded from [85] H.264/AVC reference encoder version 18.6 [84]; FFmpeg-x32 downloaded from [80] (x264 decoder only)
15	Packet losses	Generated with Gilbert–Elliot model; specific parameters for Gilbert–Elliot model: $\alpha = 0.01$, $\beta = 0.1$, $\gamma = 0.4$, $\delta = 0.01$	H.264/AVC reference encoder version 18.6 [84] Sirannon software downloaded from [34] Sirannon script proposed in [86]

(continued)

Table 9.9 (continued)

Number	Degradation	Detailed description	Tools and settings
			Error concealment and x264 decoder using FFmpeg-x64 (-ec switch set to 2)
16	2D view, H.264/AVC coding	2D left view only, QP = 44	H.264/AVC reference encoder and decoder version 18.6 downloaded from [84]
17	jpeg2000 compression	Bitrate 2 Mbps	JPEG2000 kakadu software downloaded from [87]
18	Frame-freeze	2 s long, online streaming (degraded video is same duration as original)	FFmpeg-x64 [80] and AviSynth 2.6 [82]
19	Frame-freeze	2 s long, offline streaming (degraded video is longer than original)	FFmpeg-x64 [80] and AviSynth 2.6 [82]
20	3D to 2D switching	(Left + right becomes left + left view) back to 3D, 2 s long, degraded video is same duration as original	FFmpeg-x64 [80] and AviSynth 2.6 [82]
21	3D-HEVC encoding	QP = 32: General configuration settings are based on "baseCfg_2view.cfg" in the same software, with specific setting for QP factors	HTM reference encoder and decoder 3D-HEVC version 11.0, downloaded from [88]
22	3D-HEVC encoding	QP = 44: general configuration settings are based on "baseCfg_2view.cfg" in the same software, with specific setting for QP factors	HTM reference encoder and decoder 3D-HEVC version 11.0, downloaded from [88]

ACR-HR, each original unimpaired signal is included in the experiment but not identified as such. The ratings for the original signals are removed from the scores of the associated video sequences during data processing [91]. The quality grading is done in three different dimensions, each one graded on a continuous scale from 0 to 5 with a step of 0.1, according to the [92]. The three dimensions represent picture quality, depth quality, and visual comfort. For picture quality and depth quality grade 0 represents bad, while 5 represents excellent. For visual comfort, grade 0 represents extremely uncomfortable while 5 represents very comfortable. An example of 3D subjective assessment using Generator of degradations in 3D SBS video sequences and Crowd3D application can be found in [22, 93].

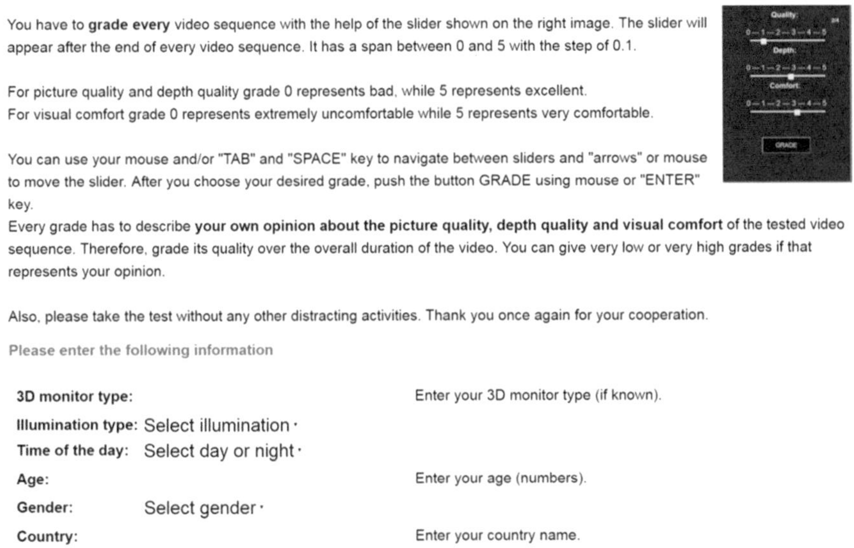
You have to **grade every** video sequence with the help of the slider shown on the right image. The slider will appear after the end of every video sequence. It has a span between 0 and 5 with the step of 0.1.

For picture quality and depth quality grade 0 represents bad, while 5 represents excellent.
For visual comfort grade 0 represents extremely uncomfortable while 5 represents very comfortable.

You can use your mouse and/or "TAB" and "SPACE" key to navigate between sliders and "arrows" or mouse to move the slider. After you choose your desired grade, push the button GRADE using mouse or "ENTER" key.
Every grade has to describe **your own opinion about the picture quality, depth quality and visual comfort** of the tested video sequence. Therefore, grade its quality over the overall duration of the video. You can give very low or very high grades if that represents your opinion.

Also, please take the test without any other distracting activities. Thank you once again for your cooperation.

Please enter the following information

3D monitor type:		Enter your 3D monitor type (if known).
Illumination type:	Select illumination	
Time of the day:	Select day or night	
Age:		Enter your age (numbers).
Gender:	Select gender	
Country:		Enter your country name.

Fig. 9.10 Start page of the application used for 3D subjective quality assessment

9.5.3 3D MOS Using DSCQS

Although many open and commercial 2D video subjective quality assessment tools exist in the community, a lower number of 3D video assessment tools have been recently proposed. In the scope of recent research studies, a new 3D and 2D video subjective quality assessment tool have been developed following the *ITU-R BT. 2021* recommendation. This tool implements the double-stimulus continuous quality scale (*DSCQS) variant I* assessment method [94]. The tool is available for free use to be downloaded from the following link: http://cones.eap.gr/?q=node/5.

The assessment consists of a number of sessions where two video sequences (signal A and B) are presented to a single observer. The assessor participating in the experiment is free to switch between the A and B signal until he/she has the mental measure of the quality associated with each signal. The projection period of signal A or B may be up to 10 s since experience suggests that longer period does not help to improve assessor's ability to measure the signal quality. Two of the video signals A and B can be the reference video sequence and the other the distorted video sequence. The characteristics/impairments are not announced or known to the assessor before the experiment. As such the assessor is not influenced and is able to perform a subjective assessment. The assessor is asked to grade both videos in terms of picture quality, depth quality, and visual discomfort. For picture and depth quality, the standard ITU continuous scale can be used with "Bad", "Poor", "Fair", "Good", and "Excellent" labels. For visual discomfort, the scale labels are replaced with "Uncomfortable", "Mildly Uncomfortable", "Comfortable", and "Very Comfortable", Fig. 9.11.

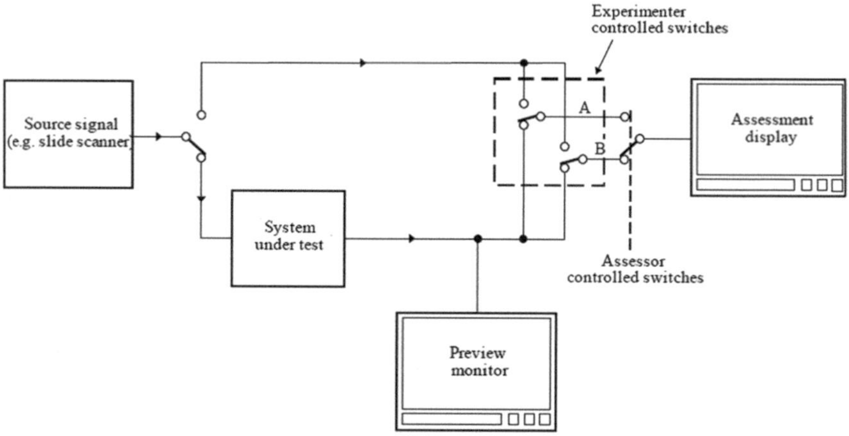

Fig. 9.11 General assessment for a test system for DSCQS method

To cope with diverse needs of different studies, the tool can render both the monoscopic version of the testing material (i.e., only the left view of the reference/ impaired sequence is presented to both eyes) as well as the stereoscopic version. This will give the opportunity to study the impact of the monoscopic parts of the material to the stereoscopic version. Moreover, the monoscopic ratings will be used as a baseline for comparing the stereoscopic material quality under the same impairments.

Technically, the tool is able to project in full-screen mode both 2D and 3D video sequences on the display by taking advantage of the *NVIDIA 3D* Vision graphics cards and API [95] and *Windows 7* and *DirectX* APIs [96]. The video sequences can be only in raw YUV format. For 3D video, the input must be the left and right part of the stereoscopic video sequences.

The experimenter can create test sessions by using test session description files. The configuration parameters are the projection type (i.e., stereoscopic or not), the input video files, characteristics of the video such as width, height, chroma sub-sampling, frame rate, and lastly the start and end frame of the test. A sample of the test case configuration is shown below.

```
Test          Objects1
Stereo        1
first_left    E:\OutLeft.yuv
first_right   E:\OutRight.yuv
second_left   E:\OutLeft_impaired.yuv
second_right  E:\OutRight_impaired.yuv
chroma        4:2:0
width         640
height        480
frame_rate    30
frame_offset  0
frame_end     300
```

On application start, the tool parses all the active configuration files and populates a dropdown box. The assessor is able to enter his/her username, expertise, and then choose and run any of the available test session. The initial tool window is shown in the following Fig. 9.12.

When a test session begins, the application reads the respective test configuration and starts the project. On stimuli projection, the application parses the input YUV traces and creates either a 2D picture or 3D picture for rendering. The sequence is rendered by the graphics card hardware and projected at the screen, Fig. 9.13.

The assessor is able to switch between the two stimuli by pressing "A" or "B" buttons. When the assessor has established its opinion on each stimulus he/she can end the test by pressing the "ESC" button. At the end of test session, a voting box is presented to the user in order to enter the picture quality for both stimuli as shown in Fig. 9.14. The score of both stimuli is logged in a file along with user and test details. The logs can be parsed later in order to gather all the scores of same tests. The assessor is free to cancel the voting session if this is needed.

For 3D viewing, the NVIDIA 3D Vision Kit includes a set of active shutter glasses and an infrared emitter. Active shutter glasses or else Liquid Crystal (LC) shutter glasses is a type of 3D glasses that uses liquid crystals in the lenses that momentarily shut off the viewing path from the eye to the screen. The emitter transmits infrared signals to the glasses, causing the left or right lens to open in synchronization with the stereo frame rendered at that moment in time. The system alternates 60 frames for the left eye with 60 frames for the right. At refresh rates of 120 frames per second (true 120 Hz), each eye sees 60 flicker-free frames per second. The advantage of LC shutter glasses is that they eliminate the ghosting perceived with polarized methods, and since colors are not being filtered out as in the anaglyph method, the viewer sees the full-color spectrum.

Fig. 9.12 Main window of the 3D assessment tool

Fig. 9.13 Sample picture of the 3D assessment tool during 3D rendering

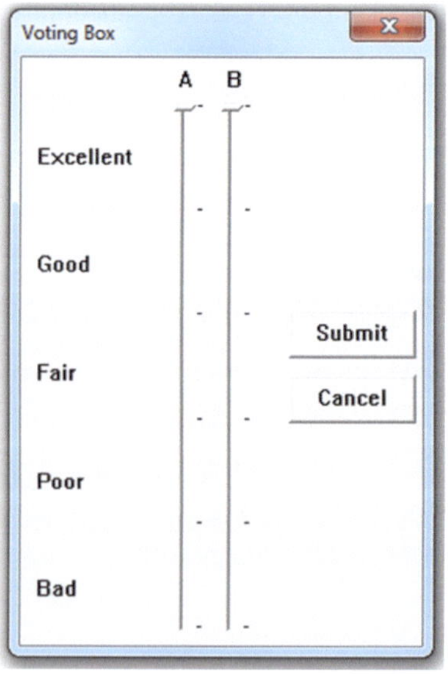

Fig. 9.14 Voting box of the 3D assessment tool

9.6 Conclusion

This chapter presented some of the existing and new 3D video tools, which find application in research and engineering of 3D media delivery systems. Several software tools integrating 3D video compression, streamers and 3D video players, network simulators, emulators, testbeds and network analysis tools and 3D video evaluation tools were described, including their main characteristics. Applications, recent works and research studies that used these tools in 3D video compression, transmission, displaying and quality evaluation were also described, highlighting their importance and usefulness in the development of new solutions and optimization of existing ones.

References

1. Wiegand, T., Sullivan, G.J., Bj ntegaard, G., Luthra, A.: Overview of the H.264/AVC video coding standard. IEEE Trans. Circuits Syst. Video Technol. **13**(7), 560–576 (2003)
2. Richardson, Ian: The H.264 Advanced Video Compression Standard, 2nd edn. Wiley, Hoboken (2010)
3. Vetro, A., Wiegand, T., Sullivan, G.J.: Overview of the stereo and multiview video coding extensions of the H.264/MPEG-4 AVC Standard. Proc. IEEE **99**(4), 626–642 (2011)
4. Stankowski, J., Domanski, M., Stankiewicz, O., Konieczny, J., Siast, J., Wegner, K.: Extensions of the HEVC technology for efficient multiview video coding. In: 2012 19th IEEE International Conference on Image Processing (ICIP), pp. 225–228, Oct 2012. https://doi.org/10.1109/icip.2012.6466836
5. Sullivan, G.J., Ohm, J.-R., Han, W.-J., Wiegand, T.: Overview of the high efficiency video coding (HEVC) standard. IEEE Trans. Circuits Syst. Video Technol. **22**(12), 1649–1668 (2012)
6. Correa, G., Assuncao, P., Agostini, L., da Silva Cruz, L.A.: Performance and computational complexity assessment of high-efficiency video encoders. IEEE Trans. Circuits Syst. Video Technol. **22**(12), 1899–1909 (2012)
7. Sullivan, G.J., Boyce, J.M., Chen, Y., Ohm, J.R., Segall, C.A., Vetro, A.: Standardized extensions of high efficiency video coding (HEVC). IEEE J. Sel. Top. Signal Process. **7**(6), 1001–1016 (2013)
8. FFmpeg website. www.FFmpeg.org
9. Zeng, H., Zhang, Z., Shi, L.: Research and implementation of video codec based on FFmpeg. In: International Conference on Network and Information Systems for Computers (ICNISC), Apr 2016, pp. 184–188. https://doi.org/10.1109/icnisc.2016.049
10. http://www.libsdl.org/
11. Xu, Y., Cao, S.: Design and implementation of a multi video transcoding queue based on MySQL and FFmpeg. In: 6th IEEE International Conference on Software Engineering and Service Science (ICSESS), Beijing, pp. 629–632 (2015). https://doi.org/10.1109/icsess.2015.7339136
12. Vulin, R., Samardžić, T., Simić, Đ., Kovačević, B.: One software solution for processing WebVTT subtitles during playback of HLS streams using FFmpeg libraries. In: 23rd Telecommunications Forum Telfor (TELFOR), Belgrade, pp. 760–763 (2015). https://doi.org/10.1109/telfor.2015.7377577
13. WebVTT: The web video text tracks format. https://www.w3.org/TR/webvtt1/
14. Apple HLS standard. https://developer.apple.com/streaming/

15. Lei, X., Jiang X, Wang, C.: Design and implementation of a real-time video stream analysis system based on FFmpeg. In: 2013 Fourth World Congress on Software Engineering, Hong Kong, pp. 212–216 (2013). https://doi.org/10.1109/wcse.2013.38
16. He, J., He, J.: Research on the synchronized transmission algorithm of embedded FFmpeg multimedia data. In: 2011 International Conference on Internet Technology and Applications, Wuhan, pp. 1–3 (2011). https://doi.org/10.1109/itap.2011.6006235
17. Tieu, D.T., Nguyen, B.C., Le, L.M., Pham, Q.M., Nguyen, Q.T., Vo, D.T.: Automatic test framework for video streaming quality assessment. In: 2015 2nd National Foundation for Science and Technology Development Conference on Information and Computer Science (NICS), Ho Chi Minh City, pp. 214–218 (2015). https://doi.org/10.1109/nics.2015.7302193
18. Ji, Q., Yu, H., Chen, H.: A smart Android based remote monitoring system. In: Third International Conference on Technological Advances in Electrical, Electronics and Computer Engineering (TAEECE), Beirut, pp. 181–184 (2015). https://doi.org/10.1109/taeece.2015.7113623
19. https://www.linuxtv.org/downloads/legacy/video4linux/API/V4L2_API/spec-single/v4l2.html
20. http://www.live555.com/mediaServer/
21. Brunnstrom, K., Sedano, I., Wang, K., Barkowsky, M., Kihl, M., Andren, B., Le Callet, P., Sjostrom, M., Aurelius, A.: 2D no-reference video quality model development and 3D video transmission quality. In: Sixth International Workshop on Video Processing and Quality Metrics for Consumer Electronics—VPQM 2012, Jan 2012, Scottsdale, Arizona, United States, pp. 1–6. https://doi.org/10.1007/s11042-015-3172-6
22. Dumic, E., Grgic, S., Sakic, K., Rocha, P.M.R., da Silva Cruz, L.A.: 3D video subjective quality: A new database and grade comparison study. Multimed Tools Appl **76**(2), 1–23 (2017). https://doi.org/10.1007/s11042-015-3172-6,2016
23. Feitor, B., Assuncao, P., Soares, J., Cruz, L., Marinheiro, R.: Objective quality prediction model for lost frames in 3D video over TS. In: IEEE International Conference on Communications Workshops (ICC), pp. 622–625, Hungary, June 2013. https://doi.org/10.1109/iccw.2013.6649308
24. Gutiérrez, J., Jaureguizar, F., Garcia, N.: Subjective comparison of consumer television technologies for 3D visualization. J. Display Technol. **11**, 967–974 (2015). https://doi.org/10.1109/JDT.2015.2448758
25. Kovacs, P.T., Nagy, Z., Barsi, A., Adhikarla, V.K., Bregovic, R.: Overview of the applicability of H.264/MVC for real-time light-field applications. In: 3DTV-Conference: The True Vision—Capture, Transmission and Display of 3D Video (3DTV-CON), pp. 1–4, July 2014. https://doi.org/10.1109/3dtv.2014.6874744
26. OpenSVC Decoder. http://sourceforge.net/projects/opensvcdecoder/
27. Blestel, M., Raulet, M: Open SVC decoder: a flexible SVC library. In: ACM Multimedia (2010)
28. Papadogiannopoulos, G., et al.: A stereo client using open SVC decoder extensions: QoE performance evaluation. In: 2014 IEEE International Conference on Image Processing (ICIP). IEEE, pp. 2482–2486, Oct 2014. https://doi.org/10.1109/icip.2014.7025502
29. Papadogiannopoulos, G., Dagiuklas, T., Politis, I., Lykourgiotis, A., Kotsopoulos, S.: Stereo video quality evaluation in heterogeneous networking conditions. In: International Conference on Telecommunications and Multimedia (TEMU), pp. 248–253, July 2014. https://doi.org/10.1109/temu.2014.6917769
30. VLC Media Player. http://www.videolan.org/vlc/
31. Han, Y., Yuan, Z., Muntean, G.-M.: An innovative no-reference metric for real-time 3D stereoscopic video quality assessment. IEEE Trans. Broadcast. **62**(3), 654–663 (2016). https://doi.org/10.1109/TBC.2016.2529294
32. Carbone, M., Rizzo, L.: Dummynet revisited. ACM SIGCOMM Comput. Commun. Rev. **40**(2), 12–20 (2010)
33. Petrovic, G., Efficient 3D video streaming. Eindhoven: Technische Universiteit Eindhoven (2013). https://doi.org/10.6100/ir754838

34. Sirannon. http://xstreamer.atlantis.ugent.be/
35. Gardikis, G., Pallis, E., Grafl, M.: Media-aware networks in future internet media. In: 3D Future Internet Media. Springer, New York, pp. 105–112, Oct 2013. https://doi.org/10.1007/978-1-4614-8373-1_7
36. Lawan, S., Sadka, A.H.: Robust multi-view video streaming through adaptive intra refresh video transcoding. Int. J. Eng. Technol. Innov. **5**(4), 209–219 (2015)
37. ETSI EN 300 744: Digital video broadcasting (DVB); framing structure, channel coding and modulation for digital terrestrial television. V.1.6.1, Sept 2008
38. Dumic, E., Sisul, G., Grgic, S.: Evaluation of transmission channel models based on simulations and measurements in real channels. Frequenz **66**(1–2), 41–54 (2012)
39. http://beam.to/datasets
40. DVB-T2: The common simulation platform, BBC R&D White Paper 196. Available at http://downloads.bbc.co.uk/rd/pubs/whp/whp-pdf-files/WHP196.pdf
41. ETSI EN 302 755: Digital video broadcasting (DVB); frame structure channel coding and modulation for a second generation digital terrestrial television broadcasting system (DVB-T2), V1.1.1, Sept 2009
42. Tralic, D., Dumic, E., Vukovic, J., Grgic, S.: Simulation and measurement of DVB-T2 system parameters. In: Proceedings of the 54th International Symposium ELMAR-2012, pp. 83–88, Sept 2012
43. Dumic, E., Grgic, S., Frank, D.: Simulating DVB-T to DVB-T2 migration opportunities in croatian TV broadcasting. In: Proceedings of 22nd International Conference on Software, Telecommunications and Computer Networks (SoftCOM 2014), Sept 2014
44. NS-2: http://www.isi.edu/nsnam/ns/
45. Evalvid: http://www.tkn.tu-berlin.de/menue/research/evalvid/
46. NS-3: https://www.nsnam.org/
47. Bi, T., Yuan, Z., Muntean, G.-M.: Network reputation-based stereoscopic 3D video delivery in heterogeneous networks. In: IEEE International Symposium on Broadband Multimedia Systems and Broadcasting (BMSB), pp. 1–7, June 2014. https://doi.org/10.1109/bmsb.2014.6873561
48. ndnSIM: http://ndnsim.net/2.1/index.html
49. Paul, A., Chilamkurti, N., Daniel, A., Rho, S.: Theory and application of vehicular networks. In: Intelligent Vehicular Networks and Communication: Fundamentals, Architectures and Solutions, Elsevier, New York (2016). https://doi.org/10.1016/b978-0-12-809266-8.00006-5
50. Zhang, L., Amin, S.O., Westphal, C.: VR video conferencing over named data networks. In: Proceedings of the Workshop on Virtual Reality and Augmented Reality Network VR/AR Network '17, pp. 7–12, Aug 2017. https://doi.org/10.1145/3097895.3097897
51. Nussbaum, L., Richard, O.: A comparative study of network link emulators. In: Proceedings of the Spring Simulation Multiconference, Article 85, Society for Computer Simulation International, San Diego, CA, USA (2009)
52. Hemminger, S.: Network emulation with NetEm. In: Proceedings of Australia's 6th National Linux Conference, Canberra, Australia, Apr 2005
53. Politis, I., Lykourgiotis, A., Dagiuklas, T.: A framework for QoE-aware 3D Video streaming optimisation over wireless networks. In: Mobile Information Systems, vol. 2016 (2016), Article ID 4913216, 18 p. https://doi.org/10.1155/2016/4913216
54. Handigol, N., Heller, B., Jeyakumar, V., Lantz, B., McKeown, N.: Reproducible network experiments using container-based emulation. CoNEXT 2012, 10–13 Dec 2012, Nice, France
55. http://mininet.org/
56. http://openvswitch.org/
57. https://omnetpp.org/
58. http://inet.omnetpp.org/
59. http://veins.car2x.org/
60. http://mixim.sf.net/
61. https://castalia.forge.nicta.com.au/index.php/en/
62. https://github.com/aarizaq/inetmanet-2.0

63. http://www.oversim.org/
64. Schwarz, H., Marpe, D., Wiegand, T.: Overview of the scalable video coding extension of the H.264/AVC standard. IEEE Trans. Circuits Syst. Video Technol. (Special Issue on Scalable Video Coding) **17**(9), 11031120 (2007)
65. Vetro, A., Pandit, P., Kimata, H., Smolic, A., Wang, Y.-K.: Joint draft 8 of multi-view video coding, Hannover, Germany, Joint Video Team (JVT) Doc. JVT-AB204, July 2008
66. PlanetLab Europe: http://www.planet-lab.eu
67. Ekmekcioglu, E., Gurler, G., Kondoz, A., Tekalp, A.M.: Adaptive multi-view video delivery using hybrid networking. IEEE Trans. Circ. Syst. Video Technol. **27**(6), 1313–1325 (2017)
68. Liotta, A.: The cognitive NET is coming. IEEE Spectr. **50**(8), 26–31 (2013)
69. Agboma, F., Liotta, A.: Quality of experience management in mobile content delivery systems. J. Telecommun. Syst. (special issue on the Quality of Experience issues in Multimedia Provision) **49**(1) (2012)
70. Torres Vega, M., Sguazzo, V., Mocanu, D.C., Liotta, A.: An experimental survey of no-reference video quality assessment methods. Int. J. Pervasive Comput. Commun. **12**(1)
71. http://packetstorm.com/packetstorm-products/hurricane-ii-software/
72. Torres Vega, M., Mocanu, D.C., Barresi, R., Fortino, G., Liotta, A.: Cognitive streaming on android devices. In: Proceedings of the IEEE/IFIP Symposium on Integrated Network and Service Management (IM'15), 11–15 May 2015, Ottawa, Canada, pp. 1316–1321. Piscataway: IEEE Service Center
73. Torres Vega, M., Zou, S., Mocanu, D.C., Tangdiongga, E., Koonen, A.M.J., Liotta, A.: End-to-end performance evaluation in high-speed wireless networks. In: Proceedings of the 2014 10th International Conference on Network and Service Management (CNSM), 17–21 Nov 2014, Rio de Janeiro, Brazil, pp. 344–347. Piscataway: IEEE Service Center
74. Mocanu, D.C., Liotta, A., Ricci, A., Torres Vega, M., Exarchakos, G.: When does lower bitrate give higher quality in modern video services? In: Proceedings of the IEEE/IFIP Network Operations and Management Symposium (NOMS'14) 5–9 May 2014, Krakow, Poland, pp. 1–5. IEEE, Piscataway
75. https://www.tue.nl/en/university/departments/electrical-engineering/research/research-groups/electro-optical-communications-eco/research/network-management-and-control/datasets/network-impaired-video-dataset/
76. Seshadrinathan, K., Soundararajan, R., Bovik, A.C., Cormack, L.K.: Study of subjective and objective quality assessment of video. Trans. Image Process. **19**(6), 1427–1441 (2010)
77. Torres Vega, M., Sguazzo, V., Mocanu, D.C., Liotta, A.: Accuracy of no-reference quality metrics in network-impaired video streams. In: 13th International Conference on Advances in Mobile Computing and Multimedia, 11–13 Dec 2015, ACM, Brussels, Belgium Brussels
78. Torres Vega, M., Giordano, E., Mocanu, D.C., Tjondronegoro, D., Liotta, A.: Cognitive no-reference video quality assessment for mobile streaming services. In: Proceedings of the 7th International Workshop on Quality of Multimedia Experience (QoMex), 26–29 May 2015, Pilos, Messinia, Greece, pp. 1–6. IEEE Service Center, Piscataway
79. https://www.wireshark.org/
80. http://FFmpeg.zeranoe.com/builds/; x64 and x32 versions from 11.12.2015
81. http://www.cl.cam.ac.uk/~mgk25/download/stirmark-1.0.tar.gz
82. http://sourceforge.net/projects/avisynth2/files/AviSynth%202.6/AviSynth%202.6.0/
83. Tiefenbehandlung, 2D-Videos in 3D abspielen: Dr. Volker Zota, Jan-Keno Janssen; c't Magazin für Computer Technik, Heise Verlag, 6, pp. 116 (2010)
84. http://iphome.hhi.de/suehring/tml/download/
85. http://avisynth.nl/index.php/Toon
86. VQEG 3DTV Group: Test plan for establishing a ground truth for quality of experience in 3D for assessment methodologies in 3D video quality assessment. GroTruQoE3D1, Draft Version 1.0 (2012)
87. http://www.kakadusoftware.com/index.php?option=com_content&task=view&id=26&Itemid=22
88. http://hevc.kw.bbc.co.uk/git/w/jctvc-3de.git/commit/d85c6b6e015c86c5c7c99ca9983304c14f8d9ad1
89. crowd3d.co.it.pt/suis3d

90. crowd3d.co.it.pt/suis3d_webm
91. ITU-T Recommendation P.910: Subjective video quality assessment methods for multimedia applications (2008)
92. ITU-R BT.2021: Subjective methods for the assessment of stereoscopic 3DTV systems. International Telecommunication Union/ITU Radiocommunication Sector (2012)
93. Mysirlidis, C., et al.: STESCAL3D: subjective evaluation of HD stereo video streaming using H.264 SVC in diverse laboratory environments. In: Quality of Multimedia Experience (QoMEX 2015), pp. 1–6 (2015)
94. http://www.itu.int/dms_pubrec/itu-r/rec/bt/R-REC-BT.2021-1-201502-I!!PDF-E.pdf
95. Internet http://www.nvidia.com/object/3d-vision-main.html
96. Internet http://msdn.microsoft.com/en-us/windows/desktop

Chapter 10
Quality of Experience and Quality of Service Metrics for 3D Content

Miguel Barreda-Ángeles, Federica Battisti, Giulia Boato, Marco Carli, Emil Dumic, Margrit Gelautz, Chaminda Hewage, Dragan Kukolj, Patrick Le-Callet, Antonio Liotta, Cecilia Pasquini, Alexandre Pereda-Baños, Christos Politis, Dragana Sandic, Murat Tekalp, María Torres-Vega and Vladimir Zlokolica

Abstract Traditionally, the quality of a multimedia system was mainly assessed through the evaluation of its Quality of Service (QoS) that is by evaluating system parameters such as bandwidth, latency, jitter, throughput, transmission delay, availability, etc. However, these metrics often failed to capture the actual end-user perceived quality, which has prompted the development of the construct of Quality of Experience (QoE), widely understood as an interaction of the technical features of multimedia systems with perceptual, and cognitive/emotional factors involved in the interpretation of those features by users. This chapter addresses the open issues

M. Barreda-Ángeles (✉) · A. Pereda-Baños
Eurecat, Barcelona, Spain
e-mail: miguel.barreda@eurecat.org

A. Pereda-Baños
e-mail: alexandre.pereda@eurecat.org

F. Battisti · M. Carli
University of Roma TRE, Rome, Italy
e-mail: federica.battisti@uniroma3.it

M. Carli
e-mail: marco.carli@uniroma3.it

G. Boato
University of Trento, Trento, Italy
e-mail: giulia.boato@unitn.it

E. Dumic
Department of Electrical Engineering, University North, Varaždin, Croatia
e-mail: dumic@gmail.com

M. Gelautz
Vienna University of Technology, Vienna, Austria
e-mail: margrit.gelautz@tuwien.ac.at

C. Hewage
Cardiff Metropolitan University, Cardiff, UK
e-mail: cthewage@gmail.com

P. A. Assunção and A. Gotchev (eds.), *3D Visual Content Creation, Coding and Delivery*, Signals and Communication Technology,
https://doi.org/10.1007/978-3-319-77842-6_10

in the field of QoS and QoE assessments. First, the perceptual characteristics of the multiview content are analyzed, and then a survey on the existing approaches for QoS and QoE estimation is performed. The analysis is then focused on the subjective aspects of QoE assessment, by describing the standard methodologies currently used and new trends based on human factors research. Finally, the chapter offers a few guidelines for future research directions in the field.

10.1 Introduction

In the last years, an increasing number of devices and complex networking infrastructures are directly affecting multimedia delivery systems [1]. In this situation, analyzing the network performance and its influence on the transmitted multimedia quality, and taking improvement measures becomes fundamental. Traditionally, quality has been measured in terms of Quality of Service (QoS) characteristics [2], which takes into account different aspects of the network services including bandwidth, latency, jitter, throughput, transmission delay, availability, etc. QoS should necessarily be considered in the new-generation networking technologies in order to support the service requirements of 3DTV over

D. Kukolj · D. Sandic · V. Zlokolica
Unviersity of Novi Sad, Novi Sad, Serbia
e-mail: Dragan.Kukolj@rt-rk.com

D. Sandic
e-mail: Dragana.Sandicj@rt-rk.com

V. Zlokolica
e-mail: vzlokolica@uns.ac.rs

P. Le-Callet
University of Nantes, Nantes, France
e-mail: patrick.lecallet@univ-nantes.fr

A. Liotta · M. Torres-Vega
Eindhoven University of Technology, Eindhoven, The Netherlands
e-mail: a.liotta@tue.nl

M. Torres-Vega
e-mail: M.Torres.Vega@tue.nl

C. Pasquini
University of Innsbruck, Innsbruck, Austria
e-mail: Cecilia.Pasquini@unitn.it

C. Politis
Kingston University London, London, UK
e-mail: C.Politis@kingston.ac.uk

M. Tekalp
Koç University, Istanbul, Turkey
e-mail: mtekalp@ku.edu.tr

IP. However, with the exponential growth of multimedia content both in terms of variety and requirements, the QoS analysis becomes insufficient, as it fails to capture the end-user perceived quality, i.e., his or her Quality of Experience (QoE) [3]. In this context, the expression QoE means "the general acceptance of an application or service" (ITU P.10/G.100), or to "the degree of delight or annoyance of the user of an application or service" [3]. QoE is described as a multidimensional process affected by influencing factors including technical and content features of systems, users' perception and cognitive processing, and context. Thus, in order to enable a comprehensive quality analysis of 3D multimedia systems, not only QoS parameters must be measured but also methods for assessing the actual end-user QoE are necessary to optimize the various technological elements comprising the delivery chain (e.g., see [3–6]) (Fig. 10.1).

Due to its subjective intrinsic nature, QoE is best assessed by means of subjective studies, i.e., using human ratings and opinion scores [7]. Depending on the application to be analyzed, the subjective study needs to be adapted to fit the needs or requirements of the application [8]. For example, measuring the end-user QoE of IPTV [9] (which requires high-resolution and quality video transmission) has to be studied in a different way than the one of a Skype call, in which the real-time bidirectional communication is at stake [10–13]. Although well aligned to human perception, subjective studies are costly, time-consuming, and prone to human bias [14]. They are fundamental to the various applications of Video Quality Assessment (VQA); yet, great effort has been directed toward mimicking subjective studies through completely automated processes and algorithms, as in objective QoE [15]. In this case, the main purpose of objective QoE metrics is to come up with an assessment that is correlated as much as possible with subjective studies and in this way to the Human Visual System (HVS) [16]. For this, Full-Reference metrics

Fig. 10.1 QoS and QoE analysis

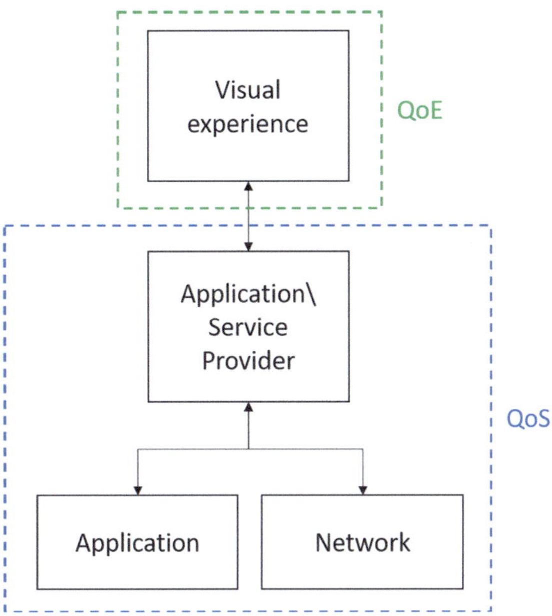

(FR) have proven to provide the best assessment [17, 18]. Algorithms such as Peak Signal-to-Noise ratio (PSNR) [19], Structural Similarity Index (SSIM) [20], or Video Quality Metric (VQM) [21], which realize a complete comparison between the original and the impaired materials, are prominent examples of these techniques. Specifically because of its high correlation with the HVS, VQM is broadly used as a valid alternative when subjective studies are not available [22]. However, the computational complexity required by these algorithms, in addition to the fact that, as FR metrics, they require both the original and the impaired material, makes them unsuited for deployments in environments where real time and low complexity are required, such as mobile [16]. In these cases, Reduced-Reference (RR) and No-Reference (NR) metrics are taking a predominant role, due to their lower requirements in terms of complexity and original material.

RR and NR metrics perform quality assessment based on features extracted from the received material (e.g., 3D video), the network and for RR metrics, also on features extracted from the original material and sent together with it on transmission. Such features provide a very fast or even real-time quality assessment while being able to be deployed in lightweight environments. Despite this fact, their accuracy in assessing video degradations and their correlation to subjective analysis are still open issues. This situation makes it particularly hard to automate the assessment of real-time streaming systems that have been subjected to network impairments [23]. A broad experimental survey on the accuracy of NR metrics to assess QoE of video delivery over networks was presented in [16, 23]. This study, applied to a 960 videos database [24] obtained from 10 original raw videos of the Live Video Quality Database [25], demonstrated that none of the NR metrics was able to perform an accurate assessment on a general base, i.e., overall video types, compressions, and network conditions. Most importantly, all metrics had low performances in lossy networks. However, it also emerged that each metric exhibited specific operational boundaries, within which the performance was accurate to the benchmark [16]. In this way, it was shown the need to use techniques that would enhance the accuracy of the metrics by means of pattern recognition methods and learning tools, i.e., through Machine Learning techniques (ML) [26].

ML has proved to be a good ally to NR and low computation video quality metrics to enhance their capabilities. Given the broad variety of ML approaches, it is important to select the most suitable learning approach to the problem to be solved. Online learning techniques provide good results in the field of NR video quality when the changing environment (i.e., number and type of users, network conditions) makes it necessary for the system to learn new conditions while the assessing is taking place. Menkovski et al. [27–35] developed several online-based methods to predict the subjective scores on IPTV services and video on demand. These methods learn from previous experience to improve their model and predicted assessment. The concept of using online learning for adaptive streaming applications, i.e., algorithms that would select the most suitable quality to the network conditions, was first introduced in [36, 37]. Adaptive streaming applications over standard WiFi networks on lightweighted devices (Android mobile devices) were tested in [38]. In this experiment, a new adaptive streaming method based on online learning techniques was developed and its performance assessed in the presence of wireless networks impaired with bandwidth, delays, or packet loss constraints (impairments generated by a network emulator).

Supervised learning, as defined as the learning task of inferring a function from labeled training data, has demonstrated to be a good approach to enhance the predicting and accuracy capabilities of objective quality of experience algorithms when the metric is based on a group of labeled data (samples labeled with their FR or subjective index). A novel, predictive No-Reference (NR) metric was presented in [39]. This method combines lightweight measurements on the video pixel and bitstream layers with artificial neural networks to achieve comparable performance to the state-of-the-art full-reference metrics (PSNR and SSIM).

Finally, when the problem at hand lacks labeled data, unsupervised learning provides the possibility of classifying it and recognizing patterns. Restricted Boltzmann machines (RBMs) have proven their capabilities as density estimators. This characteristic makes them suitable to assess the degree of degradation in multimedia quality when the content is subjected to impairments derived both from compression and network conditions. In [40], an image quality assessment algorithm based on RBMs for 2D images was presented. This work was extended to 3D images in [41].

However, despite the existence of many valuable research efforts in this direction (some of which are described below), as well as standards for subjective quality evaluation of audiovisual content [42] (e.g., ITU recommendations: 1997a, 1997b, 2000a, 2000b), quality evaluations are still very much an open research field with further work needed to correctly characterize the relation between quality and comfort with higher level constructs such as "enjoyment", "naturalness", "engagement", or "immersion".

The next sections analyze the perceptual characteristics of the multiview content that is at the base of quality assessment and summarize the state-of-the-art and future directions in the evaluation of QoE and QoS with 3D contents from heterogeneous viewpoints.

10.2 Perceptual Characteristics of Multiview Content

A systematic overview of various aspects that influence the quality of stereo and multiview images at different stages of the 3D capture and processing pipeline is given in [43]. In the following, some perceptual characteristics of synthesized multiview content that is generated by depth-image-based rendering (DIBR) techniques are analyzed. The focus is on synthesis artifacts which derive from errors in depth maps and also from individual processing steps of the view rendering pipeline, while additional error sources such as coding or transmission artifacts are neglected for now. We describe typical errors that are often present in stereo-derived depth maps and then propagate to novel views rendered from a color-plus-depth representation. Generally, the production of novel views has to cope with both geometric and radiometric challenges such as occlusions and view-dependent illumination effects, which require special attention during the rendering process. An in-depth understanding of common error sources in the synthesis chain could support the development and evaluation of metrics for objective multiview quality assessment.

The creation of novel views from color-plus-depth data using DIBR comprises two major processing steps: (a) image warping and (b) image inpainting (see, for example [44]). In step (a), the individual pixels of the original 2D view are shifted according to the corresponding depth (or disparity) map to their new position, depending on the camera viewpoint to be synthesized. Discontinuities in the depth map lead to holes—that is, pixels without projected color values—in the warped view. These so-called "disocclusions" (for simplicity, sometimes also denoted as "occlusions") usually show up as elongated regions along the borders of foreground objects, where background covered by a foreground object in the original view becomes visible under the new viewing direction. The task of the inpainting step (b) is to estimate the missing image content by appropriate propagation of color and texture information from surrounding image regions, with the goal to achieve a natural blending between the warped and newly synthesized pixel values. Artifacts introduced by the warping and inpainting procedures may interfere with each other, and nonexpert users are often not aware whether a perceived degradation in image quality is primarily due to (a) or (b). The design of evaluation procedures that explore the two effects would be beneficial to gain a deeper understanding of the visual impact of each individual step.

The result of the warping step (a) is strongly influenced by the quality of the underlying depth map. When dealing with real images, as opposed to computer-generated content, one has to cope with non-perfect depth maps that contain errors as a consequence of the employed 3D reconstruction method (e.g., stereo analysis, structured light, or time-of-flight techniques). A well-known weakness of stereo analysis is errors in homogeneously colored image regions, where a local stereo-matching algorithm may not find enough texture to identify corresponding scene points in the left and right view. While this type of error clearly affects the error statistics in quantitative evaluations of the disparity map (as, for example, provided by the well-known Middlebury stereo benchmark [45]), it is typically less disturbing in derived novel views where the effect of geometric perturbations may be smoothed out by the color homogeneity. Contrarily, reconstruction errors in heavily textured regions tend to be very apparent to the human observer of 3D views.

An important problem in both stereo reconstruction and novel view generations is reconstruction inaccuracies in the vicinity of depth discontinuities, which normally coincide with object borders. A related problem is the so-called foreground fattening effect in stereo analysis. The spatial displacement of foreground object boundaries in the depth map can lead to pronounced artifacts in the projected novel views, with patches of background texture erroneously attached to the foreground object and vice versa. Such errors are very disturbing to humans when watching 3D content, for example, on stereoscopic displays. The problem of imprecise localization of depth discontinuities becomes particularly severe for thin foreground objects, whose spatial location is oftentimes not sufficiently captured in the depth map. The uncertainties in the reconstruction of depth discontinuities can also manifest themselves in increased flickering artifacts between consecutive frames of a derived video sequence.

Besides the stereo-specific problems mentioned above, depth maps may suffer from a variety of deficiencies such as low spatial resolution, noise, and locally missing depth measurements, depending on the used sensor type. Different post-processing techniques have been suggested in the literature to enhance the quality of depth maps. This includes recent work on the suppression of inconsistencies between depth discontinuities and color edges [46] and the refinement of a low-resolution depth map in the specific context of virtual view rendering [47]. An evaluation of depth map post-processing techniques for novel view synthesis with the incorporation of both objective and subjective ratings is described in [48]. The authors investigate the effect of six different post-processing methods—including the popular edge-preserving bilateral filter and guided image filter, as well as approaches based on local image statistics—in application to stereo-derived disparity maps. The test images contain features such as thin vertical structures and fuzzy object borders, which are challenges for stereo matching. The synthesized novel views are compared against original views taken from the same viewpoint using ten objective quality metrics. Furthermore, the authors conduct a user study, in which stereoscopic views containing an original left view and rendered right views obtained from the different post-processing techniques are presented to the subjects and judged by pair comparison. The subjective evaluation shows a user preference for the bilateral filter and guided image filter, which in the case of the bilateral filter is in strong disagreement with the results from the objective quality measurement. A limited applicability of objective 2D quality metrics is also reported in an earlier study [49] with a comparable stereoscopic viewing setup. In [50], a similar test configuration containing an original and rendered virtual view is employed to assess the effectiveness of different geometric mapping options for a three-view camera system.

A variety of inpainting algorithms have been proposed in the literature (see, for example, recent work by [51, 52]), which can be used for the occlusion filling step (b). A systematic exploration of the geometric conditions leading to the appearance of holes and related inpainting strategies is presented by [44]. The authors of [53] carry out subjective and objective quality evaluations to come up with a recommendation for a new quality metric that seeks to capture rendering artifacts caused by occlusion inpainting. The study in [54] demonstrates that including depth information into a patch match based inpainting algorithm increases the subjective 3D quality impression perceived by the user.

Besides clearly visible inpainting artifacts, a less obvious point that requires attention are matting errors along object boundaries. Alpha matting refers to the occurrence of "mixed" pixels that contain color contributions from both the foreground and image background (see [55] for a survey on image and video matting).

Apart from true object transparencies, mixed pixels typically occur along object boundaries and at the location of fine structures such as hairs. In novel view generation, the fused color components vary with the viewing angle due to the geometric shift between foreground objects and image background. Negligence of this aspect can make novel views look unnatural. Even though the effect is subtle, its consideration can further improve the quality of disparity maps and derived novel views [56].

10.2.1 Previous Work on QoE and QoS Assessment

Since 3D video became a foreseeable format of interest for visual information diffusion, the question of end-user perceived quality started attracting the attention of video coding and transmission experts in academia and industry. Research works were published on the development and evaluation of stereo video quality measures aimed at predicting the final quality after the effects of compression-related degradations and transmission impairments. In any case, the large number of applications for the 3D format (e.g., 3D television, free-viewpoint television, and multiview) advocates for the definition of dedicated objective quality metrics. Although several images and 2D video quality measures with good performance have been developed, "true" 3D video quality measures have been far less researched and proposed, especially for larger combinations of different degradation types. However, similarly to what happens with image quality measures which do not perform very well on video sequences because of missing correlation between successive frames in time, 2D stereo video quality measures applied to 3D video also suffer from performance problems related to the reduced correlation between left and right views which can occur in connection to some types of degradations. In [57], authors have presented some state-of-the-art 2D objective image and video quality measures tested on new 3D Nantes (NAMA3DS1-COSPAD1) stereoscopic video database [58]. Overall results show that the performance of the measures is not very good, hinting that further study on the extension of existing image and 2D video quality measures to 3D stereoscopic video is needed.

In [59], the evaluation of DIBR-synthesized views containing synthesis artifacts (and no compression artifacts) using traditional 2D quality metrics is presented. More specifically, two use cases are addressed: visual quality assessment of still images and of video sequences in 2D context. For the still images evaluation, the 2D IQA metrics PSNR, PSNR HSVM, PSNR HSV, SSIM, MSSIM, VSNR, VIF, VIFP, UQI, IFC, NQM, and WSNR were used. It is reported that none of the tested metrics reaches 50% of human judgment. Image quality metrics used for the evaluation of still images are also used for assessing the quality of video sequences by applying these metrics to each frame separately and averaging the frames scores. For the assessment of video sequence, VQM and VSSIM metrics are used also. It is reported that in the assessment of video sequences none of these metrics reaches 50% of human judgment either. Hewage et al. presented in [60] a study on the effectiveness of several measures derived from 2D video measures applied to 3D stereo video rendered from color + depth 3D video. This study established that measures such as PSNR, SSIM, and VQM when applied to the color component or to the left and right views exhibited a reasonably high correlation with image quality and depth perception subjective scores, with room to improvement.

Based on these evidences, Benoit et al. [61], present an in-depth review of works dealing with the factors which impact perceived 3D image quality, also paying attention to the effects of display technology on final quality. Although a significant amount of the work is devoted to performance of objective 3D image quality measures and their correlations with subjective scores, this work did not extend into 3D video evaluation. In [62], Aflaki et al. look into the effects of asymmetric quality and resolution of 3D stereo video on final quality and after subjective evaluation of mixed resolution stereo video propose a logarithmic equation to model the relationship between subjective scores and downsampling ratios. More recently, Le Callet et al. proposed in [58] a new 3D stereo video dataset, which includes original content with varied characteristics and degradations together with subjective evaluation scores of the dataset components, pointing out future directions for research. Some of these suggested research topics were explored in [63, 64] based on experiments conducted by different research groups in a collaborative effort.

Based on these findings, in the state-of-the-art, several approaches have been presented to address the issue of estimating the perceived quality of 3D content. Those methods may be grouped in two macro categories: methods based on the analysis of geometric and spatial distortions, and the ones based on depth map analysis and distortion.

10.2.2 Quality Assessment Based on Geometric and Spatial Distortions

The quality assessment of 3D content based on geometric and spatial distortions is based on the extension of the paradigm used for 2D content. Full-reference objective IQA metric dedicated to view synthesis quality assessment VSQA is presented in [65]. It is reported that VSQA can be extended by any 2D IQA metric. The VSQA metric uses three visibility maps which characterize complexity in terms of textures, diversity of gradient orientations, and presence of high contrast. SSIM-based VSQA metric achieves the gain of 17.8% over SSIM in correlation with subjective measurements.

3D synthesized view IQA metric, 3DswIM [66], addresses the image quality evaluation of DIBR-synthesized views. It relies on a comparison of statistical features of wavelet subbands of the original and DIBR synthesized. Only horizontal detail subband of the first level of Haar wavelet decomposition is used for degradation measurement. A registration step is included before the comparison to ensure «shifting-resilience» property. A skin detection step weights the final quality score in order to penalize distorted blocks containing «skin-pixels» based on the assumption that a human observer is most sensitive to impairments affecting human subjects. Pearson correlation coefficient 0.76 is achieved.

Multiscale metrics based on multiresolution decompositions using morphological filters [67, 68] are proposed in order to better deal with specific geometric distortions in the DIBR-synthesized images. The nonlinear morphological filters used in the multiscale image decompositions maintain important geometric information such as edges across different resolution levels. Two morphological multiscale metrics have been proposed: morphological pyramid peak signal-to-noise ratiometric (MP-PSNR) based on morphological pyramid decomposition [68], and morphological wavelet peak signal-to-noise ratiometric (MW-PSNR) based on morphological wavelet decomposition [67] (see Fig. 10.2). The proposed metrics measure the edge distortion between the multiscale representations of the reference image and the DIBR-synthesized image using MSE. Morphological multiscale metrics achieve significantly higher correlation with human judgment compared to the state-of-the-art image quality metrics and compared to the tested metric dedicated to synthesis-related artifacts. MP-PSNR achieves improvement of 13.55% of Pearson's correlation over PSNR (Pearson 0.887 and Spearman 0.817). Morphological multiscale metrics are computationally efficient due to the morphological operators involve only integer numbers and simple operations like a min, max, and sum as well as a simple calculation of MSE.

The visual quality assessment for view synthesis in the context of 3D video delivery chain is considered in [69]. The existing 2D metrics like PSNR and SSIM are extended using depth data to assess the perceived quality of synthesized viewpoints. The proposed framework assigns more importance to regions of a synthesized frame that are more open to synthesis-related distortions. The method uses a weighting function based on depth data at the target viewpoint and a temporal consistency function to take the motion activity into account. The validation of the performances is achieved by synthesizing different viewpoints from distorted color views and distorted depth data. The performances of the proposed extended techniques are measured using correlation to video quality metric VQM metric [70]. Better performances are achieved with respect to PSNR and SSIM.

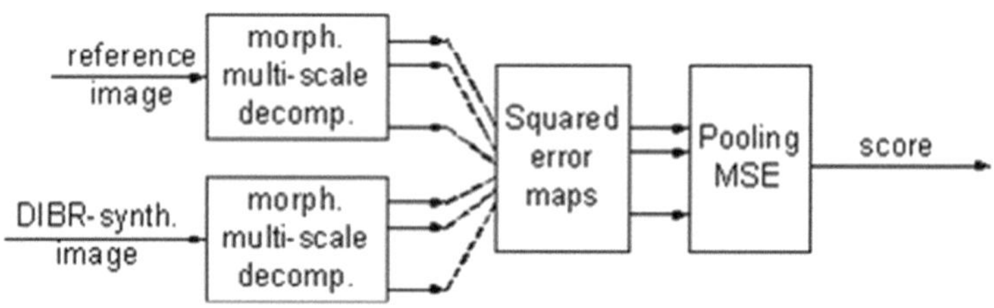

Fig. 10.2 DIBR-synthesized image quality assessment framework using morphological multiscale decomposition

Self-evaluation metric spatial PSNR (SPSNR) and temporal PSNR (TPSNR) for synthesized images is introduced in [71]. These metrics only use the synthesized view itself. The original view is unnecessary. SPSNR measures the spatial consistency by checking spatial noise caused by view synthesis. Generally, the view synthesis increases the high-frequency components since the 3D-warped images and holes have a lot of high-frequency components. Temporal consistency evaluated by TPSNR measures high-frequency components of temporal changes.

The algorithm proposed in [72] uses the original uncompressed input views to estimate the statistical characteristics of the cyclopean image by using the divisive normalization transform, which are compared to those of the synthesized image to estimate the compression and DIBR warping artifacts.

Multiview image quality measure (MIQM) [73] is created as a combination of three index measures that quantify the physical nature of multiview image distortions. The final composite measure MIQM is computed as the multiplication of the following measures: the luminance and contrast index, spatial motion index, and the edge-based structural index.

Objective quality metric of 3D video generated using DIBR technique is proposed in [74]. The metric is composed of color and sharpness of edge distortion measure. Color distortion measures the luminance loss of the rendered image and sharpness of edge distortion calculates a depth-weighted proportion of remaining edge to the original edge. The metric represents not only color artifacts but also synthesis artifacts. The metric shows significant agreement with subjective assessment.

10.2.3 Quality Based on Depth Map Analysis and Distortion

A different approach is used in the quality estimation based on depth map analysis and distortion. In this case, the depth information can be used to infer the quality of the 3D content, or it can be compressed or processed to evaluate the impact of its modification on the perceived quality.

When encoding either depth data or color sequences before performing the synthesis, compression-related artifacts are combined with synthesis artifact. The performances of commonly used objective quality metrics on free-viewpoint video FVV sequences synthesized from uncompressed texture and compressed depth data are analyzed in [75]. It is concluded that commonly used objective metrics initially designed to address video compression-related artifacts are not reliable predictors of perceived quality of video sequences synthesized from uncompressed texture and compressed depth. A measure 3VQM for objectively evaluating the quality of stereoscopic 3D videos generated by DIBR is presented in [76]. 3VQM estimates elements of the visual discomfort in DIBR-synthesized stereoscopic videos based on the ideal depth estimate. 3VQM is the combination of three measures that evaluate the temporal and spatial variation of the depth errors, fast-changing disparities, and geometric distortions. The results showed that 3VQM measure is the more accurate, coherent, and consistent compared to PSNR and SSIM objective measures.

Stereoscopic image quality assessment is presented in [77]. The database MCL-3D is created using nine image-plus-depth sources. DIBR technique is used to generate stereoscopic image pairs. At the encoder, the texture and depth maps are compressed and transmitted separately. At the decoder texture and depth, maps are decoded and a pair of stereoscopic images is rendered. The distortions are applied only on texture, only on depth map, or both on texture and depth map. The outcome of the research is that none of the existing objective quality metrics can provide satisfactory performance for stereoscopic IQ database.

A wavelet-based image quality metric for the assessment of 3D synthesized views is presented in [78]. Free-viewpoint video is generated from uncompressed color views and their compressed associated depth map. Prior to the synthesis step, the original depth maps are encoded with different coding algorithms leading to the creation of artifacts in the synthesized view. The combination of compression artifacts and synthesis artifacts is evaluated. The results show that the proposed metric has high correlation with human perception (90.1%) and better performances compared to state-of-the-art 2D quality metrics.

A quality evaluation study for assessing the distortion introduced by the view synthesis procedure and depth map compression in MVD coding systems is presented in [79]. The quality of synthesized views is assessed by PSNR metric in three cases. First, the distortion introduced only by view synthesis algorithm is calculated between virtual view video synthesized from uncompressed texture and uncompressed depth and original video. Second, the distortion introduced only by depth map compression is calculated between virtual view video synthesized from uncompressed texture and compressed depth and video synthesized from uncompressed texture and uncompressed depth. Third, the distortion introduced by both depth map compression and view synthesis algorithm is calculated between virtual view video synthesized from uncompressed texture and compressed depth and original video. It is shown that view synthesis reference software yields high distortions in terms of average luma PSNR that mask those due to depth map compression. It is concluded that reference image should be synthesized from uncompressed data when assessing codec performances.

In [80], authors analyzed the correlation between different image and video quality measures and corresponding subjective scores computed on 20 3D texture + depth video sequences from the University of Coimbra 3D (UC3D) video database. The degradations introduced to generate the impaired videos affect only the depth maps and simulate losses of packets transporting encoded depth data. For each impaired depth map sequence, a (new, degraded) synthesized view was rendered using the original texture information (and degraded depth map). Quality metrics were tested on (degraded) depth information and synthesized views. Overall results show that quality measures computed on the synthesized view correlate better with subjective (DMOS) scores than measures based on the depth information.

The authors in [81] proposed the 3DTV video quality measure as a quality measure for video-plus-depth content by measuring the quality of the virtual views that are rendered from the distorted color and depth sequences. The conventional single camera quality measures, PSNR and SSIM, were used. The undistorted reference sequence is obtained by rendering virtual views from the original color and depth maps. Clearly, the metric is sensitive to video coding effects, while geometric distortions in video sequence are not considered.

The relationship between the quality of the rendered views and different quality measures of the depth map is investigated in [82]. The depth map quality metric is proposed based on a distortion model that approximates rendering errors due to pixel errors in the depth map. The proposed metric correlates very well with the quality of the rendered views, as compared to PSNR and SSIM. The proposed depth map quality metric is incorporated at the encoding mode selection stage of a video encoder. Experimental results suggest that with the proposed encoding mode selection, scheme bit rate savings of up to 30% can be achieved compared to traditional encoding mode selection scheme based on sum of squared errors. The quality of the rendered views is measured as the PSNR between views rendered with the uncompressed and the compressed depth map.

The relationship between the overall 3D video QoE and the depth map quality measured by perceptual-based image quality metrics is investigated in [83]. Several depth map artifacts are simulated and applied to the depth map sequence corresponding to left view of a stereo video pair. The synthesized views are generated using the original views and the distorted depth map sequence. The evaluation results showed that 3D video quality depends highly on the depth map quality. Five original stereo video sequences recommended by MPEG are used in experiments. In order to simulate the compression artifacts, the depth map sequences corresponding to the left view are compressed using HEVC standard. In order to simulate packet loss effect, the H.264/AVC/based network transmission simulator is utilized. The right view is synthesized using the left view and the distorted depth map corresponding to left view. In order to get the objective evaluation of the quality of the depth maps, the reference depth map and distorted depth map videos are compared using various quality metrics: PSNR, SSIM, MS-SSIM, VQM, and VIF. VIF shows better correlations with MOS than all other metrics. The reported results confirm that overall perceived 3D video quality highly depends on the depth map quality, but also that depth map alone is not sufficient for predicting the overall 3D video quality.

The effect of compressing the data prior to rendering is evaluated in [84]. To verify the accuracy of the rendering algorithms, PSNR for the depth maps is measured since it is assumed that the accuracy of the rendering is proportional to the accuracy of a depth map for a new view. Experiments are performed using compressed video from surrounding cameras. Video-plus-depth maps of the sequences “Ballet” and “Breakdancers” were used. It has been found that the influence of H.264 compression of the video data is negligible above 1.6 Mbps joint bit rate for texture and depth. This means that for higher joint bit rates, the rendering quality is only influenced by the rendering algorithm and not by compression of the video streams from the neighboring cameras. Below 1 Mbps, the compression has a serious influence on the quality. For obtaining the highest rendering quality, depth maps should be of higher quality than texture. It was shown that the depth map consumes a significant portion of the joint bit rate (up to 50%).

10.3 Subjective Quality Evaluation

10.3.1 Standard Methods for Subjective Quality Evaluation

This interest in user evaluation stems from an obvious fact: the ultimate goal of the technological developments described above is to provide users with high-quality contents and more engaging experiences than those delivered by traditional 2D video and stereo audio displays. Standard evaluations of stereoscopic image quality often rely on the psychophysical scaling methods initially proposed in ITU recommendations 1997a, 1997b, 2000a, and 2000b for stereoscopic content, which allow quantifying specific factors such as the degree of perceived sharpness, or more general ones, such as overall image quality. An example of an application of the ITU methodology is the already mentioned assessment of the levels of binocular disparity that significantly reduce perceived quality and comfort. The ITU recommendations propose a variety of experimental designs of which the most widely employed is the Double-Stimulus Continuous Quality Scale method (DSCQS), where users evaluate the overall image quality of a series of stereoscopic images presented separately, or to pairs of images where an impaired stereo images is tested against an unimpaired stereo image acting as a reference, resulting in a set of difference scores between reference and test images. Alternative methods have been proposed for the evaluation of video sequences as well, for example, the Single-Stimulus Continuous Quality Evaluation (SSCQE), where users assess video sequences containing impairments that vary in time, such as those introduced by different coding parameters.

A relation between visual quality and immersion has often been assumed, at least implicitly. For instance, a high degree of realism in visual special effects is usually considered a key factor in engaging wide cinema audiences. But quality is an even more important issue when considering stereoscopic content. It is undisputable that stereoscopic content is a beneficial advance, for examples, it is readily given higher quality ratings when compared to an equivalent 2D counterpart; it significantly enhances perceptual resolution and figure-ground discrimination, especially in unfamiliar and noisy contexts [85]; and it lowers thresholds for visual detection in noise, as observed in the so-called "binocular unmasking" phenomenon [86]. As mentioned in ITU recommendation BT1438, regarding subjective quality assessment of stereoscopic content, in addition to factors applicable to monoscopic images (e.g., resolution, color, motion, sharpness, etc.), stereoscopic viewing entails additional factors such as depth resolution, depth motion, and size distortions (puppet and cardboard effects), and crosstalk effects (producing ghosting), to name but a few; moreover, stereoscopic content is mainly designed for presentation on large displays; in smaller screens, the parallax between the eyes reduces to the point where the depth impression might be lost and, given the depth of field of the human eye ($\pm$0.3D), even for a regular television screen the optimum viewing distance would be of about 3 meters. It is thus evident that the number of factors that can contribute to a low-quality experience is far greater than that of conventional 2D. Moreover, stereoscopic content is still far from being abundant, and this lack of content is partly being solved by transforming 2D video into 3D video by means of 2D-to-3D conversion algorithms that derive a depth map from 2D static or dynamic image; however, the result of these conversions is not always optimal [87]. It is important to realize that low quality in stereoscopic content can result in several visual discomfort symptoms. In this sense, it should be noted that in stereoscopic perception, in addition to display size and viewing distance, a key factor to take into account is the convergence/accommodation mismatch, which stems from the fact that the focus point of the eyes (accommodation) is constantly fixed on the screen, whereas the convergence point varies. Accommodation and convergence are automatically linked in natural situations, but in stereoscopic displays they can be decoupled, and if this unnatural situation is prolonged in time, it can cause visual fatigue and other symptoms such as sore eyes and headaches. Research on visual fatigue indeed indicates that one of the main factors leading to fatigue in stereoscopic displays is indeed the conflict between convergence and accommodation, although there are many others such as binocular anomalies, excessive parallax or dichoptic errors such as crosstalk, etc. [88]. Individual differences are an important factor to take into account when considering comfort and health issues. Age has to be considered as well as, for example, the ability to accommodate decreases with age so that at about 55 years of age little accommodation capabilities remain [89]; and, on the other hand, the human visual system is not fully developed until around the age of seven [90], fact that raises concerns on the use of stereoscopic 3D by children below that age. In the case of 3DTV, the literature describes visual discomfort as the primary health matter.

With the growing interest for stereoscopic 3D imaging, the Video Coding Experts Group (VCEG) and Moving Picture Experts Group (MPEG) have joined their efforts to develop new 3D video formats and coding standards. Although the history of stereoscopic video sequences dates back from the last century, the subjective quality assessment protocols that are essential to evaluate new 3D viewing systems are not standardized yet. In the next section, we discuss methodologies based on psychology/neuroscience methodologies, better suited for the study of high-level cognitive/emotional factors rather than the study of purely visual perception issues.

10.3.2 Psychology/Neuroscience-Based Methodologies

The research field known as experimental psychology, cognitive science, or more recently, cognitive neuroscience is the main framework to have placed human information processing under extensive empirical scrutiny. Psychological research characterizes human beings as characterized as "information processors". In this framework, perception is the first stage in information processing and refers to the processes whereby information from the environment coming from the senses is made available to information processing systems. In turn, cognitive processes refer to those processes by which information is manipulated (filtered, coded, compared, retrieved, etc.).

A huge battery of methods is available to measure these aspects, grounded on different methodologies depending on the type of question explored. First, to analyze conscious processes, there are well-standardized questionnaires for measuring perceptual aspects, perceived usability (IBM questionnaire), cognitive working load (e.g., NASA-TLX), or effective or emotional reactions (e.g., SAM, PANAS, Pick-a-Mood questionnaires), among others. Nevertheless, since it is unlikely that people can report information about processes over which they can have little or any awareness [91, 92], psychological research has traditionally favored methods allowing exploring unconscious or automated psychological processes, providing online moment by moment information, and not dependent on subject biases as, for example, social desirability bias. A broad classification of such methods could be (1) behavioral methods, that is, measurement of psychophysical thresholds, reaction times, motion and eye tracking, etc.; and (2) psychophysiological methods, which entail measuring physiological changes in users related to psychological stimuli. Behavioral methods (with the exception of tracking measures) often require designing specific tasks where different variables are manipulated in order to capture psychological effects in different aspects of their performance. Physiological signals, on the other hand, allow psychological measurements while users are interacting with the assessed technology, thus having been greatly favored in media research [93, 94]. When the central nervous system is the object of measurement, electrical brain activity (EEG) is the most widely used method, as it can provide information about attention and cognitive effort, and also about emotional reactions (e.g., frontal alpha asymmetry). Peripheral nervous system activity measurements are very informative when measuring emotional reactions and do also signal some attentional reactions; the main are electrodermal activity and facial electromyography, and phasic and tonic changes in heart rate, which are also indicators of diverse attentional and emotional reactions, which, as we argue in the next section, need to be accounted for in QoE research. The interest in integrating the information provided by psychological measures in modeling user's QoE is therefore obvious, and also two-sided. One of the main problems of scientific psychology has been one of external validity, psychological experiments too often test micro-hypothesis about concrete processing phenomena in tightly controlled laboratory conditions, thus making difficult its application to real-life situations. Studies of QoE provide precisely these real-life grounds in which to observe the actual validity of our models of human information processing, therefore, building bridges between QoE research and experimental psychology will clearly be a mutually beneficial endeavor for both fields.

In this framework, a substantial improvement would be represented by the possibility of monitoring physiological signals, related to the emotional status of the subject, in a nonintrusive manner. For instance, traditional methods to observe and measure the heart rate typically rely on electrocardiography (ECG), where the electrical activity of the heart is monitored by means of sensors placed on certain skin areas, while noninvasive techniques would be preferable in order to avoid additional user's discomfort that could bias the evaluation. Indeed, recent works have shown that in certain conditions video recordings allow for HR measurement through the analysis of temporal color variations of skin areas, which is performed by means of computer vision techniques. The first step in this direction has been proposed in [95], where the video-based approach is applied for the first time to the automatic evaluation of QoE with 3D video sequences. To this extent, a subjective experiment has been performed where 3D stimuli have been shown to participants and corresponding HR values have been collected. The data analysis generally shows high correlation between the video-based and sensor-based measurement of the HR, as well as a correspondence between the video-based HR measure and the 3D emotional stimuli, thus supporting ongoing research aimed at the automatic analysis of users' emotional response (see Fig. 10.3).

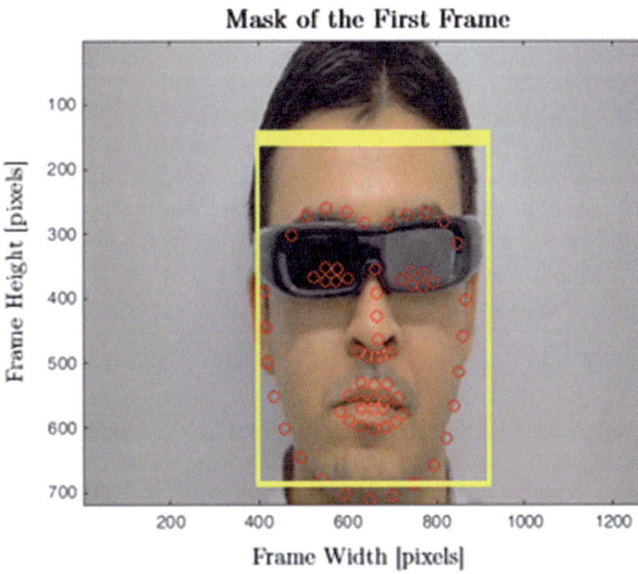

Fig. 10.3 Face detection using Viola–Jones algorithm and landmarks estimations, (red points) using discriminative response map filtering. The rectangles highlighted by the blue dashed lines are the chosen regions of interest

10.3.3 High-Level QoE Factors

Currently, a variety of QoE metrics incorporate methodologies pertaining to perceptual studies; however, there is an increasing need in the field to consider "high-level factors", that is, variables related to the aforementioned cognitive processes. The most interesting psychological variables and processes for the study of media technologies are those related to attentional and emotional phenomena. Regarding attention, cognitive science has provided large amounts of evidence that conscious information processing is mainly serial, so that when processing information in situations that require to shift the focus of attention between different tasks and/or stimuli results in an increase in the effort required to process that information [96]. In the field of basic cognitive psychology, this phenomenon has been extensively studied by means of experimental paradigms that allow to determine, for example, the degree to which performance on a given task is affected by concurrently performing a secondary task [97], or the actual performance cost of attention shifts [98]. Regarding emotions, a theoretical approach that has proven very useful in quantifying emotional reactions defines emotions as a function of two components: effective or hedonic valence, that is, if the emotion is positive or negative, and arousal, meaning the intensity of the emotion [99]. Emotions have a great impact on human behavior: they modulate perception, attention, decision-making, learning, and memory [100]. Humans constantly employ these mechanisms in their interactions with the surrounding environment. Thus, efficiency and effectiveness of media and communication applications are highly dependent on their ability to express emotions and induce emotional states in users. Previous research has shown that sounds evoke emotions [101] and provide biologically salient effective information, which may inform the design of effective auditory displays [102]. Further, more efficient auditory displays and increased sense-of-presence in virtual environments can be reached via improved auditory spatial resolution [102].

As said above, in the case of S3D, research on QoE has mainly focused on visual aspects, successfully exploring how technological features, and particularly image and video communications techniques, are related to sensorial factors, and how these determine higher order cognitive factors as visual experience. However, visual aspects do not account for all the factors that contribute to QoE. Specifically, users' emotional reactions may play a central role on the experienced QoE, as pointed out by the use of words such as "delight" and "annoyance" in the definitions described earlier [3]. This includes factors such as immersion, multimodality, and interaction. In this sense, the fundamental role of emotions in entertainment consumption is commonly taken for granted, even the concept of entertainment itself has been defined as an emotional response to mediated contents [103]. User emotions in entertainment are mainly determined by content aspects (whether it is, for example, and arousing action movie, a funny comedy, or a calmed documentary) and the interaction of these with users' needs, preferences, and circumstances. User emotions in entertainment are mainly determined by content aspects (whether it is, for example, and arousing action movie, a funny comedy, or a calmed documentary) and the interaction of these with users' needs, preferences, and circumstances. Nevertheless, formal and technological attributes can also affect viewers' emotional reactions [104]. An idea implicit to this conceptualization of emotions is that they are the result of an evaluation of a stimulus in terms of survival [105]. This evaluation not only works with real stimuli, but also with mediated stimuli that cannot have an actual incidence of survival but represent objects or events that can do. For example, people may feel fear while watching a horror film about a serial killer, although they know that there is not a real threat to them. Reeves and Nass [106] suggested that, from an evolutionary point of view, mediated messages are such a new thing that human brains find difficult to distinguish them from reality, so primarily interprets them as real events. As pointed out by Zillmann [107, 108], the human brain reacts to emotional stimuli before they have been cognitively processed in deep, so emotional responses are produced previously to the considerations about the mediated (unreal) nature of the experience. Only when the stimuli have received a cognitively deeper processing, people become aware of the unreality of the stimuli.

Some theories [109] have explained the paradoxical fact that fictional stories can produce emotions (which evolutionarily should be reserved to the appraisal of real opportunities or threats) by the existence of a dual system for the appraisal of the mediated stimuli. These theories propose the coexistence of two mental domains during the fiction contents watching: one of them, called, for example, "involvement" mode [110], processes the message as it were real, adopting a perspective from the inner of the mediated reality, while the other processes the message in the context of the broader reality of the experience, as just a symbolic representation. The switch of the attentional focus between the two domains may produce a more emotional experience (taking the represented facts as real) or limit the extent of the experienced emotions (reminding that it is only a mediated message).

In the analysis of the emotion induced by 3D, data is important to discriminate between human reaction caused by emotions and by visual discomfort or visual fatigue. As described in [111], these two signals are quite different. Fatigue can be considered as a decrease in the performance of the human visual system as a consequence of physiological strain or stress resulting from excessive exertion, while visual discomfort can be seen as the subjective counterpart of visual fatigue, partially reflecting some aspects of the QoE. Adaptation mechanisms [112] are used by the human visual system as a response to changes in the environment. These continuous adjustments in sensory processing may improve its performances while causing fatigue. Another difference is that visual discomfort is perceived instantaneously, while fatigue is induced after a discrete duration of effort. Currently, there is no clear understanding of the relation between fatigue and discomfort and how to deal with worsening and improving effects, as well as temporal aspects; furthermore, the correct methodology for measuring visual discomfort is still to be defined.

Recently, some work has been done in this direction. In [104], the effects of image distortions, which are known to trigger visual discomfort, over user's emotional reactions are investigated. The achieved results show that the main effect of visual discomfort supports the idea that the emotions are less intense and less negative. Nevertheless, in the case of emotional contents, the interactions between emotional contents and visual discomfort indicate the presence of a more intense and less positive emotion, which points out to the possibility that video discomfort elicited a negative emotion. Traditionally, basic research on human information processing (HIP) has focused on tightly controlled experimental conditions and stimuli to ensure the internal validity of the findings, which has often come at the cost of limiting their external validity. However, due in part to a surge of interest in how objective psychological metrics can address the complexity of HIP in a variety of technology-oriented contexts, contemporary research is faced with the challenge of working with increasingly complex stimuli resembling real-life conditions, while at the same time maintaining internal validity. This strand of applied research has brought to the spotlight complex constructs, such as engagement and immersion, which break the usual divide-and-conquer strategy in HIP research by tackling the interplay between different cognitive processes, such as attention, motivation, and emotion.

The extent to which a “virtual” experience is considered engaging is often referred to in the literature as immersion or presence. According to some authors (e.g., [113]), immersion should be considered as a technical feature of the system, namely, its capacity to provide a compelling, convincing and interactive virtual environment. Presence, on the other hand, as a term often attributed to Marvin Minsky, [114]) which refers to the perceptual and cognitive factors that lead the user to feel that a given experience is immersive. In other words, presence refers to the illusion of “being there”, which operationally amounts to an enhanced emotional appeal of the mediated experience. However, despite the existence of many valuable research efforts in this direction (some of which are described below), as well as of the existence of some standards for subjective quality evaluation of audiovisual content (e.g., ITU recommendations: 1997a, 1997b, 2000a, 2000b), the evaluation of quality, comfort, and immersion is, in practice, carried mainly by means of interviews and questionnaires, despite the fact that immersion is mainly operationalized by academic research as an increased emotional reaction that, again, is not always captured by self-report measures. In fact, many research labs have started to use psychophysiological indicators of emotional activation as the best operational measure of presence. More specifically, increases in the level of arousal measured by means of skin conductance and cardiac activity registries are the preferred measure. Regarding engagement, considered as a keystone of the motivations for media consumption, there is a lack of consensus on its formal definition. In the context of human–agent interaction, engagement has been defined either as the attention paid for the subject to the interaction [115], as the value attributed by the subject to the interaction [116], or as the interaction process itself. Furthermore, it accounts for both the cognitive and hedonic aspects of user interactions, and considers the characteristics of systems (e.g., usability, aesthetic appeal, interactivity), users (e.g., level of felt involvement, positive effect), as well as their interaction at the system level [117]. More specifically, the emphasis of user engagement is placed on what the user finds “innately interesting” according to their motivation for using a technology or a resource. Such diversity of interpretations is unsurprising considering that the construct of engagement lies at the very boundary between attention, emotion, and interest, all of which are considered indicators of engagement [118]. Although an extensive corpus of knowledge exists on the operationalization of attentional and emotional processes, there is not much agreement on which are the best empirical indicators of interest. In [117], the author's work provides an objective metric for one key component of this concept, namely, interest. This metric is based on electroencephalographic (EEG) registration of users' neural activity while they read sets of news pre-classified in terms of their potential interest. The authors focused on a metric that has been used as an indicator of the degree to which an item or event induces the motivation to approach or escape, the so-called Frontal Alpha Asymmetry (FAA), finding that it was indeed a good proxy for objective monitoring of interest in media contents and that, although is interpretation in terms of information processing, is yet to be fully clarified, entropy analyses show that this metric is also sensible to interest manipulation, providing results that complement traditional power spectrum analyses.

Despite the well-known importance of an adequate auditory stimulation in delivering an immersive experience [119], development of 3D technologies has traditionally conceded a preponderant role to the generation and reproduction of visual, as opposed to auditory, content. Technological developments toward 3D displays entail a demand for audio tracks of higher quality. Ideally, in a 3D audio presentation, the user should be able to perceive sound sources as localized in any point of space, regardless of the specific spatial configuration of the speaker system, but the limitations of traditional technologies make it difficult to implement adequate soundtracks, and there is currently no commercial multi-speaker system superior to 5.1, which, among other defects, does not allow to locate sound sources in every angle respect to the user, especially in the vertical dimension. The need for accurate 3D audio is undisputable, providing accurate spatial cues in virtual environments is known to significantly improve the level of immersion and the experienced pleasure of the experience [120] and, on the other hand, it has also been observed that depriving a hearing person of auditory stimulation induces a strong sensation of unreality [121]. In light of such evidence, it would be reasonable to expect realistic audio stimulation to be reflected in users' perceived quality in addition to immersiveness of the experience. The introduction of 3D sound thus introduces new possibilities for enhancing the richness of mediated experiences both in terms of quality and comfort and immersiveness. But as mentioned above, there exists a lack of dedicated methodologies for the evaluation of 3D audio, be it binaural or speaker-based, and of bimodal audiovisual 3D. Based on well-established knowledge about human multisensory information processing [122], it could be hypothesized that bimodal 3D audio content should be superior to unimodal stereoscopic content, and to stereoscopic content coupled with conventional audio, for the selected measures of quality, comfort, and immersiveness, variables which are also hypothesized, should show exhibit strong correlations. Given its novelty, whether 3D audio quality is perceived as having better quality and having more quality and being more comfortable and immersive than conventional audio systems is yet to be elucidated [123]; as well as whether the whole audiovisual 3D experience is superior to 3D audio or 3D video alone, Moreover, results on multisensory interactions allow to hypothesize that the level of quality and comfort of impaired stereoscopic visual content can be augmented through the addition of realistic 3D sound. The relation of these factors with perceived immersiveness will be also a key issue in future research.

A question that deserves attention is whether the more spatialized audio is consciously perceived by viewers, and if it is judged as having a better quality than the audio without the vertical dimension. Research using listening tests [124] showed that people rated better the audio quality when it includes the vertical dimension, but the features of the item presented had a determinant role in order to perceive the difference between conditions. Furthermore, when participants were allowed to have a reference to compare auditory quality, 3D audio rates improved. It must be taken into account that in listening tests participants are instructed to pay close attention to auditory quality, a situation that is not comparable to the real situation of fiction products consumption. Since human ears disposition make people less sensitive to the vertical distribution of sound than in the horizontal plane it is possible that, when subjects are not paying specific attention to audio quality, differences between audio with and without the vertical dimension are not so clearly perceived. Therefore, it is worth to address the question about whether the 3D audio improves the viewer's subjective perception of the audio quality in a naturalistic situation.

On the other hand, and consciously perceived or not, effects of the 3D audio might not be restricted to increased emotional reactions. Moreover, there is a tight relationship of auditory motion with vestibular perception [125], and with ocular movements [126], therefore, it is foreseeable that comfort issues might arise if presentation of fully spatialized sound does not control for these factors, which makes it all the more necessary to carry exploratory studies on the possible comfort issues in 3D auditory perception. For example, are there quality and comfort issues related to the speed of simulated auditory motion? If there are, how do these issues interact with features of the stimuli such as perceived size, location, or frequency? These questions will become even more important in the near future due to the inception of commercially viable virtual reality technologies.

10.4 Conclusions and Future Directions

Despite the active research carried out in this field, several challenges are still open and push for new solutions. In the following, a brief discussion of the main issues is reported.

10.4.1 Measurement of Different Perceptual Attributes

Even though emerging 3D quality evaluation methods accurately predict a given quality attribute, the relationship among these perception attributes has not to be thoroughly studied. The combined effect directly affects user experience and can be measured using emerging QoE indices. Therefore, the current need is to understand how 3D image processing and transmission artifacts affect the overall experience of the user, and then identify image and contextual features which can be used to quantify the overall effect on user experience. On the other hand, it is necessary to understand how the HVS perceives 3D artifacts. For instance, there could be conflicts based on whether binocular suppression or binocular rivalry is taking place based on the artifacts in question. These aspects need extended attention in order to measure the overall experience of 3D viewing. To enable unified approach to 3D objective quality subjective quality evaluation studies, standardization of these procedures is necessary. Several standardization activities are being carried out by VQEG, ITU (Recommendations: ITU-T P- and J-series), European Broadcasting Union EBU (3D-TV Group), and other Standards Developing Organizations (SDOs) in relation to 3D video subjective and objective quality evaluations. Currently, Video Quality Expert Group (VQEG), 3DTV project is being worked on creating a 3D video dataset (GroTruQoE dataset) to conduct subjective quality evaluation tests using pair comparison method. The ground truth database will then be created after extensive subjective quality evaluation to further evaluate other time-efficient subjective testing methodologies. In addition, VQEG is also active in objective quality assessment of 3D video, where they plan to evaluate 3D quality of experience in relation to visual quality, depth quality, and visual comfort dimensions. Most of these findings are reported to objective and subjective 3D video quality studies in ITU-T Study Groups (SG) 9 and 12. EBU is also working on 3D video production, formats, and sequence properties for 3D-TV broadcasting applications (e.g., EBU Recommendation R 135).

10.4.2 Lack of 3D Image/Video Databases

There are several image/video quality databases for conventional 2D image/video artifacts, although only a few have been reported for 3D image/video artifacts. This prevents developers from using a common dataset to evaluate the performance of their metrics. The amount of artifacts considered in these databases is limited. Most of them do not consider artifacts which could be introduced during transmission. The only database where artifacts introduced during transmission are considered does not report MOS values associated with the different impaired sequences, and hence it cannot be directly used for the performance evaluation of developed objective metrics. Therefore, it is a responsibility of the research community to make available the developed 3D image/video dataset publicly.

10.4.3 Visual Attention Models to Develop RR and NR Quality Metrics

The attention of users during 3D viewing can be influenced by several factors including spatial/temporal frequencies, depth cues, conflicting depth cues, etc. The studies on visual attention in 2D/3D images found out that the behavior of viewers during 2D viewing and 3D viewing is not always identical (e.g., center bias vs. depth bias). These observations are tightly linked with the way we perceive 3D video. Therefore, effective 3D video quality evaluation and 3D QoE enhancement schemes could be designed based on these observations. There are still unanswered questions such as whether quality assessment is analogous to attentional quality assessment and also how attention mechanisms could be properly integrated into design of QoE assessment methodologies. A thorough study has not been conducted to date in order to identify the relationship between 3D image/video attention models and 3D image/video quality evaluation. There are two main types of depth-integrated saliency models, namely depth-weighted 3D saliency model and depth saliency model-based methods. The depth-weighted saliency models weigh the 2D saliency map based on depth information. In-depth saliency models, the predicted 3D saliency map is derived based on the chosen weights for 2D and depth saliency maps.

10.4.4 Need for a Standard for Subjective Experiments

The International Telecommunication Union has recently released a new recommendation, ITU-R BT.2021, for the assessment of stereoscopic 3DTV systems [1]. This recommendation is mostly an extension for 3DTV of the well-known recommendation ITU-R BT.500 [2], which was established for 2D television. The recommendation includes a subset of four methods from ITU-R BT.500, namely the Single- Stimulus (SS), Double-Stimulus Continuous Quality Scale (DSCQS), Stimulus Comparison (SC), and Single Stimulus Continuous Quality Evaluation (SSCQE) methods. According to ITU-R BT.2021, the picture quality, depth quality, and visual comfort of stereoscopic imaging technologies should be assessed.

The above-mentioned recommendation does not address the specific issue of synthesized views. Therefore, subjective quality assessment of 3D contents represented in the video-plus-depth or MVD formats, and, as a consequence, of virtual synthesized views, has been conducted according to methods used for the assessment of conventional 2D contents.

In the future, new 3D objective quality measures should be researched and proposed, designed with particular care on what concerns their computation complexity and performance on stereoscopic video affected by different types of degradation.

References

1. Liotta, A.: The cognitive NET is coming. IEEE Spectr **50**(8), 26–31 (2013)
2. Torres Vega, M., Zou, S., Mocanu, D.C., Tangdiongga, E., Koonen, A.M.J., Liotta, A.: End-to-end performance evaluation in high-speed wireless networks. In: Proceedings of the 2014 10th International Conference on Network and Service Management (CNSM), 17–21 Nov 2014, Rio de Janeiro, Brazil (pp. 344–347). Piscataway: IEEE Service Center (2014)
3. Le Callet, P., Moeller, S., Perkins, A. (eds.) Qualinet white paper on definitions of quality of experience. In: COST Action IC 1003, Laussane, Switzerland (2012)
4. Agboma, F., Liotta, A.: Quality of experience management in mobile content delivery systems. J. Telecommun. Syst. (special issue on the Quality of Experience issues in Multimedia Provision) **49**(1) (2012)
5. Agboma, F., Liotta, A.: QoE-aware QoS management. In: Proceedings of the 6th International Conference on Advances in Mobile Computing and Multimedia, 24–26 Nov 2008, Linz (Austria)
6. Mocanu, D.C., Liotta, A., Ricci, A., Torres Vega, M., Exarchakos, G.: When does lower bitrate give higher quality in modern video services? In: Proceedings of the IEEE/IFIP Network Operations and Management Symposium (NOMS'14) 5–9 May 2014, Krakow, Poland, IEEE, Piscataway, pp. 1–5 (2014
7. Agboma, F., Liotta, A.: Addressing user expectations in mobile content delivery. Mobile Inf. Syst. **3**(3–4), 153–164 (2007)
8. Agboma, F., Smy, M., Liotta, A. "QoE analysis of a peer-to-peer television systems". In: Proceedings of IADIS International Conference on Telecommunications, Networks and Systems. July 2008
9. Agboma, F., Liotta, A.: Quality of experience management in mobile content delivery systems. Telecommun. Syst. (2012)
10. Exarchakos, G., Druda, L., Menkovski, V., Liotta, A.: Network analysis on Skype end-to-end video quality. Int. J. Pervasive Comput. Commun. **11**(1) (2015)
11. Exarchakos, G., Menkovski, V., Liotta, A.: Can Skype be used beyond video calling?. In: Proceedings of the 9th International Conference on Advances in Mobile Computing and Multimedia (MOMM11), ACM, 2011/12/5
12. Liotta, A., Druda, L., Menkovski, V., Exarchakos, G.: Quality of experience management for video streams: the case of Skype. In: Proceedings of the 10th International Conference on Advances in Mobile Computing & Multimedia, ACM, 2012/12/3
13. Exarchakos, G., Druda, L., Menkovski, V., Bellavista, P., Liotta, A.: Skype resilience to high motion videos. Int. J. Wavelets Multiresolution Inf. Process **11**(03) (2013)
14. Menkovski, V., Exarchakos, G., Liotta, A., Cuadra Sánchez, A.: Measuring quality of experience on a commercial mobile TV platform. In: 2010 Second International Conferences on Advances in Multimedia (MMEDIA), Athens (Greece), IEEE, 13–19 June 2010
15. Menkovski, V., Exarchakos, G., Liotta, A.: Adaptive testing for video quality assessment. In: Proceedings of Quality of Experience for Multimedia Content Sharing, Lisbon, Portugal, 29 June 2011
16. Torres Vega, M., Sguazzo, V., Mocanu, D.C., Liotta, A.: An experimental survey of no-reference video quality assessment methods. Int. J. Pervasive Comput. Commun. **12**(1) (2016)
17. Liotta, A., Mocanu, D.C., Menkovski, V., Cagnetta, L., Exarchakos, G.: Instantaneous video quality assessment for lightweight devices. In: Proceedings of International Conference on Advances in Mobile Computing & Multimedia, ACM, 2013/12/2
18. Mocanu, D.C., Santandrea, G., Cerroni, W., Callegati, F., Liotta, A.: Network performance assessment with quality of experience benchmarks. In: 2014 10th International Conference on Network and Service Management (CNSM), IEEE, 17 Nov 2014
19. Winkler, S., Mohandas, P.: The evolution of video quality measurement: from PSNR to hybrid metrics. IEEE Trans. Broadcast. **54**(3), 660–668 (2008)

20. Wang, Z., Bovik, A.C., Sheikh, H.R., Simoncelli, E.P.: Image quality assessment: from error visibility to structural similarity. IEEE Trans. Image Process. **13**(4), 600–612 (2004)
21. Pinson, M.H., Wolf, S.: A new standardized method for objectively measuring video quality. IEEE Trans. Broadcast. **50**(3), 312–322 (2004)
22. Chikkerur, S., Sundaram, V., Reisslein, M., Karam, L.J.: Objective video quality assessment methods: a classification, review, and performance comparison. TBC **57**(2), 165–182 (2011)
23. Torres Vega, M., Sguazzo, V., Mocanu, D.C., Liotta, A.: Accuracy of no-reference quality metrics in network-impaired video streams. In: 13th International Conference on Advances in Mobile Computing and Multimedia, 11–13 Dec 2015, ACM, Brussels, Belgium Brussels
24. https://www.tue.nl/en/university/departments/electrical-engineering/research/research-groups/electro-optical-communications-eco/research/network-management-and-control/datasets/network-impaired-video-dataset/
25. Seshadrinathan, K., Soundararajan, R., Bovik, A.C., Cormack, L.K.: Study of subjective and objective quality assessment of video. Trans. Image Process. **19**(6), 1427–1441 (2010)
26. Mohri, M., Rostamizadeh, A., Talwalkar, A.: Foundations of Machine Learning. The MIT Press, Cambridge (2012)
27. Menkovski, V., Oredope, A., Liotta, A., Cuadra-Sanchez, A.: Optimized online learning for QoE prediction. In: Proceedings of the 21st Benelux Conference on Artificial Intelligence, 2009/10/9
28. Menkovski, V., Exarchakos, G., Liotta, A.: Online QoE prediction. Second International Workshop on Quality of Multimedia Experience (QoMEX), IEEE, June 2010
29. Menkovski, V., Exarchakos, G., Liotta, A., Cuadra Sánchez, A.: Estimations and remedies for quality of experience in multimedia streaming. In: Third International Conference on Advances in Human-Oriented and Personalized Mechanisms, Technologies and Services (CENTRIC), IEEE, 2010/8/22
30. Menkovski, V., Exarchakos, G., Liotta, A.: Online learning for quality of experience management. In: The Annual Machine Learning Conference of Belgium and the Netherlands (2010)
31. Menkovski, V., Exarchakos, G., Liotta, A.: Machine learning approach for quality of experience aware networks. In: 2nd International Conference on Intelligent Networking and Collaborative Systems (INCOS), IEEE, 2010/11/24
32. Menkovski, V., Exarchakos, G., Liotta, A.: The value of relative quality in video delivery. J. Mobile Multimed. **7**(3), 151–162 (2011)
33. Menkovski, V., Exarchakos, G., Liotta, A., Cuadra Sánchez, A.: Quality of experience models for multimedia streaming. In: Advancing the Next-Generation of Mobile Computing: Emerging Technologies: Emerging Technologies. IGI Global (2012)
34. Menkovski, V., Exarchakos, G., Liotta, A.: Tackling the sheer scale of subjective QoE. In: International ICST Mobile Multimedia Communications Conference, vol. 29. Springer, Berlin (2012)
35. Menkovski, V., Liotta, A.: Adaptive psychometric scaling for video quality assessment. Signal Process. Image Commun. **27**(8) (2012)
36. Bertone, F., Menkovski, V., Liotta, A.: Adaptive P2P streaming. In: Streaming Media with Peer-to-Peer Networks. IGI Global (2012)
37. Menkovski, V., Liotta, A.: Intelligent control for adaptive video streaming. In: Proceedings of the International Conference on Consumer Electronics, Jan 2013
38. Torres Vega, M., Mocanu, D.C., Barresi, R., Fortino, G., Liotta, A.: Cognitive streaming on android devices. In: Proceedings of the IEEE/IFIP Symposium on Integrated Network and Service Management (IM'15), 11–15 May 2015, Ottawa, Canada, pp. 1316–1321. IEEE Service Center, Piscataway (2015)
39. Torres Vega, M., Giordano, E., Mocanu, D.C., Tjondronegoro, D., Liotta, A.: Cognitive no-reference video quality assessment for mobile streaming services. In: Proceedings of the 7th International Workshop on Quality of Multimedia Experience (QoMex), 26–29 May 2015, Pilos, Messinia, Greece, pp. 1–6. IEEE Service Center, Piscataway

40. Mocanu, D.C., Exarchakos, G., Bou Ammar, H., Liotta, A.: Reduced reference image quality assessment via Boltzmann machines. In: IFIP/IEEE International Symposium on Integrated Network Management (IM), 11 May 2015. IEEE (2015)
41. Mocanu, D.C., Exarchakos, G., Liotta, A.: Deep learning for objective quality assessment of 3D images. In: 2014 IEEE International Conference on Image Processing (ICIP). IEEE, 27 Oct 2014
42. ITU-R BT.2021, Subjective methods for the assessment of stereoscopic 3DTV systems (2012)
43. Winkler, S., Min, D.: Stereo/multiview picture quality: overview and recent advances. Sig. Process. Image Commun. **28**(10), 1358–1373 (2013)
44. Zhu, C., Li, S.: Depth image based view synthesis: new insights and perspectives on hole generation and filling. IEEE Trans. Broadcast. **62**(1), 82–93 (2016)
45. Scharstein, D., Szeliski, R.: A taxonomy and evaluation of dense two-frame stereo correspondence algorithms. Int. J. Comput. Vis. **47**(1), 7–42 (2002)
46. Liu, W., Chen, X., Yang, J., Wu, Q.: Robust color guided depth map restoration. IEEE Trans. Image Process. **26**(1), 315–327 (2017)
47. Lei, J., Li, L., Yue, H., Wu, F., Ling, N., Hou, C.: Depth map super-resolution considering view synthesis quality. IEEE Trans. Image Process. **26**(4), 1732–1745 (2017)
48. Nezveda, M., Brosch, N., Seitner, F., Gelautz, M.: Depth map post-processing for depth-image-based rendering: a user study. In: Proceedings of SPIE 9011, Stereoscopic Displays and Applications XXV, 90110K (2014). https://doi.org/10.1117/12.2039771
49. Bosc, E., Pepion, R., Le Callet, P., Pressigout, M., Morin, L.: Reliability of 2D quality assessment methods for synthesized views evaluation in stereoscopic viewing conditions. In: Proceedings of 3DTV-Con, pp. 1–4 (2012). https://doi.org/10.1109/3dtv.2012.6365457
50. Seitner, F., Nezveda, M., Gelautz, M., Braun, G., Kapeller, C., Zellinger, W., Moser, B.: Trifocal system for high-quality inter-camera mapping and virtual view synthesis. In: Proceedings of International Conference on 3D Imaging (IC3D), 1–8 (2015). https://doi.org/10.1109/ic3d.2015.7391819
51. Muddala, S., Olsson, R., Sjöström, M.: Spatio-temporal consistent depth-image-based rendering using layered depth image and inpainting. EURASIP J. Image Video Process. **9**(1), 1–19 (2016)
52. Buyssens, P., Le Meur, O., Daisy, M., Tschumperlé, D., Lézoray, O.: Depth-guided disocclusion inpainting of synthesized RGB-D images. IEEE Trans. Image Process. **26**(2), 525–538 (2017)
53. Bosc, E., Pépion, R., Le Callet, P., Köppel, M., Ndjiki-Nya, P., Pressigout, M., Morin, L.: Towards a new quality metric for 3-D synthesized view assessment. IEEE J. Sel. Top. Sign. Proces. **5**(7), 1332–1343 (2011)
54. Rittler, T., Nezveda, M., Seitner, F., Gelautz, M.: Depth-guided disocclusion inpainting for novel view synthesis. In: Proceedings of OAGM&ARW Joint Workshop Vision, Automation and Robotics, pp. 160–164 (2017). https://doi.org/10.3217/978-3-85125-524-9-34
55. Wang, J., Cohen, M.: Image and video matting: a survey. Found. Trends® Comput. Graph. Vis. **3**(2), 97–175 (2008)
56. Brosch, N., Nezveda, M., Gelautz, M., Seitner, F.: Efficient quality enhancement of disparity maps based on alpha matting. In: Proceedings of SPIE 9011, Stereoscopic Displays and Applications XXV, 90110M (2014). https://doi.org/10.1117/12.2035361
57. Dumic, E., Grgic, S., Bermejo, D.J., Cruz, L.A.S.: benchmark of state of the art objective measures for 3D stereoscopic video quality assessment on the nantes database. In: Proceedings of the 56th International Symposium ELMAR-2014, pp. 119–123 (2014)
58. Urvoy, M., Barkowsky, M., Cousseau, R., Koudota, Y., Ricorde, V., Le Callet, P., Gutierrez, J., Garcia, N.: NAMA3DS1-COSPAD1: subjective video quality assessment database on coding conditions introducing freely available high quality 3D stereoscopic sequences. In: Quality of Multimedia Experience (QoMEX), pp. 109–114 (2012)

59. Bosc, E., Le Callet, P., Morin, L., Pressigout, M.: Visual quality assessment of synthesized views in the context of 3DTV. In: Zhu, C., Zhao, Y., Yu, L., Tanimoto, M. (eds.) 3D-TV system with Depth-Image-Based Rendering, pp. 439–473. Springer, New York (2013)
60. Hewage, C.T.E.R., Worrall, S.T., Dogan, S., Kondoz, A.M.: Prediction of stereoscopic video quality using objective quality models of 2-D video. Electron. Lett. **44**(16), 963–965 (2008)
61. Benoit, A., Le Callet, P., Campisi, P., Cousseau, R.: Quality assessment of stereoscopic images. EURASIP J. Image Video Process. **2008**, Article ID 659024, 13 p (2008)
62. Aflaki, P., Hannuksela, M.M., Hakala, J., Häkkinenb, J., Gabbouj, M.: Estimation of subjective quality for mixed-resolution stereoscopic video. In: 3DTV Conference: The True Vision—Capture, Transmission and Display of 3D Video (3DTV-CON), pp. 1–4 (2011)
63. Bosc, E., Pepion, R., Le Callet, P., Pressigout, M., Morin, L.: Reliability of 2D quality assessment methods for synthesized views evaluation in stereoscopic viewing conditions. In: 3DTV-Conference: The True Vision—Capture, Transmission and Display of 3D Video (3DTV-CON), pp. 1–4 (2012)
64. Wang, K., Barkowsky, M., Brunnström, K., Sjöström, M., Cousseau, R., Le Callet, P.: Perceived 3D TV transmission quality assessment: multi-laboratory results using absolute category rating on quality of experience scale. IEEE Trans. Broadcasting **58**(4) (2012)
65. Conze, P.H., Robert, P., Morin, L.: Objective view synthesis quality assessment. In: Proceedings of SPIE, Stereoscopic Displays and Applications, Feb 2012
66. Battisti, F., Bosc, E., Carli, M., Le Callet, P., Perugia, S.: Objective image quality assessment of 3D synthesized views. Signal Process. Image Commun. **30**, 78–88 (2015)
67. Sandic-Stankovic, D., Kukolj, D., Le Callet, P.: DIBR synthesized image quality assessment based on morphological wavelets. International Workshop on Quality of Multimedia Experience QoMEX, Costa Navarino, Greece, May 2015
68. Sandic-Stankovic, D., Kukolj, D., Le Callet, P.: DIBR synthesized image quality assessment based on morphological pyramids. In: 3DTV-CON Immersive and Interactive 3D Media Experience Over Networks, Lisbon, July 2015
69. Ekmekcioglu, E., Worall, S.T., De Silva, D., Fernando, W.A.C., Kondoz, A.M.: Depth based perceptual quality assessment for synthesized camera viewpoints. In: International Conference on User Centric Media, Sept 2010
70. Pinson, M.H., Wolf, S.: A new standardized method for objectively measuring video quality. IEEE Trans. Broadcasting **50**(3), 312–322 (2004)
71. Oh, K.J., Yea, S., Vetro, A., Ho, Y.S.: Virtual view synthesis method and self evaluation metrics for free viewpoint television and 3D video. Int. J. Imaging Syst. Technol. (2010)
72. Shahid Farid, M., Lucenteforte, M., Grangetto, M.: Objective quality metric for 3D virtual views. In: International Conference on Image processing, ICIP (2015)
73. Solh, M., AlRegib, G.: MIQM: a novel multi-view images quality measure. The First International Workshop on Quality of Multimedia Experience (QOMEX), San Diego, CA, 29–31 July (2009)
74. Shao, H., Cao, X., Er, G.: Objective quality assessment of depth image based rendering in 3DTV system. 3DTV-CON (2009)
75. Hanhart, P., Bosc, E., Le Callet, P., Ebrahimi, T.: Free-viewpoint video sequences: a new challenge for objective quality metrics. International Workshop on Multimedia Signal Processing (MMSP), Jakarta, Indonesia (2014)
76. Solh, M., AlRegib, G., Bauza, J.M.: 3VQM: a vision-based quality measure for DIBR-based 3D videos. In: IEEE International Conference on Multimedia and Expo (ICME), 1–6, July 2011
77. Song, R., Ko, H., Kuo, C.C.J.: MCL-3D: a database for stereoscopic image quality assessment using 2D-image-plus-depth source. J. Vis. Commun. Image Represent. (2014)
78. Bosc, E., Battisti, F., Carli, M., Callet, P.L.: A wavelet-based image quality metric for the assessment of 3D synthesized views. In: Proceedings of SPIE 8648, Stereoscopic Displays and Applications, Mar 2013

79. El-Yamany, N., Ugur, K., Hannuksela, Gabbouj, M.: Evaluation of depth compression and view synthesis distortions in multiview-video-plus-depth coding systems. In: 2DTV-CON (2010)
80. Dumic, E., Grgic, S., Cruz, L.A.S., Assuncao, P.: Objective quality measures comparison of impaired 3D video sequences from the UC3D database. In: Proceedings of 3DTV-Conference 2014: In Pursuit of Next Generation 3D Display (2014)
81. Tikanmaki, A., Gotchev, A., Smolic, A., Miller, K.: Quality assessment of 3D video in rate allocation experiments. In: IEEE Symposium on Consumer Electronics, pp. 1–4, Apr 2008
82. Silva, D., Fernando, W., Worrall, S., Kondoz, A.: A novel depth map quality metric and its usage in depth map coding. In: 3DTV/Conference (2011)
83. Dehkordi, A.B., Pourzad, M., Nasiopoulos, P.: A study on the relationship between depth map quality and the overall 3D video quality of experience. In: 3DTV-Conference (2013)
84. Do, L., Zinger, S., de With, P.: Quality improving techniques for free-viewpoint DIBR. In: Proceedings of SPIE Stereoscopic displays and applications (2010)
85. Kooi, F.L., Toet, A.: Visual comfort of binocular and 3D displays. Displays **25**(2), 99–108 (2004)
86. Schneider, B., Moraglia, G., Jepson, A.: Binocular unmasking: An analog to binaural unmasking? Science **243**(4897), 1479–1482 (1989)
87. Meesters, L.M., IJsselsteijn, W.A., Seuntiëns, P.J.: A survey of perceptual evaluations and requirements of three-dimensional TV. IEEE Trans. Circuits Syst. Video Technol. **14**(3), 381–391 (2004)
88. Lambooij, M.T., IJsselsteijn, W.A., Heynderickx, I.: Visual discomfort in stereoscopic displays: a review. In: Proceedings of SPIE, vol. 6490 (2007)
89. Ostrin, L.A., Glasser, A.: Accommodation measurements in a prepresbyopic and presbyopic population. J. Cataract Refract. Surg. **30**(7), 1435–1444 (2004)
90. Rushton, S.K., Riddell, P.M.: Developing visual systems and exposure to virtual reality and stereo displays: some concerns and speculations about the demands on accommodation and vergence. Appl. Ergon. **30**(1), 69–78 (1999)
91. Babrow, A.S.: Theory and method in research on audience motives. J. Broadcast. Electron. Media **32**(4), 471–487 (1988)
92. Nisbett, R.E., Wilson, T.D.: Telling more than we can know: verbal reports on mental processes. Psychol. Rev. **84**(3), 231 (1977)
93. Ravaja, N.: Contributions of psychophysiology to media research: review and recommendations. Media Psychol. **6**(2), 193–235 (2004)
94. Pereda-Baños, A., Barreda-Ángeles, M.: On human information processing in information retrieval. In: NeuroIR'2015 13 Aug 2015, Santiago, Chile (2015)
95. Bonomi, M., Barreda-Ángeles, M., Battisti, F., Boato, G., Le Callet, P., Carli, M.: Towards QoE estimation of 3D contents through non-invasive methods. In: 3DTV-Conference: The True Vision-Capture, Transmission and Display of 3D Video (3DTV-CON), pp. 1–4. IEEE, July 2016
96. Kahneman, D.: Attention and Effort. Prentice-Hall, Englewood Cliffs (1973)
97. Wickens, C.D.: Processing resources in attention, dual task performance, and workload assessment. Technical Report EPL-81–3/ONR-81-3, University of Illinois at Urbana (1981)
98. Milán, E.G., González, A., Sanabria, D., Pereda, A., Hochel, M.: The nature of residual cost in regular switch response factors. Acta Physiol. **122**(1), 45–57 (2006)
99. Lang, P.J., Bradley, M.M., Cuthbert, B.N.: International affective picture system (IAPS): technical manual and affective ratings. In: NIMH Center for the Study of Emotion and Attention, pp. 39–58 (1997)
100. LeDoux, J.: Rethinking the emotional brain. Neuron **73**(4), 653–676 (2012)
101. Armony & LeDoux. In: Rees, A., Palmer, A.R. (eds.) The Oxford Handbook of Auditory Science: The Auditory Brain, vol. 2, pp. 479–505. Oxford University Press, NY (2010)
102. Larsson, P., Västfjäll, D.: Emotional and behavioural responses to auditory interfaces in commercial vehicles. Int. J. Veh. Noise Vib. **9**(1–2), 75–95 (2013)

103. Vorderer, P., Steen, F.F., Chan, E.: Motivation. In: Bryant, J., Vorderer, P. (eds.) Psychology of entertainment, pp. 3–17. Laurence Erlbaum, Mahwah, NJ (2006)
104. Döveling, K., von Scheve, C., Konijn, E.A. (eds.): The Routledge handbook of emotions and mass media. Routledge, London (2011)
105. Barreda-Ángeles, M., Pépion, R., Bosc, E., Le Callet, P., Pereda-Baños, A.: Exploring the effects of 3D visual discomfort on viewers' emotions. In: IEEE International Conference on Image Processing (ICIP), pp. 753–757. IEEE, Oct 2014
106. Bradley, M.M., Lang, P.J.: The international affective digitized sounds (IADS-2): affective ratings of sounds and instruction manual. University of Florida, Gainesville, FL, Technical Report. B-3 (2007)
107. Reeves, B., Nass, C.: How People Treat Computers, Television, and New Media Like Real People and Places. CSLI Publications and Cambridge University Press, pp. 3–18 (1996)
108. Zillmann, D.: Exemplification effects in the promotion of safety and health. J. Commun. **56** (s1) (2006)
109. Zillmann, D.: Mechanisms of emotional reactivity to media entertainments. In: The Routledge Handbook of Emotions and Mass Media, pp. 101–115 (2011)
110. Vorderer, P.A., Hartmann, T.: Entertainment and enjoyment as media effects (2009)
111. Vorderer, P.: Audience involvement and program loyalty. Poetics **22**(1–2), 89–98 (1993)
112. Urvoy, M., Barkowsky, M., Le Callet, P.: How visual fatigue and discomfort impact 3D-TV quality of experience: a comprehensive review of technological, psychophysical, and psychological factors. Ann. Telecommu. (annales des télécommunications) **68**(11–12), 641–655
113. Clifford, C.W., Webster, M.A., Stanley, G.B., Stocker, A.A., Kohn, A., Sharpee, T.O., Schwartz, O.: Visual adaptation: neural, psychological and computational aspects. Vis. Res. **47**(25), 3125–3131 (2007)
114. Sanchez-Vives, M.V., Slater, M.: From presence to consciousness through virtual reality. Nat. Rev. Neurosci. **6**(4), 332–339 (2005)
115. Minsky, M.: Telepresence (1980)
116. Peters, C., Pelachaud, C., Bevacqua, E., Mancini, M., Poggi, I.: A model of attention and interest using gaze behavior. In: Lecture Notes in Computer Science, vol. 3661, pp. 229–240 (2005)
117. Fredricks, J.A., Blumenfeld, P.C., Paris, A.H.: School engagement: potential of the concept, state of the evidence. Rev. Educ. Res. **74**(1), 59–109 (2004)
118. Mason, M.F., Norton, M.I., Van Horn, J.D., Wegner, D.M., Grafton, S.T., Macrae, C.N.: Wandering minds: the default mode network and stimulus-independent thought. Science **315** (5810), 393–395 (2007)
119. Arapakis, I., Barreda-Angeles, M., Pereda-Banos, A.: Interest as a proxy of engagement in news reading: spectral and entropy analyses of eeg activity patterns. IEEE Trans. Affect. Comput. (2017)
120. Hendrix, C., Barfield, W.: Presence within virtual environments as a function of visual display parameters. Presence Teleoperators Virtual Environ. **5**(3), 274–289 (1996)
121. Shilling, R.D., Shinn-Cunningham, B.: Virtual auditory displays. In: Handbook of Virtual Environment Technology, pp. 65–92 (2002)
122. Murray, C.D., Arnold, P., Thornton, B.: Presence accompanying induced hearing loss: Implications for immersive virtual environments. Presence Teleoperators Virtual Environ. **9**, 137–148 (2000)
123. Cengarle, G., Pereda-Baños, A.: The perception of masked sounds and reverberation in 3D versus 2D playback systems. In: Proceedings of 134th AES Convention, May 2013
124. Gorzel, M., Corrigan, D., Kearney, G., Squires, J., Boland, F.: Distance perception in virtual audio-visual environments. In: Proceedings of 25th AES UK Conference: Spatial Audio in Today's 3D World. AES, UK (2012)
125. Silzle, A., George, S., Habets, E.A.P., Bachmann, T.: Investigation on the quality of 3D sound reproduction. In: Proceedings of ICSA, p. 334 (2011)

126. Riecke, B.E., Väljamäe, A., Schulte-Pelkum, J.: Moving sounds enhance the visually-induced self-motion illusion (circular vection) in virtual reality. ACM Trans. Appl. Percept. (TAP) **6**(2), 7 (2009)

Chapter 11
3D Visual Content Datasets

Karel Fliegel, Federica Battisti, Marco Carli, Margrit Gelautz, Lukáš Krasula, Patrick Le Callet and Vladimir Zlokolica

Abstract Development and performance evaluation of efficient methods for coding, transmission, and quality assessment of 3D visual content require rich datasets of a suitable test material. The use of these databases allows a fair comparison of systems under test. Moreover, publicly available and widely used datasets are crucial for experimentation leading to reproducible research. This chapter presents an overview of 3D visual content datasets relevant to research in the field of coding, transmission, and quality assessment. Description of regular stereoscopic or multiview image and video datasets is presented. Databases created using emerging technologies, including light-field imaging, are also addressed. Moreover, there are databases of multimedia content annotated with ratings from the subjective experiment, which are a necessary resource for understanding the complex problem of quality of experience while consuming the 3D visual content.

K. Fliegel (✉)
Department of Radioelectronics, Faculty of Electrical Engineering,
Czech Technical University in Prague, Prague, Czech Republic
e-mail: fliegek@fel.cvut.cz

F. Battisti · M. Carli
University of Roma TRE, Rome, Italy
e-mail: federica.battisti@uniroma3.it

M. Carli
e-mail: marco.carli@uniroma3.it

M. Gelautz
Vienna University of Technology, Vienna, Austria
e-mail: margrit.gelautz@tuwien.ac.at

L. Krasula · P. Le Callet
University of Nantes, Nantes, France
e-mail: l.krasula@gmail.com

P. Le Callet
e-mail: patrick.lecallet@univ-nantes.fr

V. Zlokolica
Faculty of Technical Sciences, University of Novi Sad, Novi Sad, Serbia
e-mail: vzlokolica@uns.ac.rs

P. A. Assunção and A. Gotchev (eds.), *3D Visual Content Creation, Coding and Delivery*, Signals and Communication Technology,
https://doi.org/10.1007/978-3-319-77842-6_11

11.1 Introduction

The suitable test material, in this context 3D visual content, plays a crucial role in the development and performance evaluation of related coding, transmission, and quality assessment methods. Publicly available and widely used datasets are necessary for fair performance comparison and validation of systems under test and thus crucial for experimentation leading to reproducible research. Numerous research laboratories produced the relevant databases of 3D visual content. The content description is usually published in technical reports, research papers, and online resources, thus it is very scattered, and it is not easy to identify the most suitable dataset for the particular needs.

There were numerous efforts to provide overview and comparison of multimedia content datasets. Among the first published descriptions belong image and video quality resources website[1] by Stefan Winkler and related publications [1–3] providing in-depth analysis of multimedia content databases. Another notable achievement with the goal to provide rich and internationally recognized database of content of different sorts is "QUALINET Multimedia Databases Online" platform[2] created in the frame of ICT COST Action IC1003 "European Network on Quality of Experience in Multimedia Systems and Services" (QUALINET).[3] The platform, abbreviated "Qualinet Databases", is used to share the databases efficiently with other researchers and handles information on the multimedia content. The database was substantially extended to 3D visual content within the frame of ICT COST Action IC1105 "3D Content Creation, Coding and Transmission over Future Media Networks" (3D-ConTourNet).[4] As of September 2017, Qualinet Databases contains 241 registered datasets, from which about 30 datasets cover relevant 3D visual content, and there are more than 400 registered users.

3D visual content datasets relevant to research in the field of coding, transmission, and quality assessment are overviewed in this chapter. The chapter is focused mainly on selected databases available in the public domain. The databases are categorized, and then a detailed comparison of available datasets in various application domains is presented to help the users with the decision about which database is more suitable for the particular problem. For each discussed database an overview of the material is presented along with the details on how the content was created. Where available, also, experimental image acquisition setup and subjective experiment design are discussed.

The chapter has the following structure related to the fundamental categorization of 3D visual content datasets. At first, stereoscopic and multiview image and video content datasets are introduced in Sect. 11.2, with the basic description of related stereo dataset generation and multiview camera content for 3D reconstruction,

---

[1]Image and Video Quality Resources (http://stefan.winklerbros.net/resources.html).

[2]Qualinet Databases (http://dbq.multimediatech.cz/).

[3]COST Action IC1003 QUALINET (http://www.qualinet.eu/).

[4]COST Action IC1105 3D-ConTourNet (http://www.3d-contournet.eu/).

modeling, and visualization. Then, light-field content characterization and selection is addressed in Sect. 11.3 with focus on perceptual assessment. Special point-cloud and holographic content datasets are also reviewed in Sect. 11.4 with respect to image compression standardization activities. The most popular datasets annotated with ratings from subjective experiments are discussed in Sect. 11.5 and the chapter is concluded in Sect. 11.6.

11.2 Stereoscopic and Multiview Visual Content Datasets

In the following paragraphs, several stereo and multiview image and video datasets with reference depth are reviewed. These datasets have been made publicly available by the computer vision community. The creation of these datasets was primarily motivated by the need of depth ground truth to support the design and quantitative evaluation of computer vision algorithms, especially in the field of stereo matching. A particular merit of such repositories is the detailed information on how the data were created and the accuracy of their associated ground truth. Furthermore, ancillary information such as occlusion maps is often provided. Beyond their initial purpose of benchmarking computer vision algorithms, the stereo and multiview image plus depth data contained in these datasets can also give valuable support in the context of coding, transmission, and quality assessment of 3D visual content.

11.2.1 Stereo Dataset Generation for Different Scene Cases

Different approaches for generating stereo or multiview imagery with reference depth can be used to assess the quality of stereo matching or 3D reconstruction results. Similar to [4], datasets are distinguished between real scenes, laboratory scenes, and synthetic data, which are discussed in the following paragraphs.

Real Scenes

Real scenes have the advantage of rich natural texture and can faithfully represent diverse application scenarios. However, the simultaneous acquisition of depth ground truth for real outdoor videos typically requires the usage of a relatively expensive laser scanning device and techniques for coregistration between the depth measurements and RGB video. It may also result in missing depth information in areas that could not be mapped successfully by the employed 3D sensor. A well-established database that includes real stereo images and videos of traffic scenes taken by cameras mounted on a moving car is the KITTI dataset [5]. A stereo benchmark that also contains multiview data and stereo videos taken with mobile devices of real indoor and outdoor environments has been published recently [4].

Laboratory Scenes

The acquisition of depth ground truth is alleviated for indoor laboratory scenes, where structured light techniques with multiple exposure patterns can achieve high reconstruction accuracy for stationary settings. The controlled laboratory environment supports the acquisition of multiple images of the same scene taken from different viewpoints or under varying illumination conditions. The chosen spatial arrangement and surface characteristics such as material properties or texture allow to specifically address challenges such as occlusions or specular reflections, which have an impact on the quality of the stereo reconstruction and derived new views. A notable example of this group is the widely known Middlebury benchmark for stereo matching [6], which has been a driving factor for the development of stereo matching algorithms over the past 15 years.

Simulated Data

The main advantage of simulated data, as opposed to real and laboratory scenes, is that dense coregistered depth maps are generated as a byproduct of the rendering process, and video acquisitions with complex camera motions can be synthesized using freely chosen viewpoints. Related to this, certain sets of 3D models have been standardized (such as "Stanford Bunny"), which are used in virtual environments under controllable conditions to render stereo images, from which the 3D can be reconstructed and compared to the starting ground-truth 3D object. One example of such case is the Sintel dataset [7], which comprises stereo videos that were acquired with different rendering options accounting for a variety of shading and illumination effects. The obvious downside of synthetic data is their limited degree of reality, especially for natural outdoor scenes, which has traditionally limited their applicability in the design and evaluation of vision algorithms. However, there are notable recent developments which exploit high-quality renderings delivered by commercial computer games to generate synthetic data whose high degree of realism has already proven effective in the training of machine learning algorithms [8]. This shows the potential of simulation approaches to substitute real data in applications where additional information such as semantic labeling or depth maps are necessary. It is an open question whether such highly photorealistic computer images, which can be produced with a large variety of rendering options, can also support, for example, the development of improved objective quality metrics.

Description of the three selected example datasets are given in the following paragraphs.

MPI-Sintel

The dataset, which is available at the dedicated website,[5] is derived from an open source 3D animated movie that was created using the 3D creation software Blender.[6] The main purpose of the data collection is to provide ground truth for evaluation of optical flow algorithms. However, the dataset as described in the original publication [7] has been augmented over time and offers now also stereo videos with ground-truth disparity[7] and ground-truth depth maps with corresponding camera parameters[8] for download. The stereo dataset was created by simulating two parallel-viewing cameras that are placed 10 cm apart. It comprises 23 scenes consisting of up to 50 frames, captured at a resolution of 1024×436 pixels. The stereo videos are offered in two rendering options denoted as "clean" and "final", with the latter one accommodating not only illumination effects such as shading and specular reflections but also additional blurring due to camera motion or depth of field. The reference disparity maps are accompanied by masks that indicate missing stereo correspondences due to occlusions or border effects.

ETH3D

A very recent dataset containing natural multiview stereo imagery with associated depth ground truth has been released [4] at the dedicated website.[9] The data is provided along with online benchmarks that evaluate multiview reconstructions based on accuracy and completeness, and two-view results by measuring the disparity errors. The ETH3D repository includes 13 multiview stereo scenes captured with high resolution (6048×4032 pixels) by a Nikon D3X camera and five lower resolution videos (752×480 pixels) recorded by a synchronized multi-camera rig in a mobile setup with automated exposure settings. A specific design goal was to provide a broad range of indoor and outdoor scenes with both natural and man-made content, including some fine scene details such as trees or wires, which are challenging to reconstruct. The (non-dense) depth ground truth is delivered by a Faro Focus X 330 laser scanner.

KITTI

The KITTI Vision Benchmark Suite[10] was developed in the specific context of autonomous driving. It includes ground-truth datasets and evaluation tables for a variety of computer vision tasks including stereo and optical flow. The stereo cameras and a Velodyne laser scanner for capturing the color/gray-value images and ground-truth depth, respectively, were mounted on the roof of a driving vehicle. Additional multiview data are provided by consecutive frames of recorded video

---

[5]MPI-Sintel dataset (http://sintel.is.tue.mpg.de/).

[6]Blender (http://www.blender.org).

[7]MPI-Sintel stereo videos with ground truth disparity (http://sintel.is.tue.mpg.de/stereo).

[8]MPI-Sintel ground truth depth maps (http://sintel.is.tue.mpg.de/depth).

[9]ETH3D dataset (www.eth3d.net).

[10]KITTI Vision Benchmark Suite (http://www.cvlibs.net/datasets/kitti/).

sequences. The repository includes stereo images (at a resolution of approximately 0.5 Megapixels) of static environments along with semi-dense ground truth, which was released in 2012, and a dataset from 2015 comprising also dynamic scene objects. More information on the ground-truth acquisition, evaluation method, and ancillary data can be found in the literature [5, 9].

11.2.2 Multiview Camera Content for 3D Reconstruction, Modeling, and Visualization

In the following paragraphs, the existing multiview stereo (and photometric stereo) datasets that have been made available by different research groups are described. Details are provided for each dataset, including where they can be found, and their aimed application is explained. These datasets have been standardized to certain extent and can be used for benchmarking of the algorithms related to their target application, which is indicated by the reference publication (also attached to the explanation of the corresponding dataset).

Reconstructing 3D content from multiview camera images [10] has shown to be an efficient approach, which in the most general case does not require special setup and special kind of sensors. Consequently, such an approach can be performed flexibly and inexpensively in different environmental conditions, in comparison to other more advanced 3D acquisition methodologies. An additional advantage of such an approach is that the subsequent texture mapping and 3D modeling can be subsequently performed and optimized more easily for the target 3D visualization application.

However, currently, there is still lack of the standardized datasets for validation of the 3D multiview stereo reconstruction approaches as well as the availability of objective metrics for evaluating reconstructed 3D content [11, 12], in full reference or non-reference sense. One common existing method for 3D data reconstruction validation [11, 12] is to have a priori known camera configuration setup and then make images from different views, provide calibration data and also additionally provide the ground truth (obtained by some other more precise and expensive sensor technology) so that reconstruction quality could be assessed objectively, in full reference sense, in terms of accuracy and completeness.[11] The accuracy is usually computed as mean square error to the known 3D point cloud, while the completeness is determined as the number of 3D points being reconstructed.

The multiview stereo dataset from Middlebury (see footnote 11) is one of the most standardized sites in 3D community for multiview 3D reconstruction [11]. It provides two high-quality datasets: (i) Dino and (ii) Temple, which can be used for benchmarking and performance evaluation of the multiview stereo reconstruction algorithms. Each dataset is registered with a ground-truth 3D model acquired via a

---

[11]Middlebury dataset (http://vision.middlebury.edu/mview).

laser scanning process, to be used as a baseline for measuring accuracy and completeness. The ground truth cannot be downloaded directly though; usually one performs reconstruction, and then the evaluation testing is done per request. The multiview images are provided with different number of cameras and are of the 640×480 resolution. The images were captured using the Stanford Spherical Gantry[12] which enables moving a camera on a sphere to specified latitude/longitude angles. In order to obtain ground-truth model, the object was scanned from several orientations using a Cyberware Model 15 laser scanner.

An additional standardized 3D dataset is provided by the Stanford 3D Scanning Repository.[13] The purpose of this repository is to make some range data and detailed reconstructions available to the public for benchmarking. It provides different 3D models that can be used, for example, in the virtual environment to render the images and subsequently use them for 3D reconstruction purposes. The virtually generated images with known 3D model represent a valuable case scenario because it can be used for different use-cases exploration and algorithm evaluation and tunings. For generating the 3D models, different kind of 3D scanners have been used, which provide different quality of 3D viewing and 3D visualization for the target application.

On the other hand, most of the other available datasets provide only multiview images along with calibration data and silhouettes that can be used for 3D reconstruction evaluation, 3D modeling, and visualization. Selected examples of relevant datasets are listed below.

EPFL—Computer Vision Group CVLAB

The Computer Vision Group CVLAB at EPFL provides extensive datasets related to multi-camera visualization of the 3D content and 3D registration, in indoor and outdoor environment, for various applications, such as tree structure reconstruction, multiview evaluation, stereo face database, multiview stereo, multi-camera pedestrians video, multiview car dataset, deformable surface reconstruction, etc. The multiview evaluation dataset[14] represents one of the most important available dataset for evaluation of the multi-camera calibration and 3D reconstruction algorithms based on multiview imaging [12]. It consists of six multiview datasets with ground-truth 3D point cloud and rendered 3D model. Moreover, it also provides results of different structure from motion algorithms for the given data.

Next, stereo face database[15] is provided, which consists of 100 faces in eight positions captured by two cameras. These datasets are generated for the purpose of validation for the proposed approach for face modeling and face recognition from a pair of calibrated stereo cameras [13]. However, it can be used more generally for

[12]Stanford Spherical Gantry (https://graphics.stanford.edu/projects/gantry/).

[13]Stanford 3D Scanning Repository (http://graphics.stanford.edu/data/3Dscanrep/).

[14]CVLAB multiview evaluation dataset (https://cvlab.epfl.ch/).

[15]CVLAB stereo face database (http://cvlab.epfl.ch/data/stereoface).

stereo 3D reconstruction, 3D face modeling, and motion tracking. The dataset contains the camera parameter file including intrinsic matrix K, radial distortion, rotation matrixes, and translation vector. The camera images size is $640 \times 480 \times 3$, captured in controllable indoor environment.

The additional two datasets are the multiview stereo set of buildings and multiview car dataset. The multiview stereo set of buildings in outdoor environment for dense depth and 3D reconstruction[16] contains images of mid-resolution size (approx. $3000 \times 2000 \times 3$) and has been generated for validation of the proposed stereo from multiple views method [14, 15]. The original images have been compensated for radial distortion, and external and internal calibration parameters have been provided along. Additionally, initial 3D points from calibration have also been provided. The multiview car dataset[17] contains 20 sequences of cars as they rotate by 360°. There is one image approximately every 3°–4°. Using the time of capture information from the photos, it is possible to calculate the approximate rotation angle of the car. The dataset has been used for the multiview object pose estimation algorithm [16] but represents a good dataset also for general 3D reconstruction and 3D registration validation purpose.

TUM—Computer Vision Group

There are multiple datasets available capturing objects from various vantage points.[18] Each entry contains an image sequence, corresponding silhouettes, and full calibration parameters. The camera configuration setup consists of circular configuration with special lighting in indoor conditions. In this setup, five different objects were captured from various positions ("bird", "beethoven", "bunny", "head", "pig"). This dataset is specifically generated for validation of the proposed multiview camera 3D reconstruction [17], but it represents a valuable dataset to be used for 3D reconstruction benchmarking and performance comparison between different methods.

Cornell 3D Location Recognition Datasets

The 3D Location Recognition Datasets[19] contain a large amount of multiview images of Rome and Dubrovnik [18], which can be used for 2D-to-3D matching, i.e., 3D reconstruction from multiple views, based on which 3D point cloud can be obtained and used for 3D modeling and visualization evaluation.

[16]CVLAB stereo dataset of buildings (http://cvlab.epfl.ch/data/strechamvs).

[17]CVLAB multiview car dataset (http://cvlab.epfl.ch/data/pose).

[18]TUM—Computer Vision Group (http://vision.in.tum.de/data/datasets/3dreconstruction).

[19]Cornell 3D Location Recognition Datasets (http://www.cs.cornell.edu/projects/p2f).

Washington University Photo Tourism Dataset

The dataset[20] represents 715-image reconstruction of Notre Dame Cathedral in Paris, which can be used for 3D reconstruction, modeling, and visualization evaluation.

Photometric Stereo Datasets

Besides multiview stereo reconstruction algorithms, substantial progress has been made in the development of photometric stereo methodologies, which can deal with general materials and unknown illumination conditions. The main idea here is to use a single camera and capture multiple images with changeable lighting conditions, where one usually uses controllable lighting conditions. This approach is particularly valuable and important for performing fine detail 3D reconstruction that cannot be obtained with only multiview stereo correspondences. However, due to the lack of suitable benchmark data with ground-truth shapes (normals), quantitative comparison and evaluation is difficult to achieve. Related to these approaches the corresponding databases have been generated and are made available.

Photometric Harvard Stereo Dataset

Photometric Harvard Stereo Dataset[21] provides data for normal and 3D surface reconstruction. Each object in the dataset is illuminated under 20 different directional lightings, which are calibrated with two chrome spheres. The lighting strength is estimated by a simple normalization on image intensities (99 percentile) followed by a nonlinear optimization. The albedo and normal vectors of the object are solved with a least squares system, and the depth map is integrated with the Frankot-Chellappa algorithm [19]. The reconstruction error is measured by re-rendering the estimated normal map into a shading image and comparing that with the actual captured one. The data, as well as the code for normal and surface reconstruction, are provided.

"DiLiGenT" Photometric Stereo Dataset

"DiLiGenT" Photometric Stereo Dataset[22] is photometric stereo image dataset provided with calibrated directional lightings, objects of general reflectance, and "ground-truth" shapes (normals) for orthographic projection and single-view setup. In addition to the first dataset for such a purpose, a photometric stereo taxonomy is provided as well, emphasizing on non-lambertian and uncalibrated methods. Based on the dataset, state-of-the-art photometric stereo methods are quantitatively evaluated for general non-lambertian materials and unknown lightings to analyze their strengths and limitations [20].

---

[20]Washington University Photo Tourism Dataset (http://phototour.cs.washington.edu/datasets/).

[21]Photometric Harvard Stereo Dataset (http://vision.seas.harvard.edu/qsfs/Data.html).

[22]'DiLiGenT' Photometric Stereo Dataset (https://sites.google.com/site/photometricstereodata/).

Standford Computer Vision and Geometry Lab Datasets

The 3D dataset for other more advanced computer vision applications such as multiview 3D reconstruction, registration, and recognition applications are provided by Stanford Computer Vision and Geometry Lab[23]: (1) PASCAL3D + dataset [21], which is a novel and challenging dataset for 3D object detection and pose estimation. PASCAL3D + augments 12 rigid categories of the PASCAL VOC 2012 [22] with 3D annotations. This dataset represents a rich test bed to study 3D detection and pose estimation; (2) Stanford 2D-3D-Semantics Dataset, 2D-3D-S[24] [23], provides a variety of mutually registered modalities from 2D, 2.5D, and 3D domains, with instance-level semantic and geometric annotations. It covers over 6000 m^2 and contains over 70,000 RGB images, along with the corresponding depths, surface normal, semantic annotations, global XYZ images (all in forms of both regular and 360° equirectangular images) as well as camera information. It also includes registered raw and semantically annotated 3D meshes and point clouds. The dataset enables development of joint and cross-modal learning models and potentially unsupervised approaches utilizing the regularities present in large-scale indoor spaces.

11.3 Characterization and Selection of Light-Field Content for Perceptual Assessment

Many efforts have been devoted to the design of image and video quality assessment methods. In order to evaluate the quality of processed images, to compare the performance of different algorithms, or to determine the quality criteria in system optimization, the availability of test data is of primary importance. The Source Sequences (SRCs) selection is not a trivial task, especially for special content, such as light-field. In fact, the quality, the dataset cardinality, and the content of the selected SRCs may affect the performance assessment.

Concerning the content, to be as general purpose as possible, SRCs should span a wide range of content typologies. To characterize image content, low-level and high-level features can be used. In particular, low-level features, such as spatial information, color information, and brightness are considered important parameters that help in measuring the distortions suffered by data compression or transmission over a bandwidth-limited channel.

Among others, Spatial Information (SI) [24], colorfulness (CF) [25], contrast, correlation, homogeneity, brightness, hue, and saturation are related to image quality attributes and Human Visual System (HVS) characteristics [26]. In more details, SI is a perceptual indicator of spatial information of a scene, colorfulness is

[23]Stanford Computer Vision and Geometry Lab (http://cvgl.stanford.edu/resources.html).

[24]Stanford 2D-3D-Semantics Dataset 2D-3D-S (http://buildingparser.stanford.edu/dataset.html).

a perceptual attribute tied with image quality and naturalness of the images, while contrast, color information, and brightness are features strictly related to HVS features.

Several SI filters have been proposed in the literature. In [27] a method based on long edge detection is presented. Separate horizontal and vertical filters are applied, and the total edge energy is computed as Euclidean distance. Similarly, an SI filter for video is presented in [1], as a perceptual indicator of the spatial information of the scene. It measures the amount of spatial details for each frame; the SI value is higher for spatially complex scenes. SI filter has been applied to LF data in [28]. The authors show that the correlation between the SI scores estimated by using the cited methods is very high, since both the methods exploit the classical Sobel filtering. Another approach based on the ITU recommendation [29] has been adopted. The luminance component of the image is first filtered by using a Sobel filter. Then, the standard deviation over the pixels in each filtered component is computed as SI.

Colorfulness and aesthetic are important visual features having a significant impact on the perceptual quality of a scene. In literature, many efforts have been devoted to study the color impact and its assessment. A CF metric for natural images is presented in [25] based on the distribution of image pixels in CIELab color space [30]. In [31], aesthetics (e.g., "the principles of the nature and the appreciation of beauty") in photographic images is addressed by exploiting several metrics, such as light, CF, saturation, hue, and texture, to understand the human emotions with respect to the visual content.

Dealing with LF images, the inner structure of the LF must be considered. A light-field camera provides information about depth dependence and Lambertian lighting. Depth dependence implies multiple depths of semitransparent objects and the Lambertian surface reflects light with the equal intensity in all directions [32].

The depth dependence information can be exploited during coding, and the variation in depth of field information could give different compression levels at the same quality level.

Reflections and transparency are prevalent in natural images, that is, reflected and transmitted lights are superimposed on each other. The image can be modeled as a linear combination of the transmitted layer, which contains the scene of interest, and a secondary layer, which contains the reflection or transparency [33, 34]. The decomposition of the images into two layers is an ill-posed problem in the absence of additional information about the scene [35]. The light-field camera recorded information, particularly multiple views of a single scene, can be exploited to solve the problem. Therefore, in a test dataset images with transparency, reflections, and wide Depth of Field (DoF) variation are needed.

Depth Properties

One of the main properties of LF imaging is the possibility of obtaining depth information of the captured scene, offering both horizontal and vertical parallax.

As observed with 3D content, depth properties are crucial for an appropriate description and characterization of LF content.

Depth map and depth histogram: Obtaining the depth information from data captured by the acquisition systems (e.g., camera arrays, plenoptic cameras, etc.) is a challenging issue as demonstrated by the number of different approaches that are being proposed to deal with this problem. Different approaches should be considered when dealing with sparse LFs (e.g., captured by camera arrays) and dense LFs (e.g., captured by plenoptic cameras), due to the different acquisition properties, such as baseline and spatial aliasing. On one hand, depth estimations from sparse LFs can be obtained by using traditional multiview methods [36], as well as some specific techniques, for instance, based on sweeping [37] or multi-resolution matching [38]. On the other hand, for dense LF in [39], a simple technique based on computing block-wise cross-correlation is proposed. More recently, approaches taking into account multiview stereo correspondences [37, 40] have been introduced.

Disparity range: While the majority of the methods mentioned above to estimate depth provide a normalized map here only relative disparities can be obtained, the absolute disparities in terms of pixels are more important for QoE aspects, such as content characterization [41], have been presented. However, to obtain a reliable content characterization, it is required the estimate of the depth range of the scene regarding distances to the nearest and furthest objects, or the camera calibration parameters. When these data are not available, estimation algorithms, such as the multiview stereo algorithm described in [42], could return pixel disparities. This algorithm has been used (over the subaperture images of the LF images) for characterizing LF data in [28].

Occlusions are one of the most important problems to deal with in-depth estimations for LFs. However, until now, only few depth estimation algorithms specifically manage and model occlusions, i.e., the occlusion model for depth map estimation in [43]. This algorithm is applied over the LF data structure, and the amount of possible occluded pixels are computed and considered in the content characterization [28].

Refocusing Features

As aforementioned, one of the main applications of LF images is the possibility of changing the focused elements of the images. Therefore, it is important to find appropriate descriptors that could help in the characterization of LF content, providing an estimation of their possible performance in this particular use case. One alternative is to analyze the properties and shape of the disparity histogram since it provides information about the distribution in depth of the elements of the scene.

Other possibilities to deal with the use of blur metrics, such as the technique proposed in [44] taking into account the perceived image quality induced by blur. In addition, some approaches have been proposed to measure focuses specifically in LF images, such as the Multifocal Scene Defocus Quality (MSDQ) metric, which quantifies the perceptual visual quality of rendering LF images [45].

Finally, as shown in [28], the refocusing range of the LF images can be computed using the "shift and sum" algorithm that is based on the digital refocusing approach proposed in [46], which reveals that refocused images can be obtained by adding shifted subaperture images. In particular, the refocusing range is determined by the slope parameters of the algorithm used to obtain images refocused on the nearest and the furthest elements of the scene.

Selected Light-field Datasets

Several LF datasets have been proposed in the literature. The main features are reported in Table 11.1. Stanford LF Archive [47] is widely used; however, the images are captured by using a multi-camera system including gantry, microscope, etc. Nowadays, different LF cameras have been realized [48], (e.g., Lytro, Lytro Illum, and Raytrix), thus allowing the consumers to exploit such a technology. Lytro Illum is the newer version of the Lytro plenoptic camera, characterized by increased resolution and processing capabilities, while Raytrix is a so-called focused plenoptic camera. As can be noticed, the dataset [47] is not sufficient to deal with new challenges, perceptual quality evaluation, performance testing for processing algorithms, etc., which arose with the advancement of the LF technology. Other recently proposed datasets listed in Table 1 have been designed for specific purposes and the images have been acquired mostly by the Lytro plenoptic camera. In the dataset [49], the Lytro Illum camera has been used. However, most of the images have similar features and motivations behind the particular image content selection have not been reported. In [48], a LF image dataset is proposed. The dataset creation methodology using Lytro Illum, description of LF images, and analysis of LF image content is tailored. The SRCs image content selection criteria is defined, a comprehensive LF image quality dataset is proposed and made freely available to the research community, a spatial information estimation metric is exploited, an analysis of the features of the proposed dataset is provided.

11.4 Special Point-Cloud and Holographic Content Datasets

The overview of publicly available 3D visual content datasets mentioned in the previous paragraphs is far from complete since the number of relevant databases is continuously growing. For completeness, it is important to note that there are datasets used recently in the frame of development and standardization of image and video compression techniques within the Joint Photographic Experts Group (JPEG)[25] committee and the Moving Picture Experts Group (MPEG).[26]

[25]Joint Photographic Experts Group (https://jpeg.org/index.html).

[26]Moving Picture Experts Group (https://mpeg.chiariglione.org/).

Table 11.1 Overview of light-field datasets with corresponding features

Dataset	Year	Purpose	Features	Acquisition devices	Depth map
GUC light-field face and iris database [71]	2016	Face and iris recognition	Two biometric image databases collected by using a Lytro camera on multiple faces and visible iris (112 subjects for faces and 55 subjects for eye pattern)	Lytro	No
Lytro dataset [72]	2015	Light-field Reconstruction	30 images, with indoor and outdoor, motion blur, long exposure time, and flat image	Lytro	No
EPFL light-field image dataset [49]	2015	General	118 Lytro images with different categories: buildings, landscapes, people, etc.	Lytro Illum	No
LCAV-31 [73]	2014	Object recognition	Light-field images of 31 object categories captured from ordinary household objects and designed for object recognition purpose	Lytro	No
Light-field saliency Dataset (LFSD) [74]	2014	Saliency map estimation	100 light-field images with 60 indoor scenes and 40 outdoor scenes	Lytro	Yes (estimated)
Synthetic light-field archive [75]	2013	General	Artificial light-field images including images with transparencies, occlusions, and reflections	Camera (artificial light field)	No
Light-field analysis [76]	2013	Depth map estimation	Seven Blender and Six Gantry images; however, images do not cover the wide range of natural scenes	Blender Software and Gantry device	Yes
Stanford Light-Field Archive [47]	2008	General	20 light fields sampled using a camera array, a gantry, and a light-field microscope.	Gantry, light-field microscope, and camera array	No
SMART [48]	2016	General	15 Lytro Illum images with different categories	Lytro Illum	No

Notable progress is being made in the frame of JPEG Pleno[27] [50], which intends to provide a standard framework to facilitate the capture, representation, and exchange of omnidirectional, depth-enhanced, point-cloud, light-field, and holographic imaging modalities. JPEG Pleno is planned to provide an efficient compression format that will guarantee the highest quality content representation with reasonable resource requirements.

The JPEG Pleno Database[28] contains images from multiple plenoptic imaging modalities, e.g., light-field, point-cloud, and holographic imaging. There are five point-cloud datasets in the JPEG Pleno Database, one light-field dataset, and two datasets of holographic images. The light-field dataset [49] is addressed in the previous paragraph, thus only point-cloud, and holographic JPEG Pleno datasets are overviewed below. There is also one additional 3D point-cloud dataset [51] included in the overview.

11.4.1 JPEG Pleno Database: Point-Cloud Datasets

8i Voxelized Full Bodies (8iVFB v2) dataset [52] contains dynamic voxelized point cloud, i.e., sequence of frames with sets of points constrained to lie on a regular 3D grid. The dataset includes four sequences named "longdress", "loot", "redandblack", and "soldier". The human subjects' full bodies are captured by 42 RGB cameras configured in 14 clusters, at 30 fps with 10 s length. One spatial resolution is provided for each sequence: a cube of $1024 \times 1024 \times 1024$ voxels. The attributes of an occupied voxel are the red, green, and blue components of the surface color.

There are upper bodies of five subjects captured in the Microsoft Voxelized Upper Bodies dataset, named "Andrew", "David", "Phil", "Ricardo", and "Sara". The capturing was done using four frontal RGBD cameras, at 30 fps, over a 7–10 s period for each. Two spatial resolutions are provided for each sequence: a cube of $512 \times 512 \times 512$ voxels and a cube of $1024 \times 1024 \times 1024$ voxels.

ScanLAB Projects acquired and provide two datasets, namely, the Science Museum Shipping Galleries point-cloud dataset and Biplane point-cloud dataset. For the first dataset, the Shipping Galleries at the Science Museum were 3D scanned before their decommissioning in 2012 by ScanLAB Projects. A total of 256 scans were taken of the space and its exhibits to create a digital model of over two billion precisely measured points. This digital replica has been used to create a virtual flythrough of the gallery spaces providing detailed narration about the key exhibits and artefacts. The second dataset, Biplane, consists of the scan of a Handley Page Gugnunc, wooden biplane from 1920s exhibited at the Science Museum, Wroughton.

[27]JPEG Pleno (https://jpeg.org/jpegpleno/index.html).

[28]JPEG Pleno Database (https://jpeg.org/plenodb/).

The GTI-UPM Point-cloud dataset includes a directory structure consisting of several 3D models (both point clouds and naked/textured meshes) reconstructed from 2D pictures by GTI-UPM within the activities of the EU-funded research project BRIDGET (BRIDging the Gap for Enhanced broadcasT).

11.4.2 *JPEG Pleno Database: Holographic Datasets*

There are two holographic datasets available, namely ERC Interfere Holograms (data set 1) and B-com Holograms. Holography allows for recording and reproduction of wavefields of light. It is able to fully capture the three-dimensional structure of objects. Holograms represent interference patterns and their signal properties are very different from natural photography and video. The Interfere[29] database [53] contains five computer generated holograms created from 2D and 3D objects using an algorithm capable of handling self-occlusion for 3D objects. B-com Holograms [54] were synthesized using the algorithms developed by the Institute of Research & Technology (IRT) b-com.[30]

Oakland 3D Point-Cloud Dataset

This repository[31] contains labeled 3D point-cloud laser data collected from a moving platform in an urban environment. This dataset was used to produce the results presented in [51]. The data was collected using Navlab11 equipped with side-looking SICK LMS laser scanners and used in push-broom. The data was collected around CMU campus in Oakland. Data are provided in ASCII format: x y z label confidence, one point per line, space as separator. Corresponding VRML files (*.wrl) and label counts (*.stats) are also provided. The dataset is made of two subsets (part2, part3) with each its own local reference frame, where each file contains 100,000 3D points. The training/validation and testing data was filtered and labeled remapped from 44 into five labels.

11.5 Datasets Annotated with Ratings from Subjective Experiments

This section describes selected publicly available datasets that have been annotated in a subjective study. Most of the studies result in Mean Opinion Scores (MOS) quantifying the quality of each stimulus in the set. Such databases are

[29]ERC-funded Interfere project (http://www.erc-interfere.eu/).

[30]b-com hologram repository (https://hologram-repository.labs.b-com.com).

[31]Oakland Dataset (http://www.cs.cmu.edu/~vmr/datasets/oakland_3d/cvpr09/doc/).

essential for design and evaluation of objective quality metrics, described in the respective chapter in this book.

Since visual attention is of very high importance for understanding human perception in 3D applications, some effort has also been dedicated to track and record observers' gaze when exposed to the content. The datasets annotated with data from eye-tracking experiments are very useful for modeling perceptual mechanisms of the human visual system.

The described databases are further divided into image quality datasets, video quality datasets, 3D models quality datasets, and eye-tracking datasets.

11.5.1 3D Image Quality Databases

In the following paragraphs, there are six selected 3D image quality databases listed and described in detail.

IRCCyN/IVC 3D Images

The dataset[32] comprises six original stereoscopic images (with mean resolution of 512×448 pixels) and 90 distorted versions, annotated with respective Differential MOS (DMOS) values [55]. The used distortions include blur (Gaussian or downscale and upscale), JPEG, and JPEG2000 each on five different levels.

The subjective experiment was performed on 21″ Samsung SyncMaster 1100 MB display with 1024×768 pixels resolution and the frequency of 120 Hz. The viewing conditions were according to ITU-R Rec. BT.500 [56] and the viewing distance was set to four times the height of the images. The images were displayed in the center without upscaling. The observers were equipped with crystal shutter glasses.

There were 19 participants of sufficient visual acuity enrolled in the test. Their average age was 28.2. The images were evaluated using SAMVIQ [24] procedure in two sessions of 30 min per observer. The resulting DMOS scores range from 0 to 100.

LIVE 3D Image Quality Database Phase I

This database[33] contains 20 stereoscopic source images of 640×360 pixels [57]. From these scenes, 365 distorted images were created. 80 images were distorted by JPEG, 80 by JPEG2000, 80 by white noise, 80 by JPEG2000 transmitted over Rayleigh fading channel with various signal to noise ratio, and 45 by Gaussian blur. All the distortions are applied symmetrically, i.e., to both left and right image in each stereo pair.

[32]IRCCyN/IVC 3D Images dataset (http://ivc.univ-nantes.fr/en/databases/3D_Images/).

[33]LIVE 3D Image Quality Database Phase I (http://live.ece.utexas.edu/research/quality/live_3dimage_phase1.html).

A 22″ passive stereoscopic display IZ3D with the resolution set to 800 × 600 pixels was employed for subjective assessment. Each image was viewed by 17 subjects for 8 s and then assessed according to single stimulus continuous quality evaluation (SSCQE) procedure with hidden reference [56]. Two subjects were eliminated by outlier removal. The results are provided in the form of DMOS ranging from −10 to 100 (negative DMOS meaning an image evaluated better than reference).

LIVE 3D Image Quality Database Phase II

Despite the similarity in name and certain overlap in source content, this dataset [58][34] can be considered independent from the Phase I described above. Here, the distortions were applied both symmetrically and asymmetrically, and a different subjective study was conducted.

There are eight stereoscopic source images of 640 × 360 pixels and 360 distorted versions available. The applied distortions are similar to the Phase I, i.e., JPEG, JPEG2000, white Gaussian noise, Gaussian blur, and Rayleigh fading channel. From each combination of source image and distortion, three symmetrically, and six asymmetrically distorted stereo pairs were created.

The experiment was performed on 58″ Panasonic 3D television with active shutter glasses from the distance of 116 in., i.e., four times the screen height. 33 observers (22–42 years old) participated in the test which comprised of two 30 min long sessions. The procedure and data processing was the same as in case of Phase I described above.

MMSPG 3D Image Quality Assessment Database

The dataset[35] deals with the impact of distance of cameras during acquisition on the final stereoscopic image quality [59]. The set contains nine full HD (1920 × 1080) source scenes, each captured by cameras with six different distances, ranging from 10 to 60 cm.

In the test, the stereoscopic images were displayed on 46″ polarized stereoscopic full HD display Hyundai S465D. The viewing distance was three times the screen height, and the conditions were conforming to ITU-R Rec. BT.500 [56].

The content was assessed by 17 observers (22 to 53 years old, 30 on average). Single Stimulus (SS) methodology with five level discrete scale (Bad, Poor, Fair, Good, and Excellent) has been adopted. No outliers have been detected, thus the dataset provides raw scores from all of the observers, together with respective MOS and confidence intervals.

---

[34]LIVE 3D Image Quality Database Phase II (http://live.ece.utexas.edu/research/quality/live_3dimage_phase2.html).

[35]MMSPG 3D Image Quality Assessment Database (http://mmspg.epfl.ch/3diqa).

IRCCyN/IVC DIBR Images

This dataset,[36] described in detail in [60], focuses on depth image based rendering (DIBR). Three multiview sequences are considered—Book Arrival (1024 × 768, 16 cameras with 6.5 cm spacing), Lovebird1 (1024 × 768, 12 cameras with 3.5 cm spacing), and Newspaper (1024 × 768, nine cameras with 5 cm spacing). For each of them, four new viewpoints are generated using seven different algorithms thus obtaining 96 sequences in total. For the purpose of this study, a key-frame has been extracted from each of the sequences and compared to the others.

The results of two subjective experiments are available for the above-described images. In the first one, Absolute Category Rating (ACR) methodology [29] with five level discrete scale was used, while Pair Comparison (PC) procedure was adopted in the second. The conditions for the two tests were identical.

The content was displayed on a full HD TVLogic LVM401 W display. The viewing conditions were according to ITU-R Rec. BT.500 [56]. 43 subjects participated in both tests. Raw scores coming from both procedures are provided together with MOS, in case of ACR, and Thurstone-Moesteller scores [61] for PC methodology.

MCL-3D Database

The last image dataset to be described is MCL-3D [62] [37] and deals with DIBR as well. It is based on nine source scenes, provided in image plus depth form. Six of the scenes are in full HD (1920 × 1080) resolution, while the rest is in 1024 × 768. Six types of distortion, namely Gaussian blur, additive white noise, downsampling blur, JPEG, JPEG2000, and transmission errors, are applied on four different levels. Moreover, four types of rendering algorithms are employed. Overall, the dataset comprises of 693 stereoscopic pairs.

Pair comparison methodology has been adopted. The stimuli were displayed on 46.9″ LG 47LW5600 screen. The viewing distance was 3.2 meters, and each observer was given polarized glasses. The cubic function has been used to resize the images to fit the screen, and the gap between them was filled with gray pixels.

270 observers took part in the experiment in order to collect 30 opinion scores for each stimulus. The results were transformed into MOS ranging from 0 to 9.

11.5.2 3D Video Quality Databases

In the following paragraphs, there are three selected 3D video quality databases listed and described in detail.

[36]IRCCyN/IVC DIBR Images (http://ivc.univ-nantes.fr/en/databases/DIBR_Images/).

[37]MCL-3D Database (http://mcl.usc.edu/mcl-3d-database/).

IRCCyN/IVC NAMA3DS1

The video dataset described in [41][38] contains 10 progressive full HD stereo source sequences with 25 frames per second (fps) and 10 versions of each source symmetrically distorted by processing. The used algorithms include compression by H.264/AVC on three levels and by JPEG2000 on four levels, downsampling, sharpening, and a combination of downsampling and sharpening. Overall, there are 110 videos in the dataset. The duration of 99 sequences is 16 s while the other 11 videos are 11 s long.

The content was evaluated in the conditions defined by ITU-R Rec. BT.500 [56] on a 46″ full HD 50 Hz LCD Philips 46PFL9705H with shutter glasses from 172 cm which corresponds to three times the picture height.

ACR with hidden reference [29] was selected as an appropriate subjective procedure. 29 observers (12 females and 17 males of age between 18 and 63) participated in the study. One of the observers was eliminated by outlier removal. The publicly available data, therefore, include (apart from the video sequences) raw scores, MOS, and standard deviations computed from 28 observers.

MMSPG 3D Video Quality Assessment Database

Similarly to the previously described Image Quality Database [59],[39] MMSPG 3D Video Quality Database [63] studies the impact of the camera distance. The dataset comprises of six different source scenes captured by full HD cameras in six distances (10–50 cm) from each other with 25 fps.

The procedure and the conditions were similar to the experiment conducted for MMSPG 3D Image Quality Database. 20 subjects (24–37 years old, 27 on average) participated in the test, but three of them have been recognized as outliers by the post-screening procedure. The final analysis was, therefore, performed on data from 17 observers.

IRCCyN/IVC DIBR Videos

The database described in [64][40] is an extension of the previously introduced IRCCyN/IVC DIBR Image dataset [60]. The same three reference sequences have been used. The first one has 15 fps while the other two 30 fps. Apart from the three different unprocessed views, seven view interpolation algorithms were included in the test, together with three different levels of H.264/AVC compression applied on the first view. Altogether, the dataset contains 102 video sequences.

The display, the room, and the viewing conditions were the same as in the case of still images, however, only one study using ACR methodology was conducted. The resulting MOS values are obtained from 32 observers.

---

[38]IRCCyN/IVC NAMA3DS1 (http://ivc.univ-nantes.fr/en/databases/NAMA3DS1_COSPAD1/).
[39]MMSPG 3D Video Quality Assessment Database (http://mmspg.epfl.ch/cms/page-58395.html).
[40]IRCCyN/IVC DIBR Videos (http://ivc.univ-nantes.fr/en/databases/DIBR_Videos/).

11.5.3 3D Models Quality Databases

In the following paragraphs, there are two selected 3D models quality databases listed and described in detail.

LIRIS 3D Model Masking Database

The first dataset evaluating the quality of 3D models was described in [65].[41] There are four models included in the dataset. Three levels of noise and smoothing are applied to rough and intermediate areas, as well as to the whole model. The noise is also added to the smooth areas. Final set, therefore, contains four reference and 84 distorted objects.

First, the subjects were trained by showing the original and the worst cases. After that, the models (including the original) were shown sequentially, each for 20 s, and scored from 0 to 10 according to the perceived impairment (0 meaning no impairment). The observers were allowed to interact with the objects (rotation, scaling, and translation).

12 participants performed the test. The authors provide their raw scores, together with MOS and objective metrics values.

LIRIS/EPFL 3D Model General-Purpose Database

The second 3D models quality database [66][42] also include four original models, although two of them are different than in the previous dataset. Here, only three levels of noise are applied either on smooth or rough regions. This gives four original and 24 distorted objects in total.

In the subjective experiment, each reference object and all of its versions were displayed together. The observers rated each of the distorted models on the scale from 0 to 4 according to similarity to the original (4 meaning completely identical). The objects were displayed for 3 min and participants could interact with them (rotation, scaling, and translation).

The study was carried out with 11 observers. With the dataset, raw scores and MOS are made publicly available, as well as some objective perceptual metrics values [67].

11.5.4 Eye-Tracking 3D Databases

In the following paragraphs, there are three selected eye-tracking 3D databases listed and described in detail.

---

[41]LIRIS 3D Model Masking Database (http://liris.cnrs.fr/guillaume.lavoue/data/datasets.html).

[42]LIRIS/EPFL 3D Model General-Purpose Database (http://liris.cnrs.fr/guillaume.lavoue/data/datasets.html).

IRCCyN/IVC 3D Gaze

The 3D Gaze dataset [68][43] includes 18 stereo images (provided as two 2D images in png format). 10 of them are obtained from the Middlebury stereo database,[44] while the rest were obtained by the authors. The dataset focuses on the influence of content features on the visual attention deployment, therefore no distortions are added, and the observers are given a free viewing task (i.e., they were instructed to freely observe the images without any particular task).

The images were displayed on a Panasonic BT-3DL2550 polarized screen with the frequency of 60 Hz and full HD resolution. SMI Hi-Speed eye-tracker was used in binocular mode. The acquisition frequency was 500 Hz. The viewing conditions were according to the ITU-R Rec. BT.500 [56] and the viewing distance was set to three times the screen height.

35 observers between 18 and 46 years old (24.23 on average) participated in the eye-tracking experiment. Raw data from the eye-tracker for each observer are provided, along with fixation density maps, the original stereoscopic pairs, depth maps, and disparity maps. Additional information and all the files for download are publicly available.

EyeC3D: 3D Video Eye-tracking Dataset

Unlike the previous database, EyeC3D [69][45] provides the visual attention information in videos. Eight stereo sequences of 8–10 s were watched in a free viewing task. 46″ polarized stereoscopic full HD display Hyundai S465D was used together with Smart Eye Pro 5.8 eye-tracker with the accuracy less than 0.5 degrees and sampling frequency of 60 Hz.

21 subjects participated in the test (18–31 years old with average of 21.8). Each sequence was watched twice by every observer. Fixation density maps were computed for each frame. The database also provides a list of all fixation points.

IRCCyN/IVC Eye-tracking Database for Stereoscopic Videos

The last dataset to be described is also dealing with task-free visual attention in stereoscopic videos [70].[46] It is also much larger than the previously described eye-tracking datasets with 47 stereo sequences composed by two 2D videos merged on a side by side avi files. Disparity map for each frame is also provided.

Panasonic BT-3DL2550 polarized screen with frequency of 60 Hz and full HD resolution was employed along with SMI RED binocular eye-tracker with acquisition frequency of 50 Hz. The room conditions were compliant with ITU-R Rec. BT.500 [56] and the viewing distance was set to three times the screen height.

[43]IRCCyN/IVC 3D Gaze (http://ivc.univ-nantes.fr/en/databases/3D_Gaze/).

[44]Middlebury stereo database (http://vision.middlebury.edu/stereo/data/).

[45]EyeC3D: 3D Video Eye-tracking Dataset (http://mmspg.epfl.ch/eyec3d).

[46]IRCCyN/IVC Eye-tracking Database for Stereoscopic Videos (http://ivc.univ-nantes.fr/en/databases/Eyetracking_For_Stereoscopic_Videos/).

The duration of one session was approximately 20 min. 40 observers (19–44 years old with average of 26) took part in the experiment. Fixation density map for each frame is provided as a png image. Higher pixel value (i.e., more white) means higher visual saliency.

11.6 Conclusions

This chapter presents an overview of datasets, their categorization, creation, and typical applications in development and performance evaluation of methods for processing, coding, transmission, and quality assessment of 3D visual content. As for the content types, stereoscopic and multiview image and video content datasets are introduced. Then, light-field content characterization and selection is addressed. Also, selected special point-cloud and holographic content datasets are reviewed. The most popular datasets annotated with ratings from subjective experiments are presented. Development of publicly available 3D visual content datasets, recently including also special visual content, e.g., point cloud and holographic, was largely promoted also by the standardization bodies, namely JPEG and MPEG. Datasets used within selected standardization efforts are also described in this chapter.

The aim is not to provide an exhaustive listing and description of all existing 3D visual content datasets, but more to give examples of the most commonly used publicly available datasets. Any effort of this type captures the current status. However, numerous new datasets are introduced every year. It is related to the fact that novel techniques for coding, transmission, and quality assessment are being continuously developed. Description of the most recent datasets can be found in regularly updated online resources, e.g., Qualinet Databases, which were also presented in this chapter.

References

1. Winkler, S.: Analysis of public image and video databases for quality assessment. IEEE J. Sel. Top. Signal Process. **6**(6), 616–625 (2012). https://doi.org/10.1109/JSTSP.2012.2215007
2. Winkler, S., Savoy, F.M., Subramanian, R. X-Eye: a reference format for eye tracking data to facilitate analyses across databases. In: Proceedings of IS&T/SPIE Human Vision & Electronic Imaging (2014). https://doi.org/10.1117/12.2042433
3. Winkler, S., Subramanian, R. Overview of eye tracking datasets. In: Proceedings of 5th International Workshop on Quality of Multimedia Experience (QoMEX) (2013). https://doi.org/10.1109/qomex.2013.6603239
4. Schöps, T., Schönberger, J., Galliani, S., Sattler, T., Schindler, K., Pollefeys, M., Geiger, A.: A multi-view stereo benchmark with high-resolution images and multi-camera videos. In: Proceedings of IEEE Computer Conference on Computer Vision and Pattern Recognition 2538–2547 (2017). https://doi.org/10.1109/CVPR.2017.272

5. Geiger, A., Lenz, P., Urtasun, R.: Are we ready for autonomous driving? The KITTI vision benchmark suite. In: Proceedings of the IEEE Conference on Computer Vision and Pattern Recognition 3354–3361 (2012). https://doi.org/10.1109/cvpr.2012.6248074
6. Scharstein, D., Szeliski, R.: A taxonomy and evaluation of dense two-frame stereo correspondence algorithms. Int. J. Comput. Vis. **47**(1), 7–42 (2002). https://doi.org/10.1023/A:1014573219977
7. Butler, D., Wulff, J., Stanley, G., Black, M.: A naturalistic open source movie for optical flow evaluation. In: Proceedings of the European Conference on Computer Vision, pp. 611–625 (2012). https://doi.org/10.1007/978-3-642-33783-3_44
8. Johnson-Roberson, M., Barto, C., Rounak, M., Sharath, N., Ram, V.: Driving in the matrix: can virtual worlds replace human-generated annotations for real world tasks? In: Proceedings of IEEE International Conference on Robotics and Automation (2017). https://doi.org/10.1109/icra.2017.7989092
9. Menze, M., Geiger, A.: Object scene flow for autonomous vehicles. In: Proceedings of the IEEE Conference on Computer Vision and Pattern Recognition (2015). https://doi.org/10.1109/cvpr.2015.7298925
10. Furukawa, Y., Ponce, J.: Accurate, dense, and robust multi-view stereopsis. IEEE Trans. Pattern Anal. Mach. Intell. **32**(8), 1362–1376 (2010). https://doi.org/10.1109/TPAMI.2009.161
11. Seitz, S., Curless, B., Diebel, J., Scharstein, S., Szeliski, R.: A Comparison and evaluation of multi-view stereo reconstruction algorithms. In: Proceedings of Conference on Computer Vision and Pattern Recognition (CVPR) (2006). https://doi.org/10.1109/cvpr.2006.19
12. Strecha, C., von Hansen, W., Van Gool, L., Fua, P., Thoennessen, U.: On benchmarking camera calibration and multi-view stereo for high resolution imagery. In: Proc IEEE Conference on Computer Vision and Pattern Recognition (CVPR) (2008). https://doi.org/10.1109/cvpr.2008.4587706
13. Fransens, R., Strecha, C., Van Gool, L.: Parametric stereo for multi-pose face recognition and 3D-face modeling (2005). In: Proceedings of ICCV 2005 Workshop Analysis and Modeling of Faces and Gestures, vol. 3723, pp. 109–124 (2005). https://doi.org/10.1007/11564386_10
14. Strecha, C., Fransens, R., Van Gool, L.: Wide-baseline stereo from multiple views: a probabilistic account. In: Proceedings of IEEE Conference on Computer Vision and Pattern Recognition (CVPR) (2004)
15. Strecha, C., Fransens, R., Van Gool, L.: Combined depth and outlier estimation in multi-view stereo. In: Proceedings IEEE Conference on Computer Vision and Pattern Recognition (CVPR) (2006). https://doi.org/10.1109/cvpr.2006.78
16. Ozuysal, M., Lepetit, V., Fua, P.: Pose estimation for category specific multiview object localization. In: Proceedings of Conference on Computer Vision and Pattern Recognition (2009). https://doi.org/10.1109/cvprw.2009.5206633
17. Cremers, D., Kolev, K.: Multiview stereo and silhouette consistency via convex functionals over convex domains. IEEE Trans. Pattern Anal. Mach. Intell. **33**(6), 1161–1174 (2011). https://doi.org/10.1109/TPAMI.2010.174
18. Li, Y., Snavely, N., Huttenlocher, D.P.: Location recognition using prioritized feature matching. In: Proceedings of ECCV (2010). https://doi.org/10.1007/978-3-642-15552-9_57
19. Frankot, R., Chellappa, R.: A method for enforcing integrability in shape from shading algorithms. IEEE Trans. Pattern Anal. Mach. Intell. **10**(4), 439–451 (1988). https://doi.org/10.1109/34.3909
20. Shi, B., Wu, Z., Mo, Z., Duan, D., Yeung, S.-K., Tan, P.: A benchmark dataset and evaluation for non-lambertian and uncalibrated photometric stereo. In: Proceedings of IEEE Conference on Computer Vision and Pattern Recognition (CVPR) (2016)
21. Xiang, Y., Mottaghi, R., Savarese, S.: Beyond PASCAL: a benchmark for 3D object detection in the wild. In: Proceedings of IEEE Winter Conference on Applications of Computer Vision (WACV) (2014). https://doi.org/10.1109/wacv.2014.6836101

22. Everingham, M., Van Gool, L., Williams, C.K.I., Winn, J., Zisserman, A.: The pascal visual object classes (VOC) challenge. Int. J. Comput. Vis. **88**(2), 303–338 (2010). https://doi.org/10.1007/s11263-009-0275-4
23. Armeni, I., Sax, A., Zamir, A., Savarese, S. Joint 2D-3D-semantic data for indoor scene understanding. In: Computer Vision and Pattern Recognition (2017, to appear)
24. ITU-R Recommendation BT.1788: Methodology for the subjective assessment of video quality in multimedia applications, Jan 2007
25. Hasler, D., Suesstrunk, S.E.: Measuring colorfulness in natural images. Hum. Vis. Electron. Imaging VIII **2003**, 87–95 (2003). https://doi.org/10.1117/12.477378
26. Faloutsos, C., Barber, R., Flickner, M., Hafner, J., Niblack, W., Petkovic, D., Equitz, W.: Efficient and effective querying by image content. J. Intell. Inf. Syst. **3**(3–4), 231–262 (1994). https://doi.org/10.1007/BF00962238
27. Pinson, M.: Spatial information (SI) filter (2016). https://www.its.bldrdoc.gov/resources/video-quality-research/guides-and-tutorials/spatial-information-si-filter.aspx. Accessed 29 Sept 2017
28. Paudyal, P., Gutiérrez. J., Le Callet, P., Carli, M., Battisti, F.: Characterization and selection of light field content for perceptual assessment. In: Proceedings of QoMEX (2017). https://doi.org/10.1109/qomex.2017.7965635
29. ITU-T Recommendation P.910: Subjective video quality assessment methods for multimedia applications, Apr 2008
30. Tkalcic, M., Tasic, J.F.: Colour spaces: perceptual, historical and applicational background. In: Proceedings of EUROCON (2003). https://doi.org/10.1109/eurcon.2003.1248032
31. Datta, R., Joshi, D., Li, J., Wang, J.Z.: Studying aesthetics in photographic images using a computational approach. Proc. Eur. Conf. Comput. Vis. **3953**, 288–301 (2006). https://doi.org/10.1007/11744078_23
32. Bishop, T.E., Zanetti, S., Favaro, P.: Light field superresolution. In: Proceedings of IEEE International Conference on Computational Photography (ICCP 09) (2009). https://doi.org/10.1109/iccphot.2009.5559010
33. Szeliski, R., Avidan, S., Anandan, P.: Layer extraction from multiple images containing reflections and transparency. In: IEEE Conference on Computer Vision and Pattern Recognition (2000)
34. Wang, Q., Lin, H., Ma, Y., Kang, S.B., Yu, J.: Automatic layer separation using light field imaging, arXiv preprint (2015). arXiv:1506.04721
35. Levin, A., Weiss, Y.: User assisted separation of reflections from a single image using a sparsity prior. IEEE Trans. Pattern Anal. Mach. Intell. **29**(9), 1647–1654 (2007). https://doi.org/10.1109/TPAMI.2007.1106
36. Denker, K., Umlauf, G.: Accurate real-time multi-camera stereo-matching on the GPU for 3D reconstruction. J. WSCG **19**(1–3), 9–16 (2011)
37. Jeon, H.-G., Park, J., Choe, G., Park, J., Bok, Y., Tai, Y.W., Kweon, I.S.: Accurate depth map estimation from a lenslet light field camera. In: Proceedings of IEEE Conference on Computer Vision and Pattern Recognition (CVPR) (2015). https://doi.org/10.1109/cvpr.2015.7298762
38. Dabała, Ł., Ziegler, M., Didyk, P., Zilly, F., Keinert, J., Myszkowski, K., Seidel, H.-P., Rokita, P., Ritschel, T.: Efficient multi-image correspondences for on-line light field video processing. Comput. Graph. Forum **35**(7), 401–410 (2016). https://doi.org/10.1111/cgf.13037
39. Montilla, I., Puga, M., Luke, J.P., Marichal-Hernandez, J.G., Rodriguez-Ramos, J.M.: Design and laboratory results of a plenoptic objective: from 2D to 3D with a standard camera. J. Disp. Technol. **11**(1), 73–78 (2015). https://doi.org/10.1109/JDT.2014.2361257
40. Ziegler, M., Engelhardt, A., Müller, S., Keinert, J., Zilly, F., Foessel, S., Schmid, K.: Multi-camera system for depth based visual effects and compositing. In: Proceedings of European Conference on Visual Media Production (2015). https://doi.org/10.1145/2824840.2824845
41. Urvoy, M., Barkowsky, M., Cousseau, R., Koudota, Y., Ricordel, V., Le Callet, P., Gutierrez, J., Garcia, N.: NAMA3DS1-COSPAD1: subjective video quality assessment database on coding conditions introducing freely available high quality 3D stereoscopic sequences.

In: Proceedings of Fourth International Workshop on Quality of Multimedia Experience (2012). https://doi.org/10.1109/qomex.2012.6263847
42. Wang, Z.: Objective image quality assessment: facing the real-world challenges. In: Image Quality and System Performance (keynote speech paper) (2016)
43. Wang, T.-C., Efros, A.A., Ramamoorthi, R.: Depth estimation with occlusion modeling using light-field cameras. IEEE Trans. Pattern Anal. Mach. Intell. **38**(11), 2170–2181 (2016). https://doi.org/10.1109/tpami.2016.2515615
44. Liu, H., Wang, J., Redi, J., Le Callet, P., Heynderickx, I.: An efficient no-reference metric for perceived blur. In: Proceedings of European Workshop on Visual Information Processing (2011). https://doi.org/10.1109/euvip.2011.6045525
45. Wu, W., Llull, P., Tosic, I., Bedard, N., Berkner, K., Balram, N.: Content-adaptive focus configuration for near-eye multi-focal displays. In: Proceedings of IEEE International Conference on Multimedia and Expo (2016). https://doi.org/10.1109/icme.2016.7552965
46. Ng, R., Levoy, M., Brédif, M., Duval, G., Horowitz, M., Hanrahan, P.: Light field photography with a hand-held plenoptic camera. Comput. Sci. Techn. Rep. **2**(11), 1–11 (2005)
47. Vaish, V., Adams, A.: The (new) stanford light field archive. http://lightfield.stanford.edu/ (2008). Accessed 29 Sept 2017
48. Paudyal, P., Olsson, R., Sjostrom, M., Battisti, F., Carli, M.: SMART: a light field image quality dataset. In: Proceedings of International Conference on Multimedia Systems (MMSys 2016) (2016). https://doi.org/10.1145/2910017.2910623
49. Rerabek, M., Yuan, L., Authier, L.A., Ebrahimi, T.: EPFL light-field image dataset. ISO/IEC JTC 1/SC 29/WG1, Technical Report (2015)
50. Ebrahimi, T., Foessel, S., Pereira, F., Schelkens, P.: JPEG pleno: toward an efficient representation of visual reality. IEEE Multimed. **23**(4), 14–20 (2016). https://doi.org/10.1109/MMUL.2016.64
51. Munoz, D., Bagnell. J.A., Vandapel, N., Hebert, M.: Contextual classification with functional max-margin markov networks. In: Proceedings of IEEE Computer Society Conference on Computer Vision and Pattern Recognition (CVPR) (2009). https://doi.org/10.1109/cvprw.2009.5206590
52. d'Eon, E., Harrison, B., Myers, T., Chou, P.A.: 8i voxelized full bodies—a voxelized point cloud dataset. ISO/IEC JTC1/SC29 Joint WG11/WG1 (MPEG/JPEG) input document WG11M40059/WG1M74006, Geneva, January (2017)
53. Blinder, D., Ahar, A., Symeonidou, A., Xing, Y., Bruylants, T., Schretter, C., Pesquet-Popescu, B., Dufaux, F., Munteanu, A., Schelkens, P.: Open access database for experimental validations of holographic compression engines. In: 7th International Workshop on Quality of Multimedia Experience (QoMEX) (2015). https://doi.org/10.1109/qomex.2015.7148145
54. Gilles, A., Gioia, P., Cozot, R., Morin, L.: Hybrid approach for fast occlusion processing in computer-generated hologram calculation. Appl. Opt. **55**(20), 5459–5470 (2016). https://doi.org/10.1364/AO.55.005459
55. Benoit, A., Le Callet, P., Campisi, P., Cousseau, R.: Quality assessment of stereoscopic images. EURASIP J. Image Video Process. (2008). https://doi.org/10.1155/2008/659024
56. ITU-R Recommendation BT.500–13; Methodology for the subjective assessment of the quality of television pictures, Jan 2012
57. Moorthy, A.K., Su, C.-C., Mittal, A., Bovik, A.C.: Subjective evaluation of stereoscopic image quality. Signal Process. Image Commun. **28**(8), 870–883 (2013). https://doi.org/10.1016/j.image.2012.08.004
58. Chen, M.-J., Cormack, L.K., Bovik, A.C.: No-reference quality assessment of natural stereopairs. IEEE Trans. Image Process. **22**(9), 3379–3391 (2013). https://doi.org/10.1109/TIP.2013.2267393

59. Goldmann, L., De Simone, F., Ebrahimi, T.: A comprehensive database and subjective evaluation methodology for quality of experience in stereoscopic video. In: Proceedings of Electronic Imaging (EI), 3D Image Processing (3DIP) and Applications (2010). https://doi.org/10.1117/12.839438
60. Bosc, E., Pépion, R., Le Callet, P., Köppel, M., Ndjiki-Nya, P., Pressigout, M., Morin, L.: Towards a new quality metric for 3-D synthesized view assessment. IEEE J. Sel. Top. Signal Process. **6029277**, 1332–1343 (2011). https://doi.org/10.1109/JSTSP.2011.2166245
61. Thurstone, L.L.: A law of comparative judgement. Psychol. Rev. **34**(4), 273–286 (1927). https://doi.org/10.1037/h0070288
62. Song, R., Ko, H., Kuo, C.C.: MCL-3D: a database for stereoscopic image quality assessment using 2D-image-plus-depth source. J. Inf. Sci. Eng. **31**(5), 1593–1611 (2015)
63. Goldmann, L., De Simone, F., Ebrahimi, T.: Impact of acquisition distortions on the quality of stereoscopic images. In: Proceedings of 5th International Workshop on Video Processing and Quality Metrics for Consumer Electronics (VPQM) (2010)
64. Bosc, E., Le Callet, P., Morin, L., Pressigout, M.: Visual quality assessment of synthesized views in the context of 3D-TV. In: Zhu, C., Zhao, Y., Yu, L., Tanimoto, M. (eds) 3D-TV System with Depth-Image-Based Rendering Architectures, Techniques and Challenges. Springer, New York (2012). https://doi.org/10.1007/978-1-4419-9964-1_15
65. Lavoué, G., Drelie Gelasca, E., Dupont, F., Baskurt, A., Ebrahimi, T.: Perceptually driven 3D distance metrics with application to watermarking (2006). In: Proceedings of SPIE, vol. 6312. https://doi.org/10.1117/12.686964
66. Lavoué, G.: A local roughness measure for 3D meshes and its application to visual masking. ACM Trans. Appl. Percept. **5**(4) (2009). https://doi.org/10.1145/1462048.1462052
67. Lavoué, G., Corsini, M.: A comparison of perceptually-based metrics for objective evaluation of geometry processing. IEEE Trans. Multimed. **12**(7), 636–649 (2010). https://doi.org/10.1109/TMM.2010.2060475
68. Wang, J., Perreira Da Silva, M., Le Callet, P., Ricordel, V.: Computational model of stereoscopic 3D visual saliency. IEEE Trans. Image Process. **22**(6), 2151–2165 (2013). https://doi.org/10.1109/TIP.2013.2246176
69. Hanhart, P., Ebrahimi, T.: EyeC3D: 3D video eye tracking dataset. In: Proceedings of Sixth International Workshop on Quality of Multimedia Experience (QoMEX 2014) (2014). https://doi.org/10.1109/qomex.2014.6982290
70. Fang, Y., Wang, J., Li, J., Pépion, R., Le Callet, P.: An eye tracking database for stereoscopic video. In: Proceedings of Sixth International Workshop on Quality of Multimedia Experience (QoMEX 2014) (2014). https://doi.org/10.1109/qomex.2014.6982288
71. Raghavendra, R., Raja, K., Busch, C.: Exploring the usefulness of light field camera for biometrics: an empirical study on face and iris recognition. IEEE Trans. Inf. Forensics Secur. **11**(5), 922–936 (2016). https://doi.org/10.1109/tifs.2015.2512559
72. Mousnier, A., Vural, E., Guillemot, C.: Partial light field tomographic reconstruction from a fixed-camera focal stack. In: Computer Vision and Pattern Recognition, arXiv preprint (2015). arXiv:1503.01903
73. Ghasemi, A., Afonso. N., Vetterli, M.: LCAV-31: a dataset for light field object recognition. In IS&T/SPIE Electronic Imaging, pp. 902014–902014 (2014). https://doi.org/10.1117/12.2041097
74. Li, N., Ye, J., Ji, Y., Ling, H., Yu, J.: Saliency detection on light field. IEEE Trans. Pattern Anal. Mach. Intell. **39**(8), 1605–1616 (2017). https://doi.org/10.1109/TPAMI.2016.2610425
75. Wetzstein, G.: Synthetic light field archive (2016). http://web.media.mit.edu/~gordonw/SyntheticLightFields/, Accessed 29 September 2017
76. Wanner, S., Meister, S., Goldluecke, B.: Datasets and benchmarks for densely sampled 4D light fields. In: Annual Workshop on Vision, Modeling and Visualization: VMV (2013). https://doi.org/10.2312/pe.vmv.vmv13.225-226

MIX
Papier aus verantwortungsvollen Quellen
Paper from responsible sources
FSC® C105338

If you have any concerns about our products,
you can contact us on
ProductSafety@springernature.com

In case Publisher is established outside the EU,
the EU authorized representative is:
Springer Nature Customer Service Center GmbH
Europaplatz 3, 69115 Heidelberg, Germany

Printed by Libri Plureos GmbH
in Hamburg, Germany